Connectomic Medicine

Connectomic Medicine

Guide to Brain AI in Treatment Decision Planning

Michael E. Sughrue
Cingulum Health, Omniscient Neurotechnology, Phoenix, AZ, United States

Jacky T. Yeung
Department of Neurosurgery, Yale University School of Medicine, New Haven, CT, United States; Cingulum Health, Sydney, NSW, Australia

Nicholas B. Dadario
Robert Wood Johnson Medical School, Rutgers University, New Brunswick, NJ, United States

Academic Press is an imprint of Elsevier
125 London Wall, London EC2Y 5AS, United Kingdom
525 B Street, Suite 1650, San Diego, CA 92101, United States
50 Hampshire Street, 5th Floor, Cambridge, MA 02139, United States

Notices

Knowledge and best practice in this field are constantly changing. As new research and experience broaden our understanding, changes in research methods, professional practices, or medical treatment may become necessary.

Practitioners and researchers must always rely on their own experience and knowledge in evaluating and using any information, methods, compounds, or experiments described herein. In using such information or methods they should be mindful of their own safety and the safety of others, including parties for whom they have a professional responsibility.

To the fullest extent of the law, neither the Publisher nor the authors, contributors, or editors, assume any liability for any injury and/or damage to persons or property as a matter of products liability, negligence or otherwise, or from any use or operation of any methods, products, instructions, or ideas contained in the material herein.

ISBN: 978-0-443-19089-6

For information on all Academic Press publications visit our website at https://www.elsevier.com/books-and-journals

Publisher: Nikki P. Levy
Acquisitions Editor: Megan McManus
Editorial Project Manager: Tracy Tufaga
Production Project Manager: Sajana Devasi P.K
Cover Designer: Mark Rogers

Typeset by TNQ Technologies

ELSEVIER
Book Aid International
Working together
to grow libraries in
developing countries
www.elsevier.com • www.bookaid.org

This book is dedicated to my family, my wife Kanokkorn, and my daughter Isabella. - Mike Sughrue

This book is dedicated to my family who supported me in my growth, my mentors who guided me along the way, and all my patients whom I had the honor to treat. - Jacky Yeung

This book is dedicated to my family who constantly support me, my mentors who guide me, and my patients who inspire me. - Nicholas Dadario

Contents

Introduction: What is this book trying to say?

It is important to view this book for what it is intended to be.

In my lectures around the world, it has become apparent to me that one common issue in the world of evidence-based medicine is that we have deputized everyone to become absurd obstructionists. In other words, we have taught a few generations of physicians to view arguments in terms of binaries: Either an idea, diagnostic, or treatment has multiple randomized controlled trials supporting its use, or we are allowed to totally ignore the field, even if it involves denying relatively obvious, reasonable, and likely helpful treatments to patients who often lack other therapies. And we have told people they are in the right for pursuing heroic degrees of skepticism in denying therapies to patients. There are a lot of places between these extremes.

Connectomic medicine is a term we invented (at least we didn't knowingly take it from anyone else). This book is about the idea that we can understand the brain better using connectomic basic methods, and that we can use this insight to make reasonable treatment decisions to open up new options for patients who currently do not have any great alternative options. The field is new, and as such the evidence is not to the level of professional guidelines level. Having said this, those who have spent years sitting across from patients probably realize that we can only improve upon the status quo where we often accept horrific misery as something we cannot do anything about. We have written this book to promote the idea that sometimes you can do something positive for patients if you spend the time to think through their problems and to use advanced tools.

Thus, while we describe methods like TMS as treatments for problems, we are not claiming we have all the answers to the mystery of the brain, just that these kinds of approaches have been helpful for us in some patients, often where no better option was obvious to us or the other physicians treating them. The more important point we want to make is that it is possible to use AI-based tools, brain stimulation, and modifications of history and physical taking to build a new type of approach to patients. This has worked for us, and we hope others bring forward new ideas in this new way of thinking about patients.

The key points we want to leave with you after reading this book are as follows:

(1) Our understanding of the brain has radically advanced since most of us graduated from medical school, and we should take advantage of this.

(2) Brain network dynamics are repeatable ways to organizing thinking about the brain, which allow us to see through the extreme complexity of the brain in a logical way which aligns with the physiology of the brain.

(3) There are probably not real neat disease states like depression or schizophrenia, but instead by focusing on circuits as opposed to assigning names to diseases, we can begin to make these problems tractable.

(4) We must use anatomic nomenclature, like parcels, to describe brain anatomy. Otherwise, we cannot communicate with each other.

(5) We need to start with small ideas and build up to trials. There is not enough money or time in the world to run a trial on everything before we try to help the patient in front of us.

(6) The brain is too complex to be understood without machine learning and AI.

Introduction: What is this book trying to say?

SECTION

1

Foundations

CHAPTER 1

Why should you learn something new?

Introduction: One step forward or two steps back?

This book is introducing a new concept to the vast majority of physicians. It was new to us when we spent the past few years devising the principles of these approaches, and it is new to neurosurgeons, psychiatrists, neurologists, stroke doctors, and even transcranial magnetic stimulation (TMS) clinicians. We have taught people around the world our approaches: Dr. Sughrue alone has given over 300 lectures on these topics and nearly 100 courses. Over time we have developed some ways to make this new topic more digestible, but this remains difficult. We expect you to learn new things, and this is a complex topic. This is doubly true if you are a clinician who does not routinely look at brain MRIs in your existing practice. Over time, this is going to change if you apply the lessons in this book.

We all are busy: there are not enough of us, patient care is growing more complex, and hospitals are constantly redirecting us to less enlightening tasks on a nearly daily fashion. And this is one more thing being put on your plate. So, why should you bother to learn this?

The main reason to do this is that we all are not that good at our jobs, and you are tired of this and want to do better.

This is not to say everyone is incompetent, far from it. However, brain medicine in all its forms is full of diseases where the standard of care is pretty unsatisfying. Let us prove this to you with a few questions, which span a variety of specialties possibly reading this book, which will disavow you of the idea that we are not desperately in need of a better approach.

Question 1: Do you feel we have effective therapies for most patients with back and neck pain?

Given that most of the literature is aimed at determining which highly selected types of patients are helped and not harmed by surgery or other treatments aimed at the part of body where the symptoms are perceived, this shows widespread acceptance of the idea that the back is not the issue. There is a cause of psychosomatic problems, and this cause is very likely discoverable somewhere in the brain.

Connectomic Medicine. https://doi.org/10.1016/B978-0-443-19089-6.00017-3

Question 2: Do you feel that given the ~60,000 completed suicides in the United States annually suggests that our treatments are truly optimized?

This is not to say that psychotherapy, brain stimulation, and SSRIs do not work in some people. But failure rates comprising as many as half of patients treated suggests that there has to be a better way to do this.

Question 3: If you, personally, were paralyzed from a stroke, would you view a few weeks of rehab alone, and then hoping for the best to occur naturally, is the best anyone on earth could come up with?

Besides just a commentary on the understaffing our rehab services face in many nations, it is worth considering that in a world where people walked on the moon in the 1960s, and where teenagers reach thousands of people in real time from a handheld phone, that we have been taught to lower our expectations for these patients, possibly without merit.

Our biases: What we claim by writing this book

Based on the above, it should be obvious that we aim to bring the field forward with our approaches. Having said this, it is important to be clear what we do not claim. Connectomic-based approaches are not the answer to all neurologic and psychiatric diseases. Simultaneously, neither is TMS, or any other approach described in this book. We expect that things will continue to evolve, and will likely make this book look outdated. We hope this will occur, in fact, as the field progresses. We do not claim we know everything about the brain. Science progresses, and while there is a fair bit of evidence that our principle ideas are valid, and we feel we have been careful to try to get things right, inevitably, something will be shown to be wrong or a better model of cognitive brain function will be created. This book reflects our best thinking at the time of writing this book.

What we do claim is that we write this book to share our experiences, which many people seem to be interested in learning from us, and that a new philosophy for thinking through brain diseases using connectomic-based tools and data-driven personalized approaches can help better understand our patients and their problems to make better decisions and to take smarter actions, which have a better chance of working, regardless of what tools we have in the future.

Put more directly, this book claims the following three points.

Point 1: Much of our failure to successfully treat patients results from poor knowledge about how the brain works, specifically the brain of the patient in front of us

The brain is the most complex thing in the known universe, so it is not surprising that it has taken a long time to have tools which provide useful information about the brain and how it works. We spend our time with patients shrugging our shoulders when asked why a symptom is happening or its prognosis, and occasionally throw out a clinical diagnosis which very likely greatly underestimates the complexity and heterogeneity of the actual problem.

We fail to treat patients successfully because most of the time, we have little to no idea what is happening in their brains. Our mental idea of what we are trying to do when we treat a patient is the most important medical device we ever use in clinical practice. And our device is not up to the task.

Connectomic medicine is not a mature field, but improved knowledge about the nature of the disease we are treating is likely to substantially improve our ability to understand and treat these diseases. So while one or another specific approach may not work, the principle that we can only improve by reducing our ignorance about the mechanisms of brain illnesses is valid nonetheless.

Point 2: Connectomics and associated neuroimaging are ready for clinical translation

From time to time, people have heard us say this and raised objections like "there are still controversies in the field of resting state fMRI," or "we don't know what all this means." Our answer is always the same three points. First, every field of medicine has controversy or uncertainty. Yet, diseases still force us to treat people in the lengthy period while scientists are grinding out the details. People will be treated with something whether everyone has stopped arguing or not. Absolute certainty is not a mandate before doing anything as it never ever occurs. Second, it is not accurate that we have no idea of the proper methods in fMRI and DTI analysis. At the time of this writing, there are over 500,000 articles on PubMed using these techniques, in addition to over 300 NIH funded grants using resting state fMRI alone spanning over 100 million US dollars in current funding (Dadario & Sughrue, 2022). There are tens of thousands of papers using these techniques to study mental illness alone. If after 30 years of research, we still cannot draw any conclusions which are superior to educated guessing, then this is a profound betrayal of the public trust that we are continuing to not make progress in this field. Obviously, we know a lot about these techniques, best practices, and basic principles. Finally, we in medicine have never waited until 100% certainty about how an organ works to try to improve what we do. The benchmark is that the technique is thoughtful and generally better than what it is replacing. We will all be dead long before all the mysteries of the brain are discovered as will everyone we see with those diseases. It is hard to think that targeted application of improved neuroscientific principles could possibly be worse than what we do now which is basically guessing if we are being honest.

These kinds of objections are obviously holding these techniques to a higher standard than most other currently pursued diagnostic tests, as absolute certainty never occurs. These unrealistic expectations are the result of the fact that clinicians often have delegated the techniques used in these studies to Ph.D.s who in our experience, sometimes maintain unrealistic expectations about degree of certainty achieved in clinical practice, and often do not put scientific uncertainties into the context of the disease, its severity, and other options that exist. Ultimately, they are not clinicians, so we would not expect them to think like clinicians. However, we should not let them decide how we treat patients in its entirety.

Point 3: There is no such thing as "Evidence based medicine"

Obviously, this statement is not meant to imply that we should not use evidence or data to decide treatments. Instead, what we mean is that what physicians usually refer to when they use this term, and often avoid thinking about new ideas while using it, is usually poor quality evidence, even when it is a large, multicenter randomized controlled trial (RCT). Medicine will continue to learn slowly, with many hurt in its wake, if we use this archaic paradigm to tackle complex problems.

A helpful thought exercise to see why we claim this is to consider why bell-shaped curves, such as Gaussian distributions, are the mainstay of most of our statistical approaches. Gaussian-type distributions generally imply that the factor under study is affected by numerous variables not addressed in the study performed. For example, measurements of height in adults are Gaussian because hundreds of independent factors, including hundreds of genes, and the law of independent probabilities suggest that the odds of hundreds of favorable or unfavorable independent events occurring in the same person is infinitesimally low.

We see this in our clinical studies because our outcomes, for example, response to medications are impacted by thousands of variables like genes, stressors, aspects of the connectome, all of which we ignore and use people as control subjects in clinical trials for people entirely different in numerous important ways. Thus, even a good RCT is basically studying the question of whether when faced with a clinical scenario where you lack 99.99% of the relevant facts, that the proposed intervention is a good guess.

Additional issues with so-called evidence based approaches include the overestimation of the significance of studies based on the sample size and not the effect size (Zhang et al., 2013). This is the poison the use of the *P*-value has brought to our lives. For example, a large RCT of antidepressants can have a great *P*-value proving that the depression scale is dropped by 15% ... but a patient only 15% better likely would not tell you their life is a lot better, they may even still have suicidal ideations. "P" does not equal *potent*.

Finally, treating people based on Gaussian responses suggests that for many patients, the approach studied is entirely wrong and may be harmful. It is easy to dismiss the poor outcomes of outliers, even if they are not rare in some conditions, when we have no better way to address the vast complexity of the human brain (Tonelli, 1998). However, if you were hurt by someone who treated you based on a cohort you really did not belong to, the fact that this "usually works for the majority of people" does little for your symptoms.

This book thus is not a listing of trials in the field. Trials can be useful sometimes, for example, a TMS target which works over and over to reduce the depression scores is probably a good idea to consider in most people. Acknowledging the flaws in our study methods does not imply we can ignore trials and do whatever we want irrespective of them. But moving onward toward a better paradigm, forces us to think differently in three key ways. (1) We cannot assume that a trial result forces something in all cases, or that a negative trial makes something always unacceptable because this is not "evidence based." This thinking does not hold true when we have access to a greater body of relevant facts in an actionable platform. (2) When we have a greater access to the potential set of facts, it improves our ability to think through unique situations, even if there is no relevant study or trial. The lack of a trial does not render us entirely clueless about reasonable approaches. (3) Connectomic medicine is a field capable of agile learning and faster progress. If we wait for every possible permutation of brain therapy to be formally tested in expensive and long RCT's, we will fail generations of patients while bankrupting the research infrastructure running confirmatory studies on things that are likely true. This is becoming especially acute in the West, where definitive studies, even of minor interventions, can take years and tens of millions of dollars to navigate through the Byzantine processes and inertia of academic medical centers.

So what is connectomic medicine?

Put simply, connectomic medicine is the idea that by utilizing better neuroimaging, artificial intelligence (AI), and anatomically precise annotation of these data, that we can develop better treatment plans and accelerate the discovery process in ways which improve the care of patients with brain diseases. It is based on the idea that better knowledge makes better doctors, and that we need faster and larger-scale approaches to learning about how to treat these patients or it will never finish.

Connectomics involves using techniques to be able to analyze brain anatomy in a granular way and thus think through complex problems using tools of appropriate complexity (Dadario et al., 2021; Yeung et al., 2021). One such problem is shown in Fig. 1.1. When we think about white matter bundles as a single coherent structure as opposed to a collection of largely unrelated connections running together based on anatomic proximity (Fig. 1.1A), we can falsely attribute magical properties to a bundle that would be better described as a subset of this bundle (Fig. 1.1B). As this figure shows, a bundle is often a combination of numerous unrelated multimodal connections (Fig. 1.1C), and the bundle has no intrinsic meaning as a whole other than as a gross anatomic structure. Worse, if we attempt to quantify brain injury by taking metrics of entire bundles (Fig. 1.1D,E), instead of connectomic-derived subcomponents, we can come up with calculations which make no neuroanatomic sense.

Thus, we will only make progress when we begin to tackle the brain at a level of complexity necessary to really dive into its secrets, and this book is about how to use tools to think rationally at this level and to leverage these insights to make better decisions.

A key point we emphasize repeatedly in this book is an idea we term "The Fundamental Theorem of Connectomic Medicine". In order to alleviate a symptom, a therapy must make a functionally relevant circuit somewhere in the brain fire more normally. We discuss in greater detail in the upcoming Chapter 9. This idea is a powerful tool to take a complex set of interactions across many diverse diseases and express them as a simple goal: we need to find the circuit and make it work more normally.

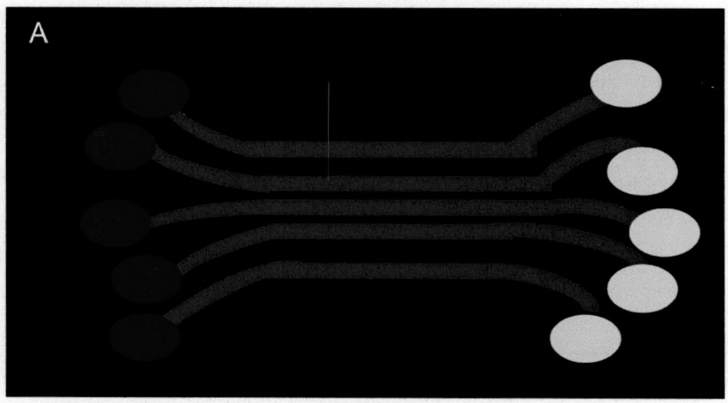

B
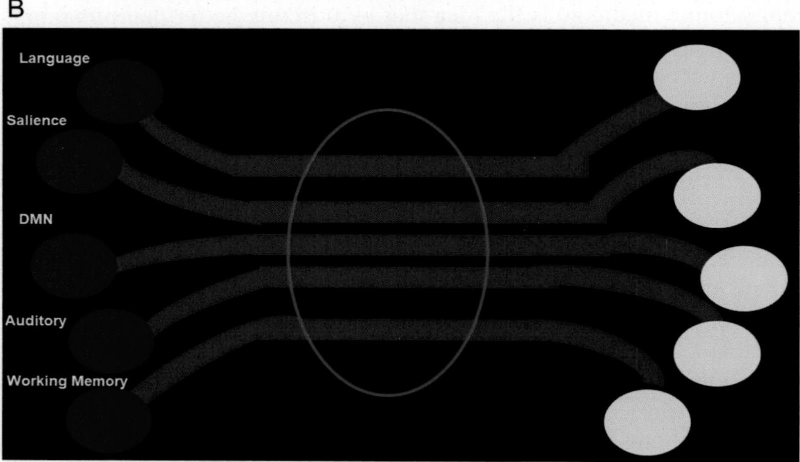

FIG. 1.1

Providing an appropriate granular understanding to white matter bundles with connectomics. In panel A, it is obvious that a single white matter bundle can contain a large volume of connections ranging from and to a variety of spatially distant regions. However, the value of the entire bundle cannot be simply understood as a single tract, but rather better understood according to its individual components often consisting of various often unrelated multimodal connections (B). An example is provided in panel C of the inferior fronto-occipital fasciculus (IFOFs) fiber bundle. The IFOF bundle is a major white matter connection likely to be involved in higher cognitive processing through multiple connectivity-related links with many networks. Relatedly, we cannot appropriately understand the value of these connections if we merely quantify integrity of the bundle as a whole as it relates to deficits, but rather must consider individual components (D,E).

The figure in panel C has been adapted from Conner et al. (2018).

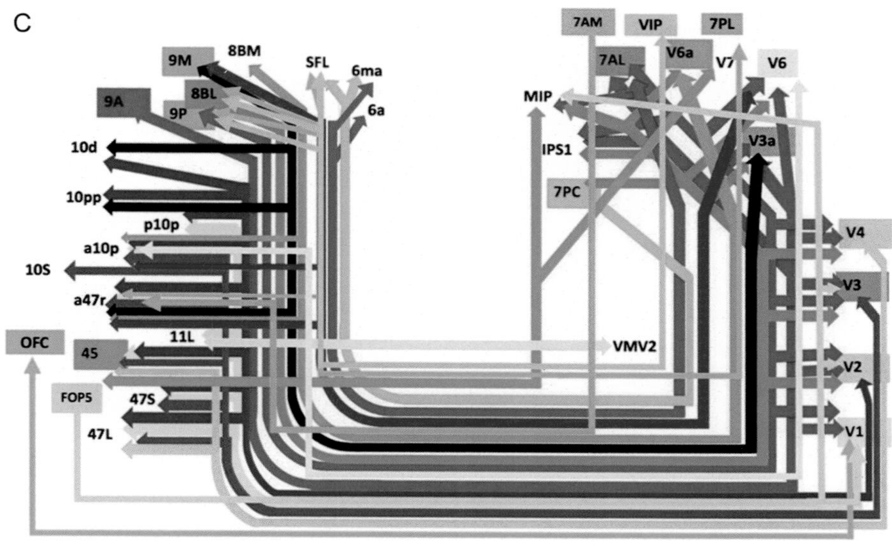
C
7AM
VIP
7PL
9M
8BM
SFL
6ma
7AL
V6a
V7
V6
9A
8BL
9P
6a
MIP
10d
IPS1
V3a
10pp
7PC
p10p
a10p
V4
10S
a47r
V3
OFC
11L
VMV2
45
V2
FOP5
47S
47L
V1

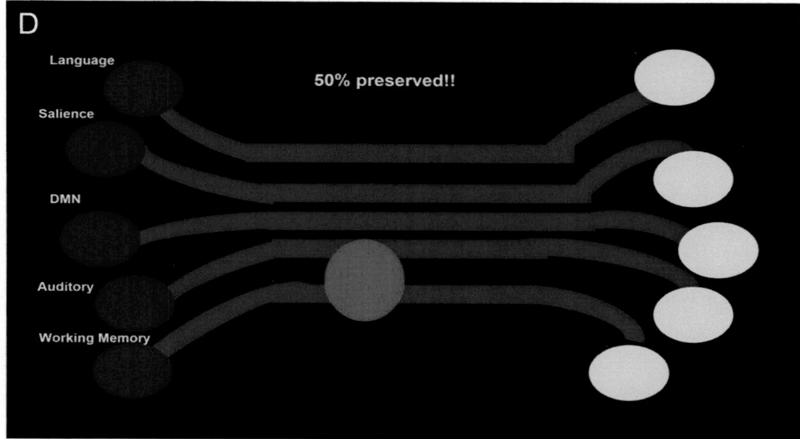
D
Language
50% preserved!!
Salience
DMN
Auditory
Working Memory

FIG. 1.1 Cont'd

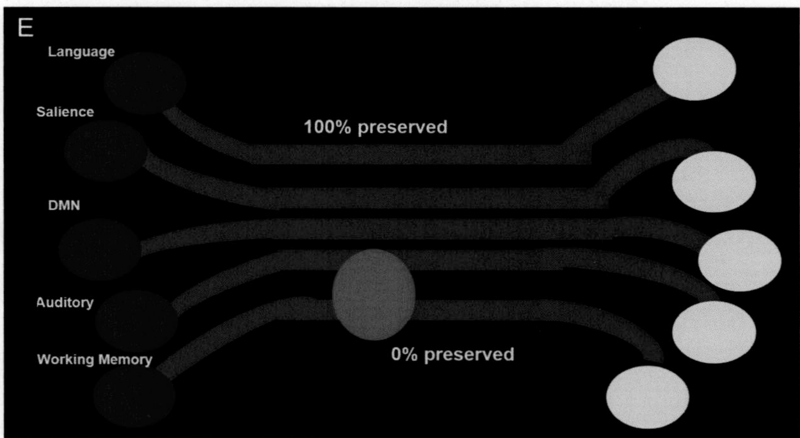

FIG. 1.1 Cont'd

Conclusions

In the upcoming chapters we dive into the various complexities of understanding the brain connectome, how to analyze these data, and how to target connectomic features to treat patients. We deconvolute these complexities into a digestible format, which will allow you to put this thinking in play with rigorous clinical practice with numerous therapies.

References

Conner, A. K., Briggs, R. G., Sali, G., Rahimi, M., Baker, C. M., Burks, J. D., … Sughrue, M. E. (2018). A connectomic Atlas of the human Cerebrum-chapter 13: Tractographic description of the inferior fronto-occipital fasciculus. *Oper Neurosurg (Hagerstown), 15*(Suppl._1), S436–s443. https://doi.org/10.1093/ons/opy267

Dadario, N. B., Brahimaj, B., Yeung, J., & Sughrue, M. E. (2021). Reducing the cognitive footprint of brain tumor surgery. *Frontiers in Neurology, 12*, 711646. https://doi.org/10.3389/fneur.2021.711646

Dadario, N. B., & Sughrue, M. E. (2022). Should neurosurgeons try to preserve non-traditional brain networks? A systematic review of the neuroscientific evidence. *Journal of Personalized Medicine, 12*(4). https://doi.org/10.3390/jpm12040587

Tonelli, M. R. (1998). The philosophical limits of evidence-based medicine. *Academic Medicine, 73*(12), 1234–1240. https://doi.org/10.1097/00001888-199812000-00011

Yeung, J. T., Taylor, H. M., Young, I. M., Nicholas, P. J., Doyen, S., & Sughrue, M. E. (2021). Unexpected hubness: A proof-of-concept study of the human connectome using pagerank centrality and implications for intracerebral neurosurgery. *Journal of Neuro-Oncology, 151*(2), 249–256. https://doi.org/10.1007/s11060-020-03659-6

Zhang, Z., Xu, X., & Ni, H. (2013). Small studies may overestimate the effect sizes in critical care meta-analyses: A meta-epidemiological study. *Critical Care, 17*(1), R2. https://doi.org/10.1186/cc11919

CHAPTER

How to map a connectome

2

What does the connectome approach add to previous neuroimaging techniques?

A connectome refers to the set of all structural connections of one brain, and all of the resulting functional interactions between these parts (Sporns, 2022). While this term can vary in the degree of focus, such as at the microscale referring to individual synapses and action potentials, in clinical practice this generally refers to white matter tracts and gross function at the level of cortical regions or gray matter nuclei. As such, throughout this chapter our use of the term "connectomics" refers to these large-scale white matter bundles and related functional connections between individual cortical regions.

It is worth emphasizing that while big data approaches to understanding the connectome, aka "connectomics," most commonly utilize advanced neuroimaging techniques; these terms are not synonyms and this distinction matters. This is important to note as neuroimaging studies provide large, complex, and rich datasets, which make them impossible to fully understand without computer assistance in noting what is relevant for a function, or how a patient differs from other people in a given domain (Doyen & Dadario, 2022; Van Horn & Toga, 2014).

Thus, a useful way to think about this is as follows: neuroimaging visualizes the connectome … connectomics is an approach to make that visualization meaningful and understandable (Fig. 2.1).

What does it mean to map the connectome?

In short, mapping a connectome requires two main steps. The first is creating an imaging dataset of structural and/or functional interactions between brain regions. This book focuses on diffusion tractography and resting state functional MRI as the ways to do this easily and in a scalable fashion. The second involves using some technique to interpret the data, and in the setting of clinical medicine, make this interpretation in a way that the information is easily linked to decision-making processes.

The latter of these two basic steps, interpretation, is a massive field of research which could dominate a much larger book if we focused on covering every approach published in this field (Doyen & Dadario, 2022). We focus in this chapter on briefly discussing what we think are the most useful ways to make the data make sense: anatomically specific clustering of the data, the use of prior information as a method for dimensionality reduction and avoidance of false discovery, and open box machine learning approaches to identify potential new ways to look at the data.

Connectomic Medicine. https://doi.org/10.1016/B978-0-443-19089-6.00007-0

FIGURE 2.1

With 100 billion neurons each having up to 15,000 connections with other neurons, how do we employ modern data analytics to make sense of these connections?

Stage 1: Methods for creating the maps

Diffusion tractography

Diffusion tractography (DTI) utilizes diffuse-weighted imaging (DWI) to determine directionality of white matter tracts in the brain (Alexander et al., 2007). It uses the basic principles of water molecule movement in DWI, commonly used for imaging cerebral ischemia, to construct white matter "highways" of the brain (Dadario & Sughrue, 2022). The principle of DTI is based on using various gradient strengths, otherwise known as B-values, in different vectors, known as B-vectors, to conjure information of directionality of white matter fibers (Soares et al., 2013). The number of directions combined with different gradient strengths are the characteristics of different diffusion protocols. Of course, the more variations in directions would equate to more information, but it comes at the tradeoff of imaging time. This would translate into fewer scans being done on the machine, making it less economically feasible for the radiology center. More raw data obtained from increased scanning directions would also necessitate more intense computing to deconvolute the crossing fibers, as will be discussed later on. A fine balance must be struck between the economy of scanning time and details one wishes to obtain in the images.

The acquisition of DTI can be commonly done on most 1.5T and 3T MRI machines. However, the issue that most users will encounter especially in the community and nonacademic settings is the processing portion of the imaging process. It is often useful to elicit the assistance of a dedicated neuroradiologist to help implement the imaging protocols. Following the acquisition, the user will then have to deal with imaging corrections. These imaging corrections include eddy current correction, distortion correction, and standard head motion correction. These can be broken down into three steps: (1) calculating the fractional anisotropy (FA) map, (2) modeling the diffusion tensors, and (3) performing tractography using the tensors (Fig. 2.1).

The simplest way to understand FA is to first know that Brownian motion for water molecules move in random directions equally (isotropically) unless there is physical impedance (restriction), such that the water molecules are forced to move in other ways (anisotropically). Anatomically and microscopically, anisotropy occurs in the form of axons, such that water molecules move in the direction of these cellular structures (Mori & Zhang, 2006). When one models these structures as an ellipsoid at every voxel (pixel resolution of the MRI) and groups them all together, one creates the FA map. The complexity occurs at the level of every voxel, which can be represented as a matrix of FA data in various sampled directions. The estimation of the FA at each voxel thus becomes the essence of DTI. One can imagine that the variation in computational method for FA mapping can "make or break" the interpretation of the raw DTI data.

What makes the processing of DTI imaging more complicated is the fact that each voxel contains multiple fibers that can be traveling in either the same, slightly different, or completely opposite directions. Most common methods assign the vector in a voxel by analysing the FA as the dominant eigenvector of that voxel. Once the vectors are modeled among the voxels, the vectors are then lined by into tracts to generate "tractography." This process may occur in two manners, probabilistic or deterministic (Soares et al., 2013). Probabilistic modeling starts by using a seed to model all possible streamlines and then weights them based on their probability of existence. Deterministic modeling lays down a seed and aligns the vectors based on that seed until a termination threshold is reached, either by defining the threshold tract angle or magnitude of change more than a threshold (Sarwar et al., 2019). Clinical-grade DTI most often employs deterministic modeling.

The benefits of crossing fiber tractography over DTI

One may already imagine that this strategy would not be good at resolving crossing fibers, present in more than 50% of voxels (Schilling et al., 2017), and inherently ignores smaller fibers that may be functionally important or tracts that run orthogonal to larger bundles (Fig. 2.2). Fig. 2.3 demonstrates various scenarios within a voxel that may occur. What also complicates matters in the real world is the presence of pathology that can alter water movements. Classic example is cerebral edema that may interfere with the mathematical process above, making false negative findings of absence tracts. This may result in an algorithm leading down a false path (Yeh et al., 2021). There are other alternative methods of modeling tractography. An example would be diffusion spectrum imaging (DSI) that creates a model with more directions and accounts for more eigenvectors per voxel (Wedeen et al., 2008). DSI is great as a research tool, but practically difficult to implement in clinical practice with imaging times well over an hour. Another example would be Q-ball imaging that resolves multiple intravoxel fiber orientations using high angular resolution diffusion imaging (Tuch, 2004). It is outside of the scope of this book to go into the various methods and attempts to overcome these barriers, but we will briefly describe the use of constrained spherical deconvolutions (CSD) as an alternative to typical DTI processing. We summarize these concepts in Fig. 2.4.

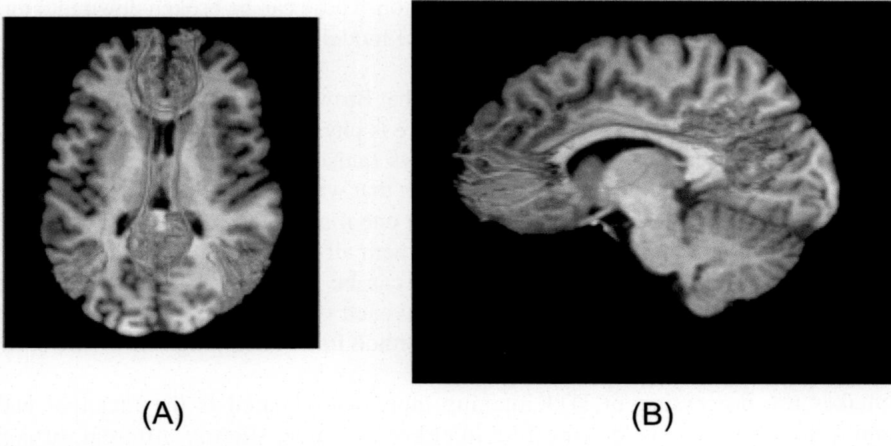

FIGURE 2.2

This schematic demonstrates various fiber tracts containing various fiber orientations which running rostral-caudal on axial (A) and (B) sagittal planes.

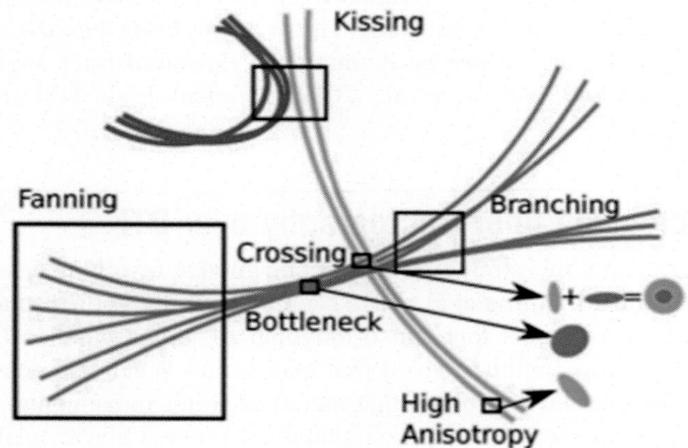

FIGURE 2.3

Various types of brain tract conflicts that need to be accounted for in postprocessing.

CSD fits the raw data using prestructured models of anisotropic ellipsoid (Morez et al., 2021). Fiber orientation distribution function (fODF) is an alternative model to DTI which accommodates the issue of "crossing fibers" that is supported by CSD. CSD calculates multiple tensors in heterogeneous regions and is thus able to support crossing fibers and areas of multiple fiber orientations. It captures the primary eigenvector signal that is generated in the DTI model, but can additionally differentiate the

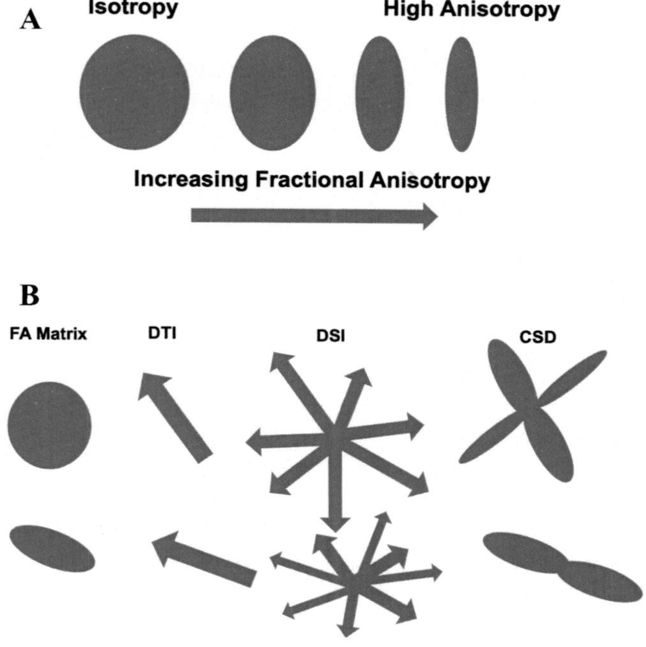

FIGURE 2.4

This schematic demonstrates in brief the steps of creating a diffusion tensor imaging (DTI) tractographic image. (A) In panel A we show the basic concept behind anisotropy as a matrix describing the basic shape of anisotropy. Although, this is a high-dimensional vector set. (B) In panel B we highlight the different methods of modeling a set of vectors from the FA matrix. Differences between DTI, DSI and CSD in handling these vectors are as follows: DTI models this as the dominant eigenvector, diffusion spectrum imaging (DSI) models multiple vectors, and constrained spherical deconvolution (CSD) uses a collection of predetermined ellipsoids to model the vector set.

presence of nonprimary crossing fibers. This enables CSD to accurately reconstruct white matter structures in regions such as the corticospinal tract.

The CSD algorithm calculates the white matter fODF using diffusion weighted imaging data. This process involves estimating the fiber response function and using the fiber response function in the context of constrained spherical deconvolution to determine the fODF. Once the fODF has been calculated, whole brain tractography can be conducted by pairing fiber direction data with a set of seed points to define start points and stopping criteria in the process offiber tracking. With these criteria and input parameters defined a deterministic fiber tracking algorithm can generate individual white matter fiber bundles called streamlines. The extension of individual streamlines to the entire brain is called whole-brain tractography.

Functional MRI

Functional MRI (fMRI) is a well-known technique to neurosurgeons and to research neuroscientists (Logothetis, 2003), but many do not fully comprehend its capabilities and limitations, especially for clinical application. This type of imaging measures blood oxygen level-dependent (BOLD) signal changes during the activation of a brain region as a result of vasodilation and the associated change in hemoglobin oxygen level. These signal changes are measured as differences from the sequential acquisition of an echo planar image (EPI).

The successful capture of fMRI images requires postprocessing that accounts for several factors (Glover, 2011). First, the patient's head motions need to be corrected and realigned in order for calculated differences within a voxel to be accurate during the time period of measurement. Second, cardiac and respiratory motion variations can produce noise in the BOLD signals and need to be accounted. Third, anatomic considerations, such as cerebrospinal fluid spaces and large vessel pulsations need to be corrected during the measured time periods. Similarly, the skull and meninges will need to be stripped from the processed images.

Most clinicians, especially neurosurgeons, are familiar with task-based fMRI for its ability to discern areas of the brain that are involved in motor and language functions (Fig. 2.5). This makes sense as task-based fMRI involves comparing the BOLD signal between a baseline condition and the execution of a task, such as moving one's hands or legs or performing language functions. This type of fMRI technique is limited by the need of an able and willing participant and a trained technician who can administer the task. These are reasons why task-based fMRIs are usually performed at resourceful academic centers. While task-based fMRIs are great for the localization of archetypal functions, such as limb movements, sensation, and language sidedness, it is limited in the study of more complex functions. How do we localize accurately the ability to empathize? How do we localize our ability to process highly complex emotions? What about our abilities to pay attention to active or inactive events or objects?

Resting-state fMRI (rs-fMRI) is different from its task-based counterpart in that it relies on large-scale brain networks to coactivate each other even during periods of rest, when we think that we are not

FIGURE 2.5

Task-based fMRI measures changes in BOLD signals upon the execution of a task, most commonly concerning language or sensorimotor functions.

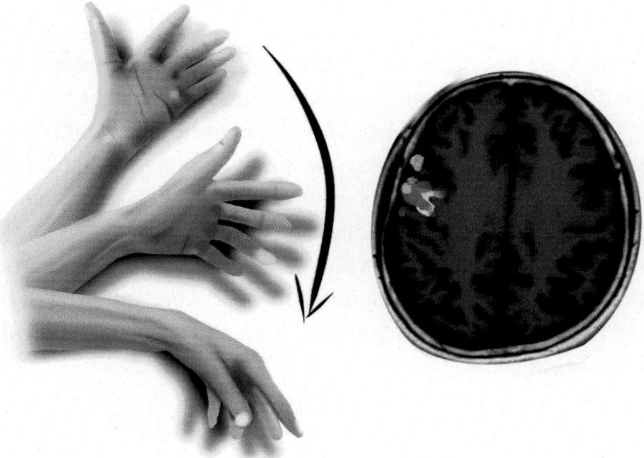

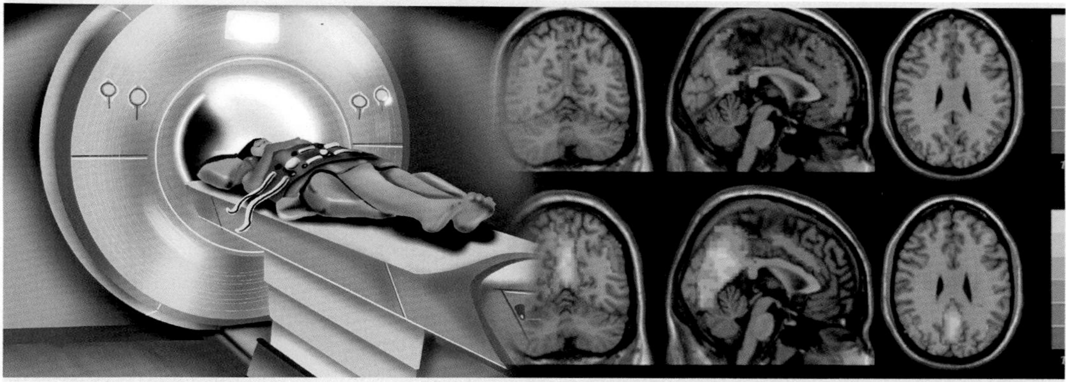

FIGURE 2.6

Resting-state fMRI is conducted for brain mapping to evaluate regional interactions that occur in a task-negative state, when a subject is not instructed to do anything in an awake state.

active (*despite this not entirely being completely accurate*) (Fig. 2.6). We can take advantage of established knowledge on the anatomical locations of large-scale brain networks to map them out simultaneously within 10 min of scanning time, thus highlighting the functional network architecture across the entire brain rather than one specific region with a single task as in task-based fMRI. The patient or subject is required to lie still in the MRI scanner, while avoiding falling asleep, with no technicians required, unlike task-based scans. In essence, this can be done at any community scanners. Once the rsfMRI is obtained, postimage processing is required and can be performed using two general methods. The first method is seed-based, where areas within the signal time course will be correlated with a specifically chosen region. While it is simple, it is subject to human bias and errors as the seed is manually placed. The second method is more complicated but can avoid human bias. This method utilizes independent component analysis, which takes into account the cross-talk among areas and identifies the one that behaves independently of other areas and statistically accounts for the greatest variance within the data (McKeown et al., 2003). The beauty of rs-fMRI is the ability to minimize resource utilization and the fact that we can leverage existing knowledge on large-scale brain networks that have been thoroughly studied for their involvement in complex brain functions that cannot be easily assessed during task-based fMRI.

Phase 2: Making the maps make sense

Parcellation: Both an alphabet for describing cortical anatomy and a method for dimensionality reduction

For decades, we have used gyral and sulcal anatomy as guideposts for determining where functions are located in the human brain. This has led to only minimal progress for two reasons: (1) the brain is not organized this way, and has important areas which cross sulcal boundaries and others which sit entirely within sulci, and (2) these designations are vague, hard to replicate, and hard to communicate to others. Nonreproducible nomenclatures are well-known recipes for failure in science.

Parcellating the cortex refers to making a division of the cortical surface into regions which serve different functional roles based on cytoarchitecture, imaging, or other bases. This provides a valuable way to simplify the complexity of the cortex to something understandable, and to provide a scalable system for communicating finding and applying them in a reproducible manner.

Clinical application of a parcellation scheme to a set of neuroimaging data is almost always done using computers, and a variety of approaches have been described (Eickhoff et al., 2018). Defining the functional organization of the human neocortex has long been the central focus of clinical neurology and neurosurgery, due to the importance of linking anatomic location to clinical observations, as well as creating strategies to avoid neurologic deficits during surgical procedures of the supratentorial brain. The traditional mechanism of defining regions/partitions (or parcellations, as they are called in the case of the Human Connectome Project atlas (HCP-MMP1)) in neuroimaging analyses is to align (or register) an atlas (a set of three-dimensional points or voxels assigning identity in a standard coordinate space to various parcellations) to the brain after warping it into a standard coordinate space, such as the Montreal Neurological Institute (MNI) space (Chau & McIntosh, 2005), or a surface-based coordinate system.

Limitations of traditional registration-based parcellation

While useful, it can be time-consuming and/or inaccurate due to topological variance (e.g., brain shape, presence of tumor tissue), and assumes that coordinates in space, not specific circuits in the brain, are the defining unit of organization of brain function. Pure anatomic-based techniques for atlas registration entirely underperform largely when applied to patients with structurally abnormal brains, such as those with brain tumors, stroke, hydrocephalus, traumatic brain injury, and atrophy. The most obvious reason is that the gyri and sulci in these cases are distorted and/or displaced.

Topography-based mapping

Topography-based mapping is a “top-down” approach that calibrates the HCP-MMP1 template based on the physical shape of a subject’s cortex formed by the attributes of their gyri and sulci. Because of the emphasis on topography, this approach has high gray-matter coverage, and is particularly suitable for fMRI analyses as it prioritizes the capture of relevant BOLD signal. However, as it relies on topographic data, it is also less adaptable to gross structural abnormalities. The topography-based map is most useful for analysing data derived from the cortex, such as fMRI BOLD signals, and delineating the boundaries of cortical parcels.

Connectivity-based mapping

Connectivity-based mapping is a “bottom-up” approach that uses a machine learning (ML) algorithm to define and calibrate the HCP-MMP1 template to a subject based on their subcortical white matter connectivity (i.e., tracts) (Doyen et al., 2022). The algorithm leverages the known structural aspects of the white matter to offer an additional source of information that is used to improve the localization and contouring of parcellations in the brain. A key advantage of this is that it refines the HCP-MMP1 atlas to the individual’s white matter structure. In addition, the use of white matter structure ensures that no parcellations are placed in the CSF or ventricles as there is no white matter in these areas. Because of the emphasis on connectivity, the connectivity-based map is able to adapt around structural abnormalities and robustly identify the nodal cores of each parcel.

Using prior information

One problem with using a data-driven approach to mapping the connectome in isolation is that it assumes we know nothing about the brain, its functions, or its diseases. Not only does this substantially increase statistical noisiness and reduce reproducibility, but it makes clinical correlation next to impossible and makes false discovery basically certain.

A simple statistical explanation highlights the latter point. If we search randomly in the structural and functional connectome for abnormal connectivity features to explain a given patient's function, we are guaranteed to find many false positives with $P < .05$. A simple calculation of the false discovery rate of looking at a structural and functional 379×379 connectivity matrix (basically the 360 HCP parcellations with 19 subcortical regions) would be expected to produce 7182 features with false positive $P = .05$ based on simple probability alone. By only looking at parts of the brain which are known to be linked to the networks or parts of the brain which relate to the symptom being explored, this issue drastically reduces the severity of this problem both by reducing the search space dramatically, and by improving the pretest probability and thus the posttest probability of a positive result. It turns out that AI-based tools still require us to take a careful history and think like doctors.

At this point, prior information can be brought into the interpretation of brain connectome data by linking published values from fMRI and other studies to specific brain regions in a parcellation scheme or a subcortical region (Dadario et al., 2021). This highlights the intense need for the field to converge on a single scheme for parcellating the cortex as opposed to everyone using a difference scheme of choice: quite simply put, there is no path toward helping patients with this research without a common language by which we teach doctors and communicate results. We and others have done a great deal of work with the HCP scheme and this seems like a solid option.

Using machine learning approaches

There is an entire chapter on this topic in this book, so we will be brief. Machine learning provides new insights into where to look for answers to our questions. The importance of this issue should be obvious if you read the previous section.

Conclusion

In this chapter we introduced a number of important concepts regarding the brain connectome, the methodological and mathematical approaches to highlighting these connections, and the various ways to begin to try to understand cerebral structure and function from connectomic-based mapping. It is clear that spatially distinct regions can be functionally connected to support similar specific functions, and that the large white matter bundles which exist and often connect these regions can be identified in a variety of ways, but with important considerations discussed in this chapter. In the next chapter, we focus on highlighting the specific anatomy with which these techniques can identify, specifically as it relates to, large-scale brain networks.

References

Alexander, A. L., Lee, J. E., Lazar, M., & Field, A. S. (2007). Diffusion tensor imaging of the brain. *Neurotherapeutics, 4*(3), 316–329. https://doi.org/10.1016/j.nurt.2007.05.011

Chau, W., & McIntosh, A. R. (2005). The Talairach coordinate of a point in the MNI space: How to interpret it. *NeuroImage, 25*(2), 408–416. https://doi.org/10.1016/j.neuroimage.2004.12.007

Dadario, N. B., Brahimaj, B., Yeung, J., & Sughrue, M. E. (2021). Reducing the cognitive footprint of brain tumor surgery. *Frontiers in Neurology, 1342*.

Dadario, N. B., & Sughrue, M. E. (2022). Advanced neuroimaging of the subcortical space: Connectomics in brain surgery. In *Subcortical neurosurgery* (pp. 29–47). Cham: Springer.

Doyen, S., & Dadario, N. B. (2022). 12 Plagues of AI in Healthcare: A practical guide to current issues with using machine learning in a medical context. *Frontiers in Digital Health, 4*.

Doyen, S., Nicholas, P., Poologaindran, A., Crawford, L., Young, I. M., Romero-Garcia, R., & Sughrue, M. E. (2022). Connectivity-based parcellation of normal and anatomically distorted human cerebral cortex. *Human Brain Mapping, 43*(4), 1358–1369. https://doi.org/10.1002/hbm.25728

Eickhoff, S. B., Yeo, B. T. T., & Genon, S. (2018). Imaging-based parcellations of the human brain. *Nature Reviews Neuroscience, 19*(11), 672–686. https://doi.org/10.1038/s41583-018-0071-7

Glover, G. H. (2011). Overview of functional magnetic resonance imaging. *Neurosurgery Clinics of North America, 22*(2), 133–139. https://doi.org/10.1016/j.nec.2010.11.001. vii.

Logothetis, N. K. (2003). The underpinnings of the BOLD functional magnetic resonance imaging signal. *Journal of Neuroscience, 23*(10), 3963. https://doi.org/10.1523/JNEUROSCI.23-10-03963.2003

McKeown, M. J., Hansen, L. K., & Sejnowsk, T. J. (2003). Independent component analysis of functional MRI: What is signal and what is noise? *Current Opinion in Neurobiology, 13*(5), 620–629. https://doi.org/10.1016/j.conb.2003.09.012

Morez, J., Sijbers, J., Vanhevel, F., & Jeurissen, B. (2021). Constrained spherical deconvolution of nonspherically sampled diffusion MRI data. *Human Brain Mapping, 42*(2), 521–538. https://doi.org/10.1002/hbm.25241

Mori, S., & Zhang, J. (2006). Principles of diffusion tensor imaging and its applications to basic neuroscience research. *Neuron, 51*(5), 527–539. https://doi.org/10.1016/j.neuron.2006.08.012

Sarwar, T., Ramamohanarao, K., & Zalesky, A. (2019). Mapping connectomes with diffusion MRI: Deterministic or probabilistic tractography? *Magnetic Resonance in Medicine, 81*(2), 1368–1384. https://doi.org/10.1002/mrm.27471

Schilling, K., Gao, Y., Janve, V., Stepniewska, I., Landman, B. A., & Anderson, A. W. (2017). Can increased spatial resolution solve the crossing fiber problem for diffusion MRI? *NMR in Biomedicine, 30*(12). https://doi.org/10.1002/nbm.3787

Soares, J. M., Marques, P., Alves, V., & Sousa, N. (2013). A hitchhiker's guide to diffusion tensor imaging. *Frontiers in Neuroscience, 7*, 31. https://doi.org/10.3389/fnins.2013.00031

Sporns, O. (2022). The complex brain: Connectivity, dynamics, information. *Trends in Cognitive Sciences, 26*(12), 1066–1067. https://doi.org/10.1016/j.tics.2022.08.002

Tuch, D. S. (2004). Q-ball imaging. *Magnetic Resonance in Medicine, 52*(6), 1358–1372. https://doi.org/10.1002/mrm.20279

Van Horn, J. D., & Toga, A. W. (2014). Human neuroimaging as a "big data" science. *Brain Imaging Behavior, 8*(2), 323–331. https://doi.org/10.1007/s11682-013-9255-y

Wedeen, V. J., Wang, R. P., Schmahmann, J. D., Benner, T., Tseng, W. Y., Dai, G., ... de Crespigny, A. J. (2008). Diffusion spectrum magnetic resonance imaging (DSI) tractography of crossing fibers. *NeuroImage, 41*(4), 1267–1277. https://doi.org/10.1016/j.neuroimage.2008.03.036

Yeh, F. C., Irimia, A., Bastos, D. C. A., & Golby, A. J. (2021). Tractography methods and findings in brain tumors and traumatic brain injury. *NeuroImage, 245*, 118651. https://doi.org/10.1016/j.neuroimage.2021.118651

CHAPTER

The anatomy of human brain networks

3

A better anatomically specific nomenclature for structure and function

This chapter outlines the anatomy of the human brain, focusing on the concept of large-scale brain networks. Brain networks are a useful way to describe cognitive brain function and they are a learnable way to get started doing connectome-based thinking without a tremendous investment in time. As described in later chapters, all aspects of brain function cannot be explained entirely by merely knowing the brain networks. However, what is helpful about brain networks is they provide a basic nomenclature for explaining the brain functions that we previously lack anatomic frameworks for intelligently localizing in the brain. When in doubt, thinking about patients in terms of a brain network or two is not a terrible idea.

When we first engaged with the medical literature on brain networks, two key problems immediately became obvious. The first was a lack of anatomic specificity: *papers often called something DLPFC or parietal lobe which are huge and complex areas*. Without anatomic detail, we are not doing medicine, we are doing something else entirely which does not require attention to detail. Second, as is common in many fields, there is a lack of clarity or consensus in taxonomy. We have worked hard to address these issues, and in this chapter, we hope to make brain network anatomy accessible and clear.

What do we mean by a brain network?

Neuroscience has repeatedly demonstrated, using a variety of methods, that parts of the brain which are not immediately adjacent to each other, have strongly correlated patterns of coactivation (Fig. 3.1). This does not mean that they are completely synchronous, this would imply they are possibly seizing. However, strongly correlated areas have been shown to share many common functions, and we and many others have shown that usually (but not always) these statistical relationships are accompanied by underlying structural connections. Together, these observations highlight the well-known idea: *regions that fire together are wired together.*

So, a brain network is a statistically defined set of brain regions whose activity patterns are highly similar. What may be obvious to you is that statistical arguments have some inherent challenges. The main one is that it is not clear how to define this term in a way which is consistent between people, and between time points. We would not expect the same person to be thinking the same set of thoughts or be in identical brain states if they hopped into the scanner two different times, let alone expect different people to do this. So where is the cut-off of *normal*?

Connectomic Medicine. https://doi.org/10.1016/B978-0-443-19089-6.00015-X

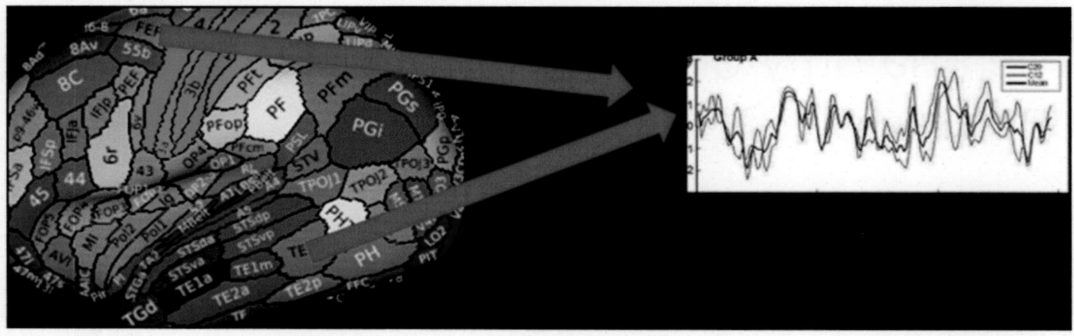

FIGURE 3.1

Spatially distant areas (left) which demonstrate similar patterns of co-activation (right) are functionally connected.

Interestingly, extensive studies of resting state imaging data, normally resting state fMRI, has shown us that despite these challenges, a few fortunate things are fairly clear. (1) Despite the fact that the functional relationships are not identical between scan time points, they are similar. (2) Individual patterns of network connectivity are fairly consistent across subjects. They are not identical, some enclaves of one network (Fig. 3.2) might lie within another in a way unique to that person, however, they are not random and these differences between connectivity patterns are modest and manageable. (3) We can define an expected distribution of connectivity features for a healthy population, and expect most people to have connectivity between areas which usually lie within this range. (4) Disease states cause connectivity features which lie outside this normal range, in other words, regional interactions in the brains of these patients are more or less correlated than seen in normal patients.

In other words, we can expect much of network function to be consistent in normal people, and disease states cause this network function to become abnormal, specifically in a way we can capture (Shahab et al., 2022). These findings are what makes connectomic medicine possible. If brain network function is random between people, not reproducible at all between time points, or the same between diseased brains and normal ones, then there would be no reason to use these techniques to study our patients (Dadario & Sughrue, 2022). While network functional abnormalities are not 100% sensitive or specific (*as nothing is*), they generally perform reasonably well if you have a solid basis on what they do and do not tell us, and you know where to look.

The basic anatomy of the human brain

We will not start out by describing the gyri and sulci as many books do. These structures, though visible to the naked eye, are mainly the result of physical processes created by tension bands around white matter connections as the cerebral cortex thickens during neuronal migration. It is well known that brain functions are only poorly referable to these areas, and thus we will not try to salvage this hopelessly flawed paradigm, and will instead discuss the brain the way the best available data suggest it is organized in the current era of thinking. Although, we will occasionally use gyral/sulcal nomenclature to simplify descriptions in the same way we use other terms like dorsal/ventral.

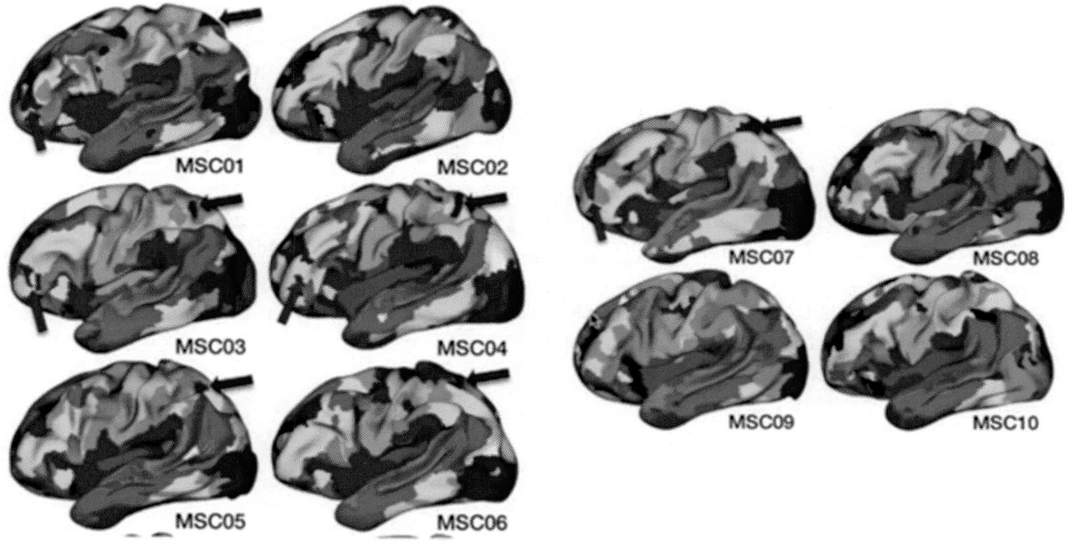

FIGURE 3.2

Unique differences in network connectivity between individuals. Despite most patterns of network connectivity being similar between most individuals, specific aspects of a network may have a unique orientation or architecture specific to that person, but often in a manageable and understandable way. This figure highlights these differences between individual subjects as described in an example from the literature.

Figure adapted from Gordon et al. (2017).

The cortical parcels

The fact that different regions of the cortical sheet do different things has been known for well over a century. Annotating the connectome from its raw data to an understandable representation of reality, must involve making these anatomical distinctions in a way which is neurologically meaningful and practically accessible. In other words, it is necessary to deconvolute the brain into regions and study the interaction of these parts. The process of dividing the cortical sheet into distinct areas based on some neurobiologically relevant criteria is called parcellation, and the regions which these criteria determine are distinct and cannot be meaningfully subdivided further are called parcels (Eickhoff et al., 2018). Thus, in the connectome, parcels play the role of the atom in elementary chemistry. Taking this analogy further, there are ways to subdivide both atoms and parcels in ways which make sense at smaller scales, for parcels this might involve smaller cortical columns or individual neurons, but this is awkward at the macroscale our applications work, and thus a parcel is the best unit of working for clinical medicine in our opinion.

People have been parcellating the brain for a long time. Brodmann used cytoarchitecture to outline his 47 areas. This was a landmark approach for the time, but we know his maps were not entirely accurate. It is not surprising that a work done on a single brain in 1908 might not have definitively settled the organization of the neocortex. Many efforts using other cytoarchitectural approaches followed in subsequent decades: some subdivided a Brodmann area, others defined boundaries differently. Still others used electrophysiologic recording data, usually from macaques, especially notable in defining visual areas and intraparietal sulcal regions.

More recent efforts have utilized neuroimaging (Glasser et al., 2016; Eickhoff et al., 2018). These approaches have the benefit of being based on some measure of functional difference between regions, which by definition has functional significance, as opposed to cytoarchitectural differences which can only be assumed to have a functional meaning. A good example of how these could differ is Brodmann area 8 (Dadario, Tanglay, & Sughrue, 2023). Despite the fact that the cytoarchitecture is similar throughout this strip of the posterior dorsal frontal cortex, different parts of this strip subserve different functions suggesting that while a similar calculation may occur at different parts of area 8, because these areas are wired differently, these calculation may occur in a different context, and thus lead to a different functional role.

We personally strongly recommend some familiarity with the Human Connectome Project (HCP) parcellation scheme (Fig. 3.3), though memorizing all areas from the start is not usually necessary (Glasser et al., 2016). However, the scheme was rigorously determined (the paper including

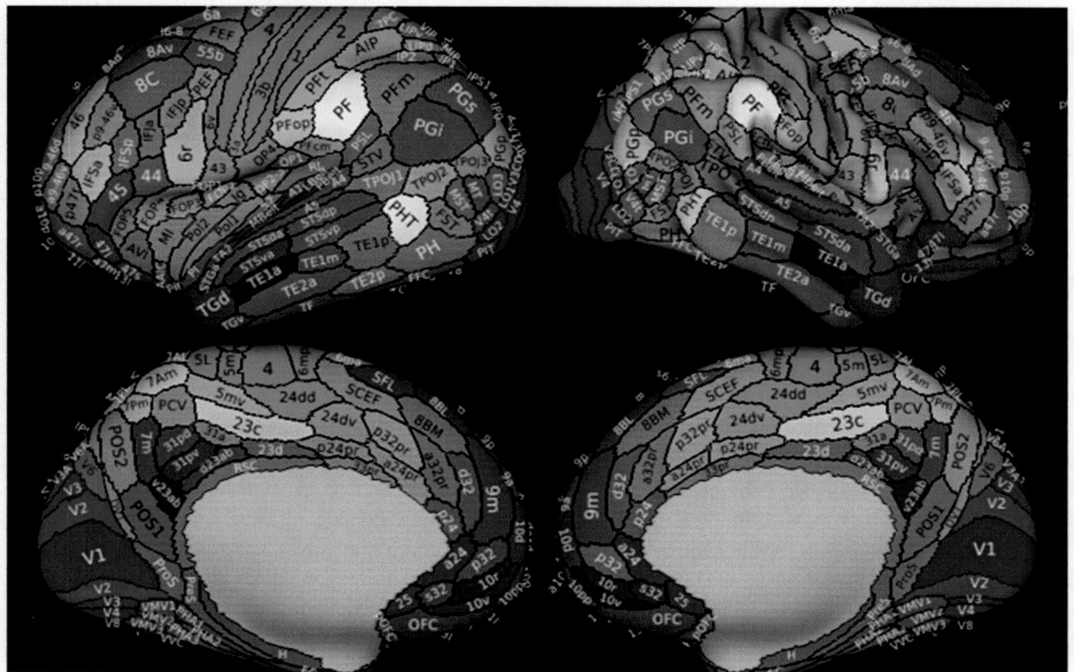

FIGURE 3.3

Human Connectome Project (HCP) atlas as described by Glasser et al. (2016) which has been updated here to include overlaid individual parcellations.

supplemental materials covers around 500 printed pages) and importantly, a deep effort to rectify numerous previous parcel nomenclature helps make it simple to learn as brain regions are named with atleast some learnable rationale for the names of regions. By using such a scheme, we can maximize the use of the existing neuroscientific knowledge base to infer some things about the brain, we can share results using a valid and reproducible alphabet, and as we describe in other places in this book, we can learn more from machine learning approaches when we use a scheme for dimensionality reduction which is not only based on scientific reality as opposed to statistical techniques alone, but also in our experience gets the modeling technique to a sweet spot balance between explainability and avoiding the curve of dimensionality (Dadario, Brahimaj, et al., 2021; Elam et al., 2021).

We have written an entire 500 page atlas, and dozens of published papers on the anatomic and functional nature of the regions in the HCP parcellation, if you wish to dive that deep into the minutia, they are available for you to read (Baker, Burks, Briggs, Conner, et al., 2018; Briggs et al., 2018). For all others, a few observations are worth noting. First, the brain does not care that gyri and sulci are easy for us to understand, as many areas are tucked into clefts of the insula, cross sulcal-gyral boundaries, or share a gyrus with other parcels. Function is located everywhere on the cortical sheet.

Second, Brodmann seems to have been basically right about a few areas: the sensorimotor cortices are basically the same, the lateral frontal lobe has many similar areas, such as areas 44, 45, and 46, as well as areas like 8, 9, 10, and 11 which are similar boundaries but made up of several distinct subdivisions. Other areas which basically are the same include the superior parietal lobule (areas 5 and 7 were subdivided not redrawn) (Lin et al., 2021), the precuneus (Dadario & Sughrue, 2023; Tanglay et al., 2022), and the cingulate gyrus (the latter two retaining areas 23, 24, 25, 31, and 31 in subdivided form). The temporal lobe regions are new however (often named with T parcel subdivided as TA, TE, TF, and TG subdivisions) (Briggs, Tanglay, et al., 2021), as are the inferior parietal lobule (named with P parcels, such as PF, PG, and PH subdivisions and IP subdivisions which refer to positions in the banks of the inferior parietal lobule which are often analogous to known subdivisions of this sulcus from macaque electrophysiologic studies), the occipital lobe (mostly named V subdivisions for hierarchically organized visual processing areas began by the Hubel and Weisel work), and insular cortex (with the undersurfaces of the opercular cortices given names with FOP and OP for subdivisions of the frontal and parietal opercula, respectively, or for their position on the insular cortex proper, such as anterior ventral insula (AVI), middle insula (MI), etc). This basic list does not cover every area obviously, but it should make this look less daunting.

Which brain networks should we study?

As with literally every topic we have ever seen in the literature that has not been codified in text book form yet, the literature on this uses different terms, different definitions, and is needlessly confusing if you are a clinician who just wants to have a list to learn. So this is your basic list.

As stated above, networks are defined as statistical relationships between activity patterns seen in different brain regions. In other words, areas that fire with other areas in that network more frequently than the rest of the brain. Basically, areas with strong functional connectivity. However, given that no two areas are 100% synchronized and no network is completely independent of the rest of the brain, it should be obvious that depending on the time point, an area will frequently be cross-communicating with other areas, and we know some parts of the network are more often doing this than others. Therefore depending on how stringent you set your statistical threshold, you will get a network containing different numbers of areas, or conversely, you will find there are different total numbers of networks in the brain. While the concept of a parcel being the minimum area below which further subdivisions do not improve meaningful explanatory power suggests that there is a maximum number of networks you can divide the brain into (most modern studies estimate the number of parcels at between 180 and 240 per hemisphere with fairly similarly sized cortical regions, suggesting that there is a limit at least when you are using neuroimaging to do this division). Ultimately, we know parcels do not work completely in isolation, and that some grouping of these into networks seems useful for some applications.

A highly cited study which aimed to determine this basic concept was published by Yeo and colleagues (Yeo et al., 2011). In this study, they took resting state fMRI data and aimed to cluster the brain voxels at an average person level into a specific number of networks. In other words, they assigned every voxel in the brain to a cluster of other regions that shared the closest brain activity patterns. They found that there are natural cut-points in the data suggesting solutions involving 7 or 17 networks fit the data best (something roughly similar has been noted by multiple other similar studies so this approach seems as reasonable as any other). A way to think of this distinction is as follows. There are around 17 main brain networks which are made up of around 360 total subunits called parcels. Some of these networks work largely independent of others while other networks are loosely more related to other networks, suggesting that there is a degree of hierarchy, with some of these sister networks likely driving the organization of other networks and their function. These sister networks thus can be combined to divide the brain into seven main axes of brain organization. Fig. 3.4 shows the basic organization of these hierarchies. As an example, a DMN subnetwork, is more similar in activity to other DMN subnetworks than it is to sensorimotor, visual, or other subnetworks, but its subunits (aka parcels) are more similar to each other in activity than they are to the other networks of the brain (aka, its more like a subnetwork of the DMN rather than the CEN).

The anatomy of the seven main brain network axes

Below, we characterize the seven main large-scale brain networks along with their relevant connectomic architecture. These networks are the Default Mode Network (DMN) (Andrews-Hanna et al., 2010; Sandhu et al., 2021), Central Executive Network (CEN) (Niendam et al., 2012), Salience Network (SN) (Briggs et al., 2022), Dorsal Attention Network (DAN) (Allan et al., 2019), Sensorimotor Network, Limbic Network, and Visual Network. Structural and functional diagrams of these large-scale brain networks are summarized in Fig. 3.5.

The cognitive control networks

Three canonical resting state networks which are helpful to first understand include the DMN, CEN, and SN. These networks are believed to likely sit on the top of the network hierarchy described above, and upon which the other networks align themselves to subserve complex human cognition

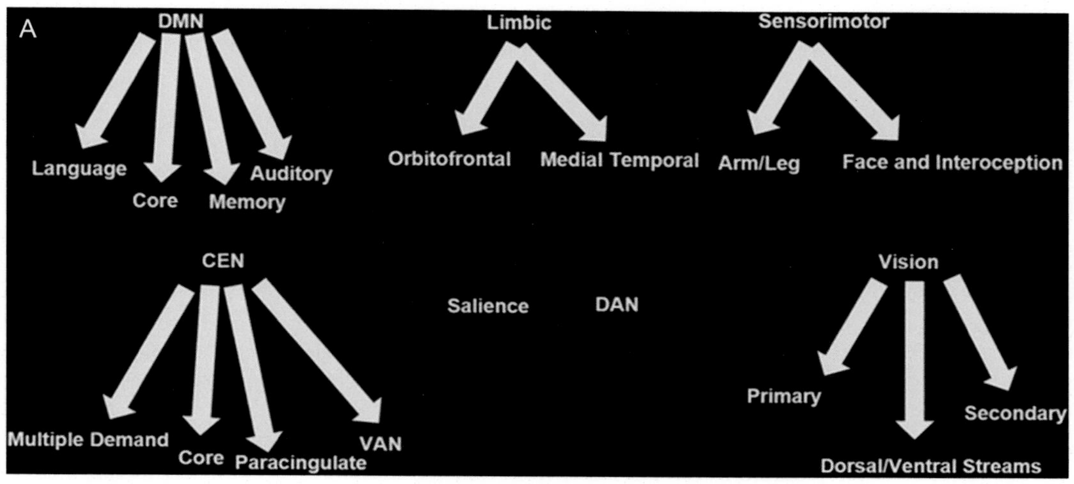

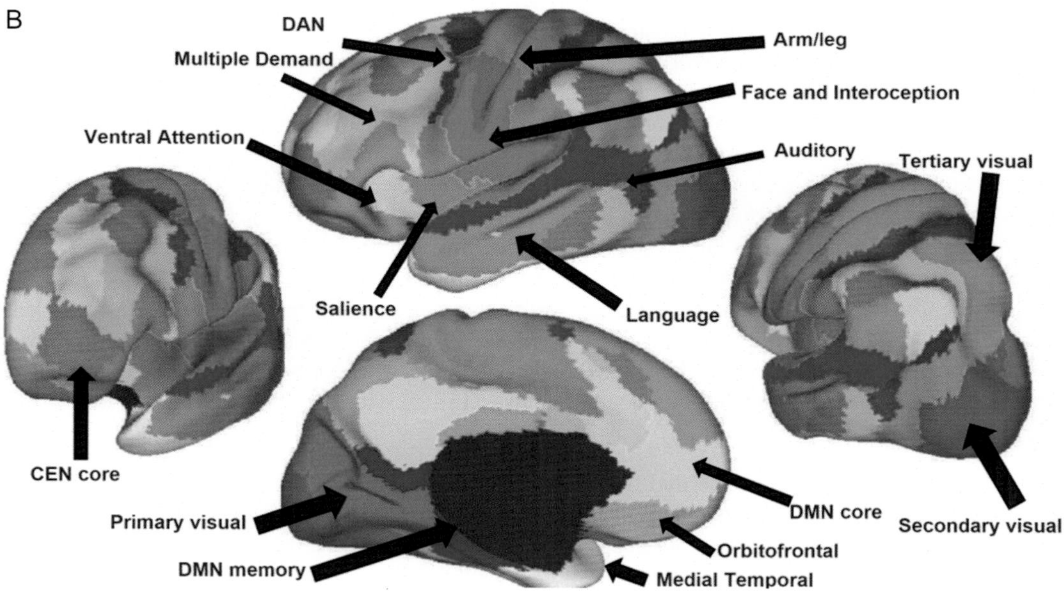

FIGURE 3.4

7 large-scale brain networks and their subnetwork hierarchy. Networks and subnetworks are shown in stick figure (A) and (B) superimposed on the Yeo et al. (2011) network model (Yeo et al., 2011). An additional recently discovered CEN subnetwork, which we coined "the para-cingulate network", is described elsewhere (Dadario & Sughrue, 2023) and not mentioned for simplicity.

and related functions (Dadario, Brahimaj, et al., 2021). Therefore, we term these three networks the "cognitive control networks." Damage to the cognitive control networks has been significantly represented in the literature. A recent review of the neuroscientific literature conducted by our team identified that damage or dysfunction of the nontraditional large-scale brain networks (e.g., not language, motor, or visual networks) almost always causes a neurological, cognitive, or emotional deficit. Most of these deficits (33%) are specifically the result of disruption in a combination of these three cognitive control networks. As such, we introduce the cognitive control networks first below.

DMN axis

The DMN is largely the most documented higher-order network in the literature. It is believed to represent the internal mind that is at work when an individual is at a resting state, not actively engaged in *externally* oriented tasks or attentional processing, but also not entirely "at rest" as once previously believed. During this time at rest, the DMN is not stagnant but its activity increases during internal thought and passive sensory processing (Raichle et al., 2001).

The core of the DMN includes three clusters which can be found in the posterior cingulate cortex, inferior parietal lobule (especially the angular gyrus) and the anterior cingulate and medial frontal regions (Mazoyer et al., 2001; Raichle & Snyder, 2007). Its main structural connections is the cingulum bundle which links these clusters.

There are three additional networks which loosely affiliate with the DMN. One we call the language system, as it contains Broca's area, the posterior temporal semantic areas include Wernicke's area, and other areas commonly linked to language function. This however is not a completely satisfying name, as it has a symmetrical right side area, and it includes areas which have not been strongly linked to the language system, like large parts of the medial superior frontal gyrus. But presently, we lack a better term to refer to this network. Also note, this is not a speech system, it is a cognitive system which controls language and is the cognitive control of speech motor function similar to praxis (O'Neal et al., 2021) and tool use is to hand motor function.

Additionally, the auditory system (specifically the later auditory processing areas like A4 which are higher areas) (note A1 and neighboring areas seem to be strongly connected to the face and interoception sensorimotor network which is interesting), and the memory network (which contains parts of isthmus of the cingulate gyrus and hippocampus) also affiliate with the DMN axis.

CEN axis

While the DMN represents the internal mind, the CEN is the external mind that is turned on during active tasks and external thinking involving working memory (Dosenbach et al., 2007). These two networks, CEN and DMN, work in anticorrelation, but the CEN works in correlation with the dorsal attention network (DAN) for attention processing, as well as visual spatial planning (Fox et al., 2005; Niendam et al., 2012).

In brief, the CEN consists of a three-part network with clusters in the anterior cingulate cortex, the inferior parietal lobe, and the posterior most portions of the middle and inferior temporal gyri (Boschin et al., 2015; Bilevicius et al., 2018; O'Neill et al., 2019). Structurally, its main fiber bundle is the superior longitudinal fasciculus (SLF), which is a major fiber bundle that links a number of other networks as well, which are involved in attentional processing (Briggs et al., 2018).

Compared to the DMN in large review studies, damage or dysfunction in the CEN was associated with additional motor and emotional deficits (Dadario & Sughrue, 2022). Furthermore, dysfunctional CEN connectivity with other networks, especially with the DMN, has been implicated in various psychiatric disorders, such as schizophrenia and posttraumatic stress disorder (Daniels et al., 2010; Dong et al., 2018).

The CEN axis has four affiliated networks. In brief, these include its core component, a multiple demand (MD) network, a novel para-cingulate network (Dadario & Sughrue, 2023), and the ventral attention network (VAN). The MD network is a diverse network of frontoparietal regions responsible actively for the organization and control of diverse cognitive operations (Cope et al., 2022). Specifically, MD encompasses aspects of the lateral prefrontal cortex (LPFC) in inferior fontal sulcus (IFS), the anterior insula/frontal operculum (AI/FO), the dorsal anterior cingulate/pre-supplementary motor area (ACC/pre-SMA), a small region in the anterior frontal cortex (AFC), and the intraparietal sulcus (IPS) (Duncan, 2010). Other studies have furthered this definition suggestion an extended multiple demand network demonstrated including the areas bilateral pre-SMA/MCC, aINS, and MFG/IFG for higher cognitive functions (Camilleri et al., 2018).

More recently, when parcellating the cerebral cortex into a finer network organization (e.g., from 7 networks to 17), an additional and less previously explained network seems to separate from the CEN, which can be seen flanking the cingulate cortex and core DMN (Dadario & Sughrue, 2023). We recently termed this network the para-cingulate network, which has likely significant roles in self-referential processes and related psychiatric symptoms based on available studies in this region. While not explicitly explained in great detail yet in the literature, it is important to note that there is a great deal of data converging on the presence of a para-cingulate network using a variety of analytical methods, including parcellation studies (Akiki & Abdallah, 2019), graph theoretical work (Power et al., 2011), and independent component analyses (Doucet et al., 2011; Power et al., 2011; Yeo et al., 2011). Anatomically, the para-cingulate network can be seen as a two-part system with its ventral component beginning at the isthmus of the cingulate gyrus bordering the parahippocampal gyrus and then extending to the middle cingulate gyrus, and its more superior-dorsal component at the opposite side of the core DMN beginning at the most superior portion of the precuneus and then extends in an inverse "V" fashion. This anatomy has been highlighted by the authors extensively, and is instrumental in understanding different precuneus subnetworks and related functions (Dadario & Sughrue, 2023).

The ventral attention network (VAN) is discussed later along with the dorsal attention network. In brief, it is a unilateral, right-hemispheric network (Allan et al., 2020), which reorients attention to behaviorally relevant stimuli.

SN axis

The SN constitutes a particularly important network. It is believed to serve as a key intermediary between the DMN and CEN through controlling the allocation of resources between these two networks (Menon, 2011). Specifically, the SN is thought to process external stimuli from the outside world and modulates how the different networks view the information (Menon & Uddin, 2010).

Anatomically, it includes an anterior insula-operculum cluster, a middle cingulate cluster, and the middle portion of the dorsolateral prefrontal cortex. The main fiber bundle linking the SN is the frontal aslant tract (FAT) (Briggs et al., 2022).

As the SN is in charge of processing of information from the external world, its hyperactivity can lead to neuroticism or anxiety (heightened sensitivity to outside stimuli) and hypoactivity can be a hallmark of autism (lack of sensitivity to social cues) (Fox et al., 2013; Uddin et al., 2013). More recently, the SN has become of popular interest in the neurosurgical community due to its major bundle the FAT, which has been recently characterized as a prominent feature implicated in supplementary motor area (SMA) syndrome (Dadario, Tabor, et al., 2021; Palmisciano et al., 2022). Damage to the FAT can cause SMA syndrome, while a network-based approach which spares these fibers results in significantly reduced SMA syndrome outcomes (Briggs, Allan, et al., 2021). Simultaneously, transcallosal FAT connections which link premotor regions and the SMA through the corpus callosum can facilitate recovery from SMA syndrome (Baker, Burks, Briggs, Smitherman, et al., 2018; Tuncer et al., 2022).

Traditional networks expanded

Visual axis

The visual system has been obviously well discussed in the traditional literature and is believed to be situated mostly in the occipital lobe, consisting of two major streams, dorsal and ventral. Connections with the parietal lobe in the dorsal stream facilitates the guidance of actions and recognition of objects in space, while the ventral stream is associated with object recognition and form representation (Dadario, Brahimaj, et al., 2021; Norman, 2002). It has strong connections to the medial temporal lobe via the basal, tentorial service (Baker, Burks, Briggs, Stafford, et al., 2018). We have previously shown in detailed anatomic studies that despite these many separate visual functions are thought to be housed separately in specific parcellations within this network, these two streams are interconnected via the vertical occipital fasciculus and may participate in more interconnected functions than previously understood (Baker, Burks, Briggs, Stafford, et al., 2018; Dadario, Tanglay, Stafford, et al., 2023; Palejwala et al., 2021).

Sensorimotor axis

Like the visual system, the sensorimotor network is likely the next or equally most studied network in history, along with the limbic network discussed next. Most classic works describe the sensorimotor network as including the primary motor cortex, cingulate cortex, premotor cortex, supplementary motor area, sensory cortices in the parietal lobe (Chenji et al., 2016). The sensorimotor network allows us to produce an appropriate motor response as a response to both external and/or internal stimuli. However, what has been improved with recent connectomic work is its dynamic interactions with other major networks to integrate relevant information for an appropriate response to execute. This may include transducing information related to the visual system for sight or perhaps information from the CEN for task processing.

Limbic axis

The limbic network is a multilobar network which was traditionally believed to have a primary role in controlling emotional processing. However, further study has improved our understanding of this region suggesting additional roles of wide scope, including social cognition, emotional learning and memory, and olfaction, among others (RajMohan & Mohandas, 2007; Rolls, 2019). It is an extremely widespread network including the amygdala, thalamus, hypothalamus, hippocampus, and many paralimbic structures (Koikegami et al., 1967). Its anatomic architecture basically includes the prefrontal-limbic system, anterior cingulate cortex, medial temporal network, parahippocampal gyrus, olfactory lobe, and the ventral tegmental area. Damage to this network, unsurprisingly, is linked to a variety of psychiatric disorders, including anxiety, bipolar disorder, schizophrenia, and autism (Bogerts et al., 1985; Haznedar et al., 2000; Cannistraro & Rauch, 2003; Brambilla et al., 2008; Dadario, Brahimaj, et al., 2021).

Dorsal attention network

The DAN is an important network responsible for goal-directed, top-down processing of an object and goal-related stimuli (Allan et al., 2019). In comparison to the ventral attention network (VAN), which reorients attention to behaviorally relevant stimuli, the DAN controls more voluntary attentional processing and works closely with the CEN to execute appropriate responses. It is a bilateral network located in the frontal eye fields, intraparietal sulcus, superior parietal lobe, and visual cortex. Similar to the CEN, it is mainly structurally connected by superior longitudinal fasciculus (SLF) fibers. Damage or dysfunction in the DAN can induce severe cognitive deficits, but has mostly been studied in the context of other networks like the CEN in unison (Dadario & Sughrue, 2022).

The connectomic initiation axis

An important point we would like to highlight is the importance of considering these networks as facilitating independent functions, but most commonly also through dynamic multi-network interactions to subserve complex human functioning. A particularly important example can be seen with the networks which underlie the functions of spontaneous, internally guided behavior. Multiple lines of evidence converge on a likely prefrontal cognitive initiation axis, where the DMN (linked by the cingulum) and the SN (linked by the FAT) form a strip across the medial frontal lobe extending up until the SMA (Baker, Burks, Briggs, Stafford, et al., 2018; Darby et al., 2018; Briggs, Allan, et al., 2021; Dadario, Brahimaj, et al., 2021). Through the interactions of these networks, the initiation of spontaneous, internally motivated actions may be facilitated. In fact, the salience network and pre-SMA even share a parcellation, area SCEF, thus presenting an likely important region for the outflow of information from higher order networks to the motor network in order to execute decisions accordingly based on internal motivations aggregated from saliency processing by the SN network. Simultaneously, damage to this axis may result in abulia and akinetic mutism (Briggs, Allan, et al., 2021; Darby et al., 2018).

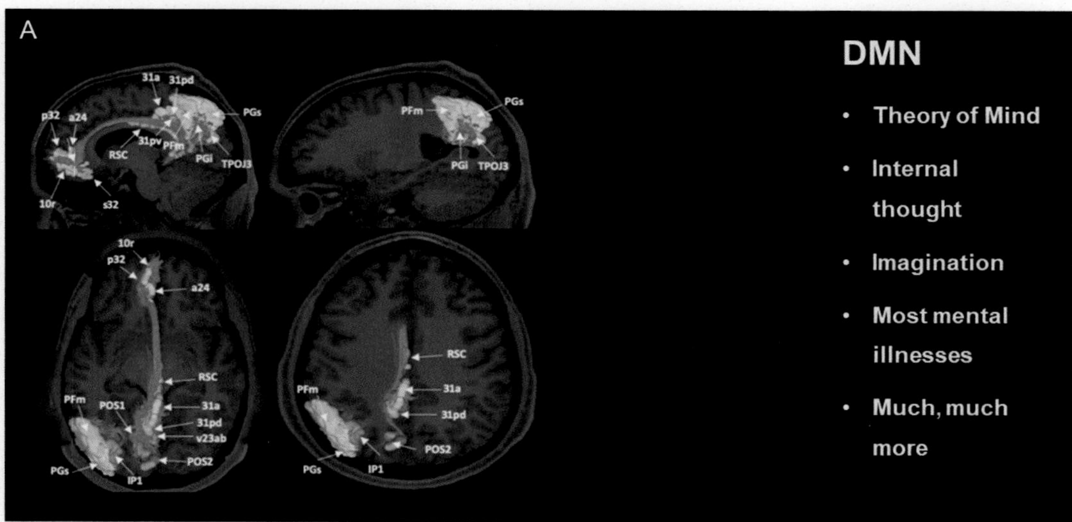

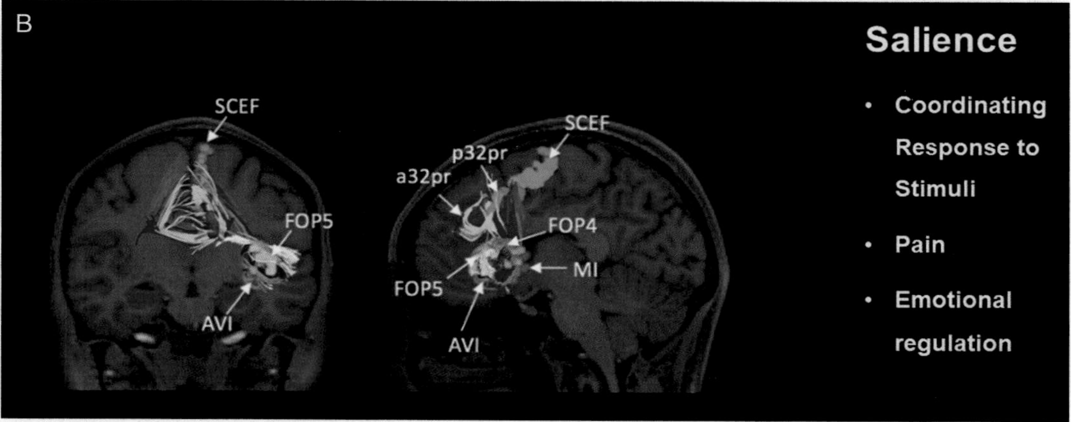

FIGURE 3.5

The functions of each network are listed along with key regions in panels A–K.

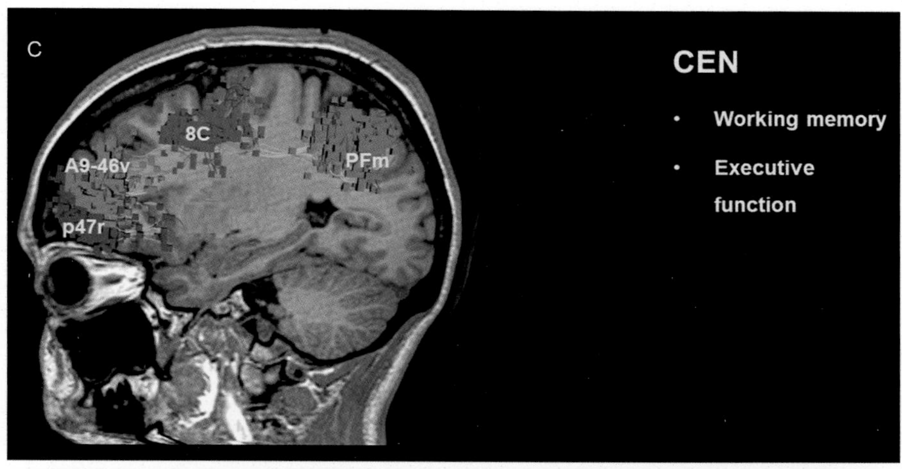

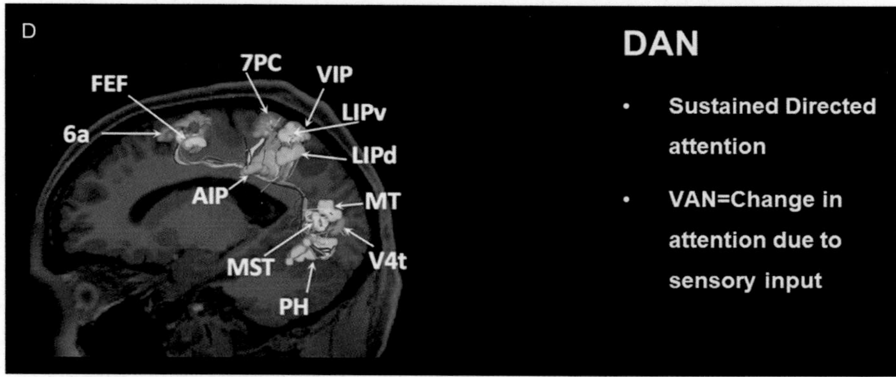

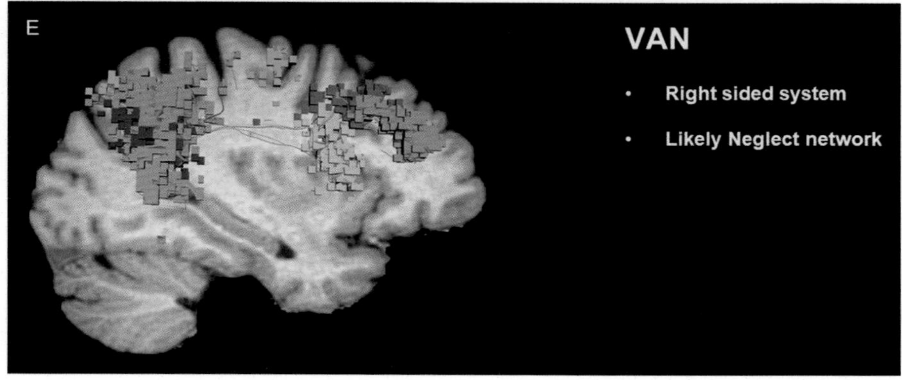

FIGURE 3.5 Cont'd

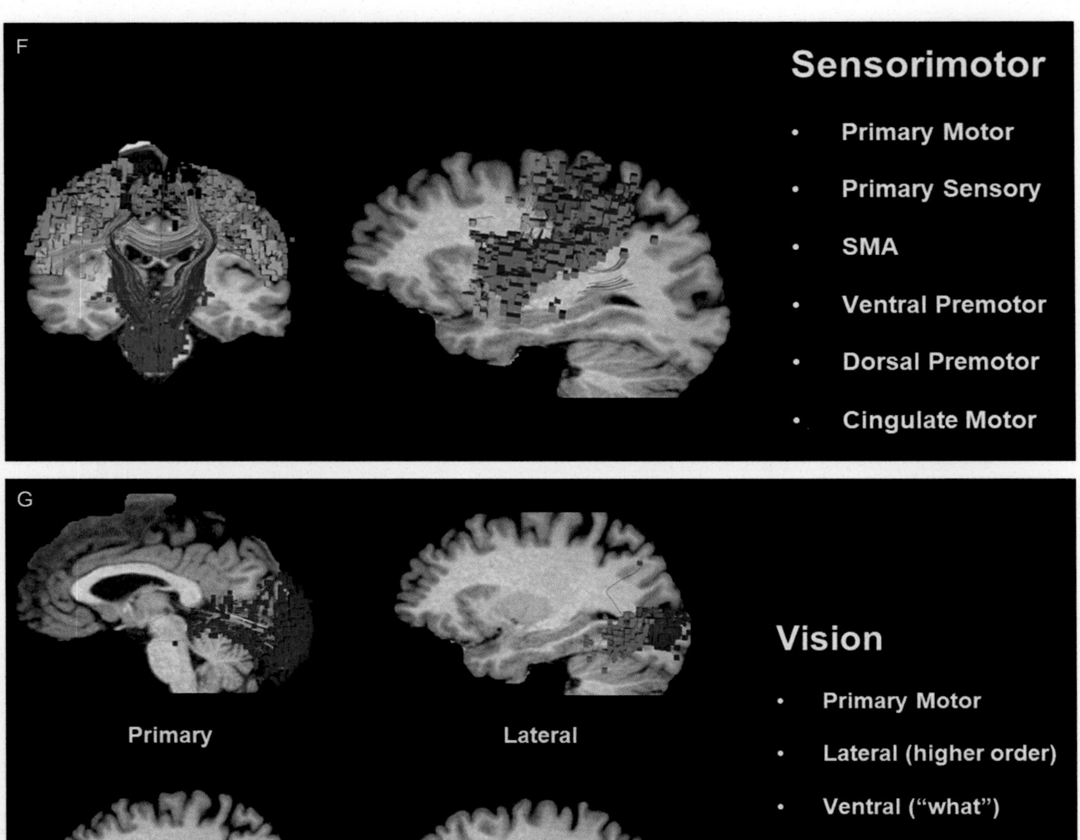

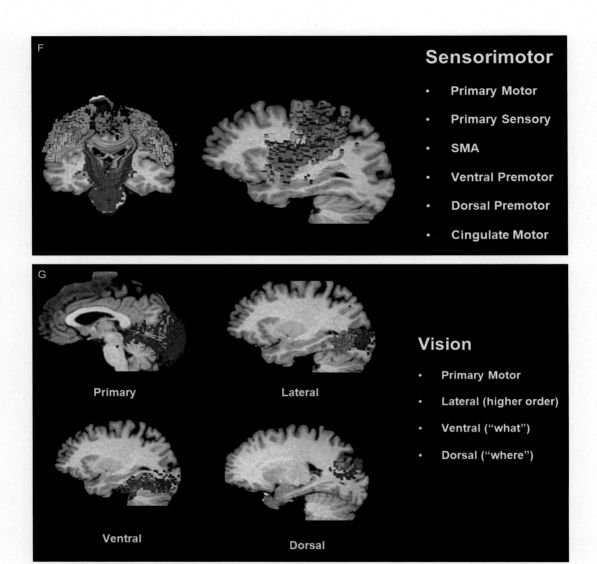

FIGURE 3.5 Cont'd

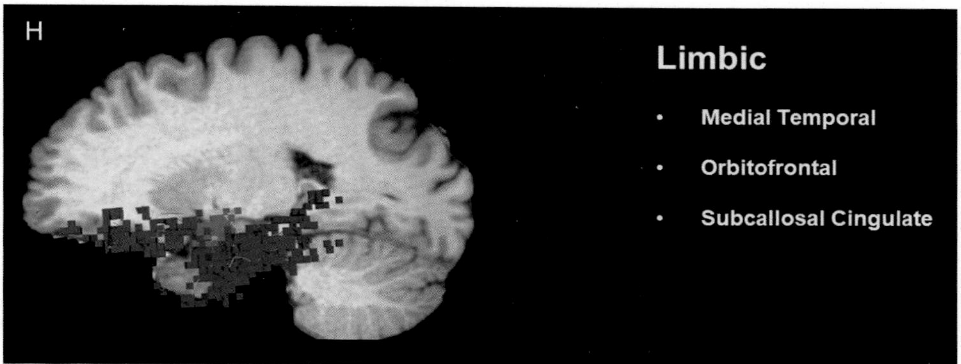

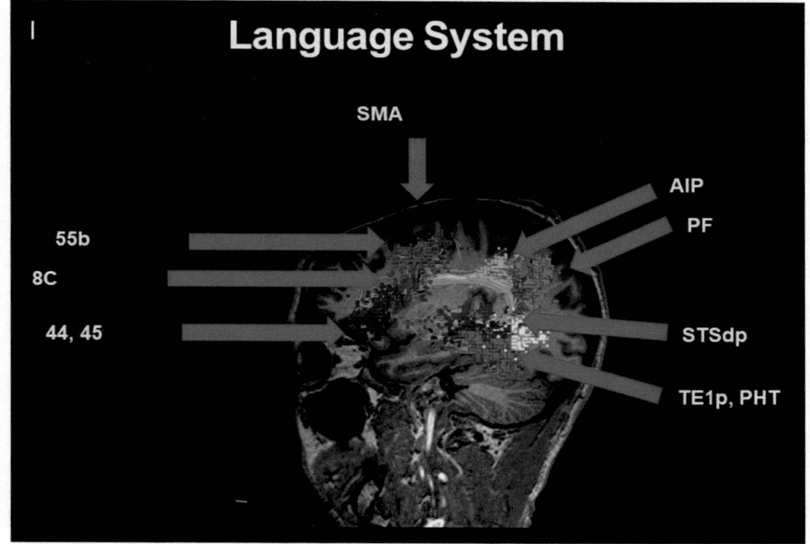

FIGURE 3.5 Cont'd

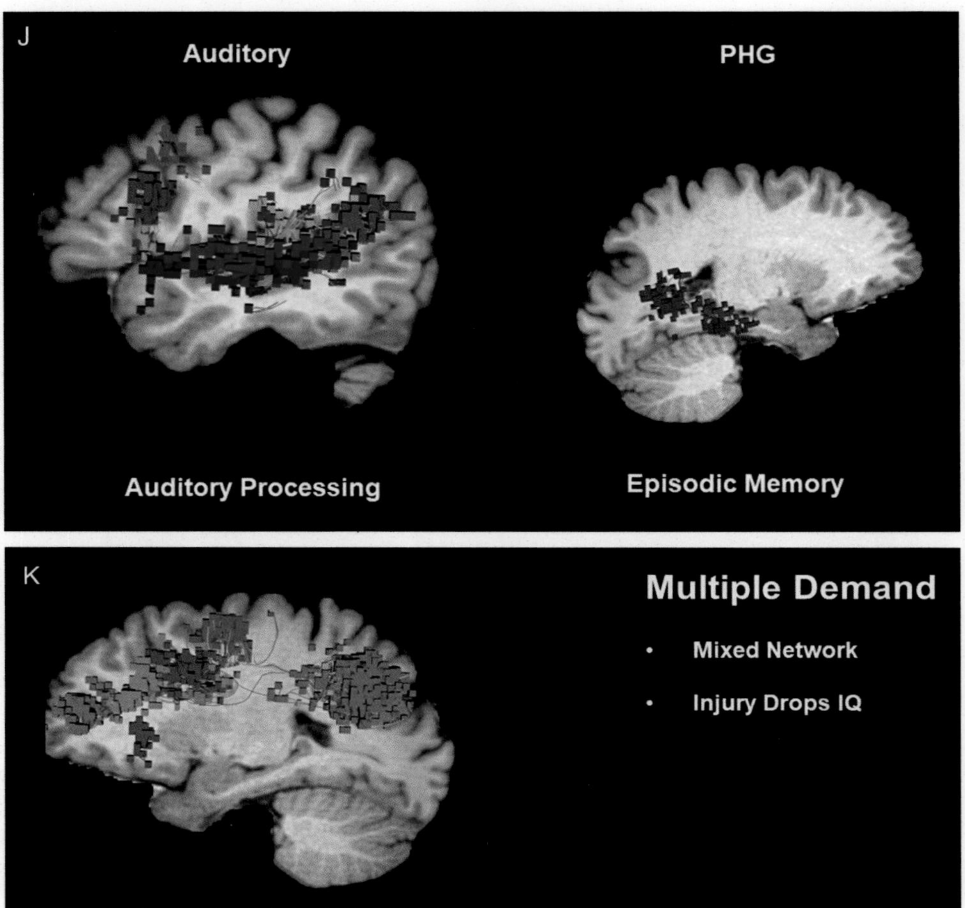

FIGURE 3.5 Cont'd

Conclusion

In this chapter, we introduce an incredibly important concept and improvement in thinking regarding the structure and function of the human cerebrum—that is, *large-scale brain networks*. Specifically, we characterized the seven main large-scale brain networks along with their relevant connectomic architecture and subnetworks, including the Default Mode Network (DMN), Central Executive Network (CEN), Salience Network (SN), Dorsal Attention Network (DAN), Sensorimotor Network, Limbic Network, and Visual Network. In the next chapter, we introduce how this anatomy relates to the traditional Brodmann's atlas.

References

Akiki, T. J., & Abdallah, C. G. (2019). Determining the hierarchical architecture of the human brain using subject-level clustering of functional networks. *Scientific Reports, 9*(1), 19290. https://doi.org/10.1038/s41598-019-55738-y

Allan, P. G., Briggs, R. G., Conner, A. K., O'Neal, C. M., Bonney, P. A., Maxwell, B. D., et al. (2019). Parcellation-based tractographic modeling of the dorsal attention network. *Brain and Behavior, 9*(10), e01365. https://doi.org/10.1002/brb3.1365

Allan, P. G., Briggs, R. G., Conner, A. K., O'Neal, C. M., Bonney, P. A., Maxwell, B. D., et al. (2020). Parcellation-based tractographic modeling of the ventral attention network. *Journal of the Neurological Sciences, 408*, 116548. https://doi.org/10.1016/j.jns.2019.116548

Andrews-Hanna, J. R., Reidler, J. S., Sepulcre, J., Poulin, R., & Buckner, R. L. (2010). Functional-anatomic fractionation of the brain's default network. *Neuron, 65*(4), 550–562. https://doi.org/10.1016/j.neuron.2010.02.005

Baker, C. M., Burks, J. D., Briggs, R. G., Conner, A. K., Glenn, C. A., Sali, G., et al. (2018). A connectomic atlas of the human cerebrum-chapter 1: Introduction, methods, and significance. *Operative Neurosurgery (Hagerstown), 15*(Suppl. l_1), S1–s9. https://doi.org/10.1093/ons/opy253

Baker, C. M., Burks, J. D., Briggs, R. G., Smitherman, A. D., Glenn, C. A., Conner, A. K., et al. (2018). The crossed frontal aslant tract: A possible pathway involved in the recovery of supplementary motor area syndrome. *Brain and Behavior, 8*(3), e00926. https://doi.org/10.1002/brb3.926

Baker, C. M., Burks, J. D., Briggs, R. G., Stafford, J., Conner, A. K., Glenn, C. A., et al. (2018). A connectomic atlas of the human cerebrum-chapter 9: The occipital lobe. *Operative Neurosurgery (Hagerstown), 15*(Suppl. l_1), S372–s406. https://doi.org/10.1093/ons/opy263

Bilevicius, E., Kolesar, T. A., Smith, S. D., Trapnell, P. D., & Kornelsen, J. (2018). Trait emotional empathy and resting state functional connectivity in default mode, salience, and central executive networks. *Brain Sciences, 8*(7). https://doi.org/10.3390/brainsci8070128

Bogerts, B., Meertz, E., & Schönfeldt-Bausch, R. (1985). Basal ganglia and limbic system pathology in schizophrenia. A morphometric study of brain volume and shrinkage. *Archives of General Psychiatry, 42*(8), 784–791. https://doi.org/10.1001/archpsyc.1985.01790310046006

Boschin, E. A., Piekema, C., & Buckley, M. J. (2015). Essential functions of primate frontopolar cortex in cognition. *Proceedings of the National Academy of Sciences of the U S A, 112*(9), E1020–E1027. https://doi.org/10.1073/pnas.1419649112

Brambilla, P., Hatch, J. P., & Soares, J. C. (2008). Limbic changes identified by imaging in bipolar patients. *Current Psychiatry Reports, 10*(6), 505–509. https://doi.org/10.1007/s11920-008-0080-8

Briggs, R. G., Allan, P. G., Poologaindran, A., Dadario, N. B., Young, I. M., Ahsan, S. A., et al. (2021). The frontal aslant tract and supplementary motor area syndrome: Moving towards a connectomic initiation Axis. *Cancers, 13*(5). https://doi.org/10.3390/cancers13051116

Briggs, R. G., Conner, A. K., Baker, C. M., Burks, J. D., Glenn, C. A., Sali, G., et al. (2018). A connectomic atlas of the human cerebrum-chapter 18: The connectional anatomy of human brain networks. *Operative Neurosurgery (Hagerstown), 15*(Suppl. l_1), S470–s480. https://doi.org/10.1093/ons/opy272

Briggs, R. G., Tanglay, O., Dadario, N. B., Young, I. M., Fonseka, R. D., Hormovas, J., et al. (2021). The unique fiber anatomy of middle temporal gyrus default mode connectivity. *Operative Neurosurgery, 21*(1), E8–E14.

Briggs, R. G., Young, I. M., Dadario, N. B., Fonseka, R. D., Hormovas, J., Allan, P., et al. (2022). Parcellation-based tractographic modeling of the salience network through meta-analysis. *Brain and Behavior, 12*(7), e2646. https://doi.org/10.1002/brb3.2646

Camilleri, J. A., Müller, V. I., Fox, P., Laird, A. R., Hoffstaedter, F., Kalenscher, T., et al. (2018). Definition and characterization of an extended multiple-demand network. *NeuroImage, 165*, 138–147. https://doi.org/10.1016/j.neuroimage.2017.10.020

Cannistraro, P. A., & Rauch, S. L. (2003). Neural circuitry of anxiety: Evidence from structural and functional neuroimaging studies. *Psychopharmacology Bulletin, 37*(4), 8–25.

Chenji, S., Jha, S., Lee, D., Brown, M., Seres, P., Mah, D., et al. (2016). Investigating default mode and sensorimotor network connectivity in amyotrophic lateral sclerosis. *PLoS One, 11*(6), e0157443. https://doi.org/10.1371/journal.pone.0157443

Cope, T. E., Hughes, L. E., Phillips, H. N., Adams, N. E., Jafarian, A., Nesbitt, D., et al. (2022). Causal evidence for the multiple demand network in change detection: Auditory mismatch magnetoencephalography across focal neurodegenerative diseases. *Journal of Neuroscience, 42*(15), 3197. https://doi.org/10.1523/JNEUROSCI.1622-21.2022

Dadario, N. B., Brahimaj, B., Yeung, J., & Sughrue, M. E. (2021). Reducing the cognitive footprint of brain tumor surgery. *Frontiers in Neurology, 12*, 711646. https://doi.org/10.3389/fneur.2021.711646

Dadario, N. B., & Sughrue, M. E. (2022). Should neurosurgeons try to preserve non-traditional brain networks? A systematic review of the neuroscientific evidence. *Journal of Personalized Medicine, 12*(4), 587.

Dadario, N. B., & Sughrue, M. E. (2023). The functional role of the precuneus. *Brain: A Journal of Neurology.* awad181. Advance online publication https://doi.org/10.1093/brain/awad181.

Dadario, N. B., Tabor, J. K., Silverstein, J., Sun, X. R., & RS, D. A. (2021). Postoperative focal lower extremity supplementary motor area syndrome: Case report and review of the literature. *The Neurodiagnostic Journal, 61*(4), 169–185. https://doi.org/10.1080/21646821.2021.1991716

Dadario, N. B., Tanglay, O., Stafford, J. F., Davis, E. J., Young, I. M., Fonseka, R. D., et al. (2023). Topology of the lateral visual system: The fundus of the superior temporal sulcus and parietal area H connect nonvisual cerebrum to the lateral occipital lobe. *Brain and Behavior, 13*(4), e2945. https://doi.org/10.1002/brb3.2945

Dadario, N. B., Tanglay, O., & Sughrue, M. E. (2023). Deconvoluting human Brodmann area 8 based on its unique structural and functional connectivity. *Frontiers in Neuroanatomy, 17*, 1127143. https://doi.org/10.3389/fnana.2023.1127143

Daniels, J. K., McFarlane, A. C., Bluhm, R. L., Moores, K. A., Clark, C. R., Shaw, M. E., et al. (2010). Switching between executive and default mode networks in posttraumatic stress disorder: Alterations in functional connectivity. *Journal of Psychiatry & Neuroscience, 35*(4), 258–266. https://doi.org/10.1503/jpn.090175

Darby, R. R., Joutsa, J., Burke, M. J., & Fox, M. D. (2018). Lesion network localization of free will. *Proceedings of the National Academy of Sciences, 115*(42), 10792–10797. https://doi.org/10.1073/pnas.1814117115

Dong, D., Wang, Y., Chang, X., Luo, C., & Yao, D. (2018). Dysfunction of large-scale brain networks in schizophrenia: A meta-analysis of resting-state functional connectivity. *Schizophrenia Bulletin, 44*(1), 168–181. https://doi.org/10.1093/schbul/sbx034

Dosenbach, N. U., Fair, D. A., Miezin, F. M., Cohen, A. L., Wenger, K. K., Dosenbach, R. A., et al. (2007). Distinct brain networks for adaptive and stable task control in humans. *Proceedings of the National Academy of Sciences of the U S A, 104*(26), 11073–11078. https://doi.org/10.1073/pnas.0704320104

Doucet, G., Naveau, M., Petit, L., Delcroix, N., Zago, L., Crivello, F., et al. (2011). Brain activity at rest: A multiscale hierarchical functional organization. *Journal of Neurophysiology, 105*(6), 2753–2763. https://doi.org/10.1152/jn.00895.2010

Duncan, J. (2010). The multiple-demand (MD) system of the primate brain: Mental programs for intelligent behaviour. *Trends in Cognitive Sciences, 14*(4), 172–179. https://doi.org/10.1016/j.tics.2010.01.004

Eickhoff, S. B., Yeo, B. T. T., & Genon, S. (2018). Imaging-based parcellations of the human brain. *Nature Reviews Neuroscience, 19*(11), 672–686. https://doi.org/10.1038/s41583-018-0071-7

Elam, J. S., Glasser, M. F., Harms, M. P., Sotiropoulos, S. N., Andersson, J. L. R., Burgess, G. C., et al. (2021). The human connectome project: A retrospective. *NeuroImage, 244*, 118543. https://doi.org/10.1016/j.neuroimage.2021.118543

Fox, M. D., Liu, H., & Pascual-Leone, A. (2013). Identification of reproducible individualized targets for treatment of depression with TMS based on intrinsic connectivity. *NeuroImage, 66*, 151–160. https://doi.org/10.1016/j.neuroimage.2012.10.082

Fox, M. D., Snyder, A. Z., Vincent, J. L., Corbetta, M., Van Essen, D. C., & Raichle, M. E. (2005). The human brain is intrinsically organized into dynamic, anticorrelated functional networks. *Proceedings of the National Academy of Sciences of the U S A, 102*(27), 9673–9678. https://doi.org/10.1073/pnas.0504136102

Glasser, M. F., Coalson, T. S., Robinson, E. C., Hacker, C. D., Harwell, J., Yacoub, E., et al. (2016). A multi-modal parcellation of human cerebral cortex. *Nature, 536*(7615), 171–178. https://doi.org/10.1038/nature18933

Gordon, E. M., Laumann, T. O., Gilmore, A. W., Newbold, D. J., Greene, D. J., Berg, J. J., et al. (2017). Precision functional mapping of individual human brains. *Neuron, 95*(4), 791–807. https://doi.org/10.1016/j.neuron.2017.07.011. e797.

Haznedar, M. M., Buchsbaum, M. S., Wei, T. C., Hof, P. R., Cartwright, C., Bienstock, C. A., et al. (2000). Limbic circuitry in patients with autism spectrum disorders studied with positron emission tomography and magnetic resonance imaging. *American Journal of Psychiatry, 157*(12), 1994–2001. https://doi.org/10.1176/appi.ajp.157.12.1994

Koikegami, H., Hirata, Y., & Oguma, J. (1967). Studies on the paralimbic brain structures. *Psychiatry and Clinical Neurosciences, 21*(3), 151–180. https://doi.org/10.1111/j.1440-1819.1967.tb01290.x

Lin, Y.-H., Dadario, N. B., Hormovas, J., Young, I. M., Briggs, R. G., MacKenzie, A. E., et al. (2021). Anatomy and white matter connections of the superior parietal lobule. *Operative Neurosurgery, 21*(3), E199–E214.

Mazoyer, B., Zago, L., Mellet, E., Bricogne, S., Etard, O., Houdé, O., et al. (2001). Cortical networks for working memory and executive functions sustain the conscious resting state in man. *Brain Research Bulletin, 54*(3), 287–298. https://doi.org/10.1016/s0361-9230(00)00437-8

Menon, V. (2011). Large-scale brain networks and psychopathology: A unifying triple network model. *Trends in Cognitive Sciences, 15*(10), 483–506. https://doi.org/10.1016/j.tics.2011.08.003

Menon, V., & Uddin, L. Q. (2010). Saliency, switching, attention and control: A network model of insula function. *Brain Structure and Function, 214*(5–6), 655–667. https://doi.org/10.1007/s00429-010-0262-0

Niendam, T. A., Laird, A. R., Ray, K. L., Dean, Y. M., Glahn, D. C., & Carter, C. S. (2012). Meta-analytic evidence for a superordinate cognitive control network subserving diverse executive functions. *Cognitive, Affective, & Behavioral Neuroscience, 12*(2), 241–268. https://doi.org/10.3758/s13415-011-0083-5

Norman, J. (2002). Two visual systems and two theories of perception: An attempt to reconcile the constructivist and ecological approaches. *Behavioral and Brain Sciences, 25*(1), 73–96. https://doi.org/10.1017/s0140525x0200002x. discussion 96-144.

O'Neal, C. M., Ahsan, S. A., Dadario, N. B., Fonseka, R. D., Young, I. M., Parker, A., et al. (2021). A connectivity model of the anatomic substrates underlying ideomotor apraxia: A meta-analysis of functional neuroimaging studies. *Clinical Neurology and Neurosurgery, 207*, 106765. https://doi.org/10.1016/j.clineuro.2021.106765

O'Neill, A., Mechelli, A., & Bhattacharyya, S. (2019). Dysconnectivity of large-scale functional networks in early psychosis: A meta-analysis. *Schizophrenia Bulletin, 45*(3), 579–590. https://doi.org/10.1093/schbul/sby094

Palejwala, A. H., Dadario, N. B., Young, I. M., O'Connor, K., Briggs, R. G., Conner, A. K., et al. (2021). Anatomy and white matter connections of the lingual gyrus and cuneus. *World Neurosurgery, 151*, e426–e437.

Palmisciano, P., Haider, A. S., Balasubramanian, K., Dadario, N. B., Robertson, F. C., Silverstein, J. W., et al. (2022). Supplementary motor area syndrome after brain tumor surgery: A systematic review. *World Neurosurg, 165*, 160–171. https://doi.org/10.1016/j.wneu.2022.06.080. e162.

Power, Jonathan D., Cohen, Alexander L., Nelson, Steven M., Wig, Gagan S., Barnes, Kelly A., Church, Jessica A., et al. (2011). Functional network organization of the human brain. *Neuron, 72*(4), 665–678. https://doi.org/10.1016/j.neuron.2011.09.006

Raichle, M. E., MacLeod, A. M., Snyder, A. Z., Powers, W. J., Gusnard, D. A., & Shulman, G. L. (2001). A default mode of brain function. *Proceedings of the National Academy of Sciences of the U S A, 98*(2), 676–682. https://doi.org/10.1073/pnas.98.2.676

Raichle, M. E., & Snyder, A. Z. (2007). A default mode of brain function: A brief history of an evolving idea. *NeuroImage, 37*(4), 1083–1090. https://doi.org/10.1016/j.neuroimage.2007.02.041. discussion 1097-1089.

RajMohan, V., & Mohandas, E. (2007). The limbic system. *Indian Journal of Psychiatry, 49*(2), 132–139. https://doi.org/10.4103/0019-5545.33264

Rolls, E. T. (2019). The cingulate cortex and limbic systems for emotion, action, and memory. *Brain Structure and Function, 224*(9), 3001–3018. https://doi.org/10.1007/s00429-019-01945-2

Sandhu, Z., Tanglay, O., Young, I. M., Briggs, R. G., Bai, M. Y., Larsen, M. L., et al. (2021). Parcellation-based anatomic modeling of the default mode network. *Brain and Behavior, 11*(2), e01976. https://doi.org/10.1002/brb3.1976

Shahab, Q. S., Young, I. M., Dadario, N. B., Tanglay, O., Nicholas, P. J., Lin, Y.-H., et al. (2022). A connectivity model of the anatomic substrates underlying Gerstmann syndrome. *Brain Communications, 4*(3), fcac140.

Tanglay, O., Young, I. M., Dadario, N. B., Briggs, R. G., Fonseka, R. D., Dhanaraj, V., et al. (2022). Anatomy and white-matter connections of the precuneus. *Brain Imaging and Behavior, 16*(2), 574–586.

Tuncer, M. S., Fekonja, L. S., Ott, S., Pfnür, A., Karbe, A.-G., Engelhardt, M., et al. (2022). Role of interhemispheric connectivity in recovery from postoperative supplementary motor area syndrome in glioma patients. *Journal of Neurosurgery*, 1–10. https://doi.org/10.3171/2022.10.JNS221303

Uddin, L. Q., Supekar, K., Lynch, C. J., Khouzam, A., Phillips, J., Feinstein, C., et al. (2013). Salience network-based classification and prediction of symptom severity in children with autism. *JAMA Psychiatry, 70*(8), 869–879. https://doi.org/10.1001/jamapsychiatry.2013.104

Yeo, B. T., Krienen, F. M., Sepulcre, J., Sabuncu, M. R., Lashkari, D., Hollinshead, M., et al. (2011). The organization of the human cerebral cortex estimated by intrinsic functional connectivity. *Journal of Neurophysiology, 106*(3), 1125–1165. https://doi.org/10.1152/jn.00338.2011

CHAPTER

4 Updating the traditional Brodmann's Atlas based on structural and functional connectivity

Introduction

The human cerebral cortex has been divided into a number of different cortical maps over previous decades. These atlases have been established using a variety of different analytical methods to identify isolated brain regions, originally according to histological differences between regions in the context of their speculative functions. The most traditionally used atlas was outlined by Brodmann, and characterizes the cerebral cortex into approximately 52 regions according to cytoarchitectural-based regional differences in cell and laminar structures (Amunts & Zilles, 2015). Several other cyto- (Bailey, 1951; Petrides et al., 2012; von Economo & Koskinas, 1925; Sarkissov et al., 1955) and myleoarchitectural-based studies (Vogt & Vogt, 1919) have also further provided a number of anatomical-based organizational schemes; however, despite utilizing similar methodology of anatomical delineations, all of these maps vary in meaningful ways and are heterogenous in nature (Zilles & Amunts, 2010). One reason for this heterogeneity is they are purely *anatomical*-based schemes, and therefore are largely hindered by their single unit of neurobiological property, mostly cytoarchitectonic, and not to mention their often limited sample size which further increases variability.

Brodmann's Atlas has long stood as one of the most commonly used cortical schemes to organize brain anatomy, with related terms used to guide clinical discussions and treatments as well as communicate research findings. However, with advanced neuroimaging capabilities and improved techniques for structural and functional imaging there has been a recent growing interest in many for an improved characterization of Brodmann areas. One of the largest recent advancements in the neuroimaging community has been that of the Human Connectome Project (HCP) given their creation of a multimodal atlas of the human cerebral cortex—combining architectural, functional, neural connectivity, and topographical differences between cortical regions in healthy individual brains. According to the Glasser et al. HCP atlas, a total of 180 anatomically fine cortical parcellations were identified between hemispheres according to these various neurobiological properties.

The work by the HCP authors has given us a significant body of data on structural-functional relationships in the human brain according to a more anatomically specific and neurobiologically diverse parcellated atlas. However, this work has predominantly explained an atlas using unfamiliar and nonanatomic-based maps (e.g., flat maps which do not explain gyri and sulci in depth). This makes communication and translation of findings using this map in a clinical setting, such as for psychiatry, neurology, and neurosurgery, rather difficult. Therefore, we have recently built off of this work and

Connectomic Medicine. https://doi.org/10.1016/B978-0-443-19089-6.00018-5

published an atlas describing all 180 HCP parcellations in each hemisphere according to the surrounding cortical anatomy, functional connectivity, and structural connectivity (Baker et al., 2018a).

In this chapter, we attempt to briefly bridge information on the structural and functional connectivity of the brain connectome with Brodmann's Atlas. Specifically, we discuss recent work by our lab and the recent literature on the structural-functional connectivity of HCP regions across the human cerebral cortex, and how this information provides an improved understanding of various Brodmann areas.

The new anatomy of Brodmann areas—the basic anatomical and structural-functional connectivity patterns

Each Brodmann area is discussed below in the context of its (1) anatomical boundaries followed by their (2) structural and (3) functional connections with other HCP regions. We outline how the chapter will break up each Brodmann area in Table 4.1. Additional details provided on each HCP region and its connectivity can be found in our atlas published in *Operative Neurosurgery*.

Table 4.1 HCP regions included in each Brodmann area.

Brodmann area	HCP region	Brodmann area	HCP region
1	1	27	PreS
2	2	28	PEEC
3	3a, 3b	29	RSC
4	4	30	RSC
5	5m 5L 5mv	31	31a 31pd 31pv
6	6ma 6mp 6a 6d 6v 6r FEF PEF 55b	32	p32 s32 d32 a32pr p32pr
7	7Am 7AL 7Pm 7PL 7pc 7m VIP AIP MIP PCV LIPv LIPd	33	PFm PGs PGi IP1 IP0 33pr
8	SCEF SFL 8AV 8C 8AD 8BL s6-8 i6-8	26	RSC
9	9a 9p 9m 9-46d IFJa IFJp	34	EC
10	10pp 10d a10p p10p 10v 10r	35	PHA1 PHA2 PHA3
11	11L	37	TPOJ1 TPOJ2 TE2a FFC VVC VMV1 VMV2 VMV3
12	12	38	TGd
13	13L	39	PGS PGp
14	OFC POFC	40	PFop PFt PFcm PF IP2
17	V1 ProS	41	A1, Lbelt, Mbelt
18	V2 V3	42	Pbelt
19	V4 V6 V7 V3a V3b V4t IPS1 MIP LO1 LO2 LO3 V8 FST PH MST MT POS2 PIT POS1 TPOJ3 DVT V6A V3CD	43	43
20	TE2a TGV TE2p TF	44	44
21	TE1a TE1m TE1p PHT STSda STSva	45	45
22	A4 A5 STGa STSvp STSdp PSL STV TA2	46	46 p9-46v a9-46v IFSp IFSa
23	23d 23c d23ab v23ab	47	a47r p47r 47L 47m 47s
24	24a 24p a24pr p24pr 24dd 24dv	48	Hippocampus
25	25	52	PI, 52

Brodmann areas 1, 2, and 3 (BA1, BA2, BA3)

We group Brodmann areas 1, 2, and 3 into a single section given their similarities in structure and function. Brodmann areas 1 and 2 each only include one respective HCP parcellation. Area 3, part of the primary sensory area, can be split into two regions: area 3a and area 3b.

Area 1

Area 1 is located on the visible surface of the postcentral gyrus and forms most of the postcentral operculum. It further extends up to the midline without folding onto the medial face. Area 1 shares its anterior convexity border with area 3b and its posterior border with area 2. Meanwhile, its inferior end is bordered by OP4 and PFop, while its superior end is fully encompassed by areas 3b and 2.

Area 1 demonstrates functional connectivity with various brain regions. It is connected to both the sensory strip (areas 2 and 3a) and motor strip (area 4) with further motor involvement with connection to the premotor regions (areas SCEF, 6mp, 6d, and 6v). Additionally, area 1 is connected to the middle cingulate regions (areas 24dd, 24dv, 5m, and 5L) and the superior insula opercular regions (areas 43, OP1, OP2-3, OP4, IG, PFcm, and FOP2). Area 1 is functionally connected to the lower opercula and Heschl's gyrus regions (areas A4, A5, RI, 52, MBelt, LBelt, PBelt, TA2, and STV) and the parietal lobe (areas LIPv, VIP, IPS1, 7AL, and 7PC). It establishes visual involvement by having functional connections to both the medial occipital lobe (areas V2, V3, and V4) and the lateral occipital lobe (areas PH, TPOJ1, FST, V4t, MST, MT, and LO3). It strengthens its visual involvement with connections to the dorsal visual stream (areas V6, V6a, and V7) and the ventral visual stream (area FFC).

Area 1, working in conjunction with area 3b, plays a role in processing sensory information related to touch. More specifically, area 1 functions as the secondary activation point following the initial reception of tactile stimuli in area 3b (Ploner et al., 2000). Additionally, area 1 and area 2 work together to process information related to the sensation of touch in both hands following bilateral tactile stimulation (Martuzzi et al., 2015).

Area 1 is structurally connected to the brainstem via the pyramidal tracts, the thalamus via thalamocortical projections, and area PFm in the parietal lobe. Its connections to pyramidal tracts descend through the posterior limb of the internal capsule and cerebral peduncle to eventually terminate in the brainstem. The thalamocortical tracts course medial to pyramidal projections to establish connection with the thalamus. The parietal projections that connect area 1 to PFm are portions of the SLF. Additionally, local short association fibers connect area 1 with surrounding parcellation that include areas 1, 2, 3a, 3b, 4, and 6. However, it is worth noting that white matter tracts of area 1 in the right hemisphere may have less consistent projections to surrounding parcellations.

Area 2

Area 2 is located on the anterior bank of the postcentral sulcus and occupies most of its surface. It is limited by the Sylvian fissure on its lateral border and does not reach the midline. The main anterior border of area 2 is with area 1, while it does make minor contact anteriorly with area 3b superiorly. Inferiorly, it borders PFop. Area 51 constitutes the superior border of area 2, while area 7AL is its posterosuperior border. Its long posterior border contacts area 7PC, AIP, and PFt from superior to inferior, respectively.

Area 2 demonstrates functional connectivity with a vast number of brain regions. It is connected to both the sensory strip (areas 1, 3a, and 3b) and the motor strip (area 4). It furthers it motor involvement by having functional connections to the premotor regions (areas SCEF, FEF, 6a, 6mp, 6d, and 6v). It is also functionally connected to the middle cingulate regions (areas 24dd, 24dv, p32prime, 5mv, 5m, and 5L) and the superior insula opercular regions (areas 43, OP1, OP2-3, OP4, IG, PFcm, FOP1, and FOP2). Area 2 has numerous functional connections to the lower opercula and Heschl's gyrus regions (areas PoI1, PoI2, A4, A5, RI, 52, A1, MBelt, LBelt, PBelt, TA2, and STV) and the parietal lobe (areas AIP, VIP, LIPv, PFop, PFt, IPS1, 7AL, and 7PC). Area 2 establishes visual functional connections with the medial occipital lobe (areas V2 and V3), the lateral occipital lobe (areas PH, TPOJ1, TPOJ2, FST, V4t, MST, MT, and LO3), the dorsal visual stream (area V6), and the ventral visual stream (area FFC).

Area 2 is integral in the role of deep tissue sensations, such as through the processing of proprioceptive and kinesthetic sensory information, as well as the perception of movement and force (Hyvarinen & Poranen, 1978; Ploner et al., 2000). It has been shown to be activated by information related to pressure, joint position, and complex touch (Choi et al., 2015). Furthermore, it is also involved in the integration of tactile information from both hands, as evidenced by its activation during bilateral tactile stimulation of the hands (Martuzzi et al., 2015).

Area 2 is structurally connected to the brainstem via the pyramidal tracts, the thalamus via thalamocortical projections, areas PFm, IP1, and IP2 in the parietal lobe, and parcellation 7am in the contralateral lobe. Its connections to pyramidal tracts descend through the posterior limb of the internal capsule and cerebral peduncle to eventually terminate in the brainstem. The thalamocortical tracts course medial to pyramidal projections to establish connection with the thalamus. The parietal projections that connect Area 1 to PFm, IP1, and IP2 are portions of the SLF. Area 2 is structurally connected to the parcellation 7am in the contralateral hemisphere through tracts that travel through the body of corpus callosum. Additionally, local short association fibers connect area 2 with surrounding parcellation that include areas 6r, 7Pc, AIP, 1, 3a, 3b, and 4.

Area 3a

Located deep within the central sulcus, area 3a extends up to the midline but does not fold onto the hemisphere's medial surface. Its long anterior border is adjacent to area 2, while its posterior border is area 3b. Area 23 lies inferior to area 3a, while area 5m is situated superiorly. Near the Sylvian fissure, area 3a has a minor region of posterior contact with area 2.

Area 3a has functional connectivity to both motor and sensory strip with connections to areas 1, 2, 4, and 3b. In addition to the motor strip, area 3a connects to the premotor regions (areas SCEF, 55b, 6d, 6v, and 6mp). Area 3a also has connections to the areas 24dd, 24dv, 5m, and 5L in the middle cingulate regions. Area 3a connects to both the superior insula opercular regions (areas 43, OP1, OP2-3, OP4, IG, and FOP2) and lower opercula and Heschl's gyrus (areas PoI2, 52, A4, A5, RI, A1, MBelt, PBelt, LBelt, TA2, and STV). It connects to the parietal lobe through areas PFcm, 7AL, and 7PC. It is connected to the medial occipital lobe (areas V2, V3, and V4) and the lateral occipital lobe (areas PH, TPOJ1, FST, V4t, MST, MT, and LO3). Area 3a furthers visual involvement with connections to the dorsal visual stream (areas V6, V6a, V3a, and V7) and the ventral visual stream (area FFC).

Area 3a has functions for both proprioception and deep body tissue sensation (Huffman & Krubitzer, 2001; Hyvarinen & Poranen, 1978; Ploner et al., 2000; Vierck et al., 2013). More recently, area

3a has been noted to be reception of slow burning pain generated from the deep somatic tissue (Vierck et al., 2013). Further studies suggest that abnormal processing within area 3a may contribute to clinical pain conditions (Whitsel et al., 2019).

Area 3a is structurally connected to the brainstem through the pyramidal tracts. The pyramidal tracts travel through the posterior limb of the internal capsule and cerebral peduncle to end in the brainstem. Area 3a is structurally connected to the thalamus through the thalamocortical tracts which run medial to the pyramidal projections. Area 3a is connected to the parcellations of 3b and 4 in the contralateral hemisphere through the corpus callosum. Area 3a connects to the parietal lobe through the portion of the SLF which connects to PFm. Area 3a is connected to areas 4, 3b, 2, 1, and 6 through short association fibers.

Area 3b

Area 3b constitutes the entire anterior bank of the postcentral gyrus, not extending to the Sylvian fissure or folding onto the medial face of the hemisphere. Its anterior convexity border is adjacent to area 3a, and area 2 is its long posterior convexity border. The inferior end of area 3b terminates above the Sylvian fissure, with areas 3a and 2 making contact beneath it, forming its inferior boundary. Superomedially, area 3b connects with areas 5m and 5l, while also sharing a small posterosuperior border with area 2.

Area 3b is functionally connected both to the motor strip (area 4) and the sensory strip (areas 1, 2, 3a). It is also associated with the premotor regions as it is functionally connected to areas SCEF, 6d, and 6v. It is connected to the middle cingulate region through areas 24dd, 24dv, 5m, and 5L. It is connected to both the superior insula opercular regions (areas 43, OP1, OP2-3, OP4, IG, PFcm, and FOP2) and the lower opercula and Heschl's gyrus regions (areas A4, A5, RI, 52, A1, MBelt, LBelt, PBelt, TA2, and STV). It is connected to the parietal lobe through areas 7AL and 7PC. It has functional involvement in both the medical occipital lobe (areas V2, V3, and V4) and lateral occipital lobe (areas TPOJ1, FST, V4t, MST, MT, and LO3). Further functional visual interactions occur with the dorsal visual stream (areas V6, V6a, V3a, and V7) and ventral visual stream (area FFC).

Area 3b has shown to be activated primarily to tactile stimulation. When there is a tactile stimulus area 3b is activated before areas 1 and 2 (Ploner et al., 2000). Area 3b provides more specific localizations of a stimulus as compared to areas 1 and 2 (Martuzzi et al., 2015). The example commonly used to illustrate this difference is that area 3b can discern which finger is being touched while area 1 and 2 provide information regarding the general area of the tactile stimulation. There are also studies that have shown that area 3b is activated when people observe someone else's fingers being touched and are not actually mechanically stimulated themselves (Kuehn et al., 2018).

Area 3b is structurally connected to the brainstem through the pyramidal tracts. The pyramidal tracts travel through the posterior limb of the internal capsule and cerebral peduncle to end in the brainstem. Area 3b is structurally connected to the thalamus through the thalamocortical tracts which run medial to the pyramidal projections. Area 3b is connected to the parcellations of 3a, 3b, and 4 in the contralateral hemisphere through the corpus callosum. Area 3a connects to the parietal lobe through the portion of the SLF, which connect to IP1 and IP2. Area 3a is connected to areas 1, 2, 3a, and 4 through short association fibers.

Brodmann area 4 (BA4)

Area 4

Area 4, situated in the precentral gyrus, primarily occupies the posterior portion of the gyrus and constitutes the anterior border of the central sulcus. Medially, it expands to cover a larger part of the paracentral lobule, which corresponds to the leg motor region. The well-established somatotopic organization of this area also includes a dedicated region for eye control, located adjacent to the frontal eye fields. Area 4 shares a posterior convexity border with area 3a. Its inferior end encompasses area 43, and at its superior end, it borders area 5m posteriorly, area 44dd inferiorly, and area 6mp anteriorly on the medial surface (paracentral lobule). Area 4's long anterior border contacts several areas, including area 6d, FEF, area 55b, premotor eye field (PEF), and area 6v, mostly contacting it on the anterior half of the precentral gyrus.

Area 4 demonstrates functional connectivity with a diverse set of brain regions. In the sensory strip, it is connected to areas 1, 2, 3a, and 3b. It has motor involvement with functional connections to the premotor regions (areas SCEF, 55b, 6d, 6v, and 6mp). Area 4 also has connections with the middle cingulate regions (areas 24dd, 24dv, 5m, and 5L) and the superior insula opercular regions (areas 43, OP1, OP2-3, OP4, IG, and FOP2). Additionally, Area 4 is connected to the lower opercula and Heschl's gyrus regions (areas A4, A5, RI, PBelt, LBelt, TA2, and STV) and the parietal lobe (areas VIP, IPS1, LIPv, and 7PC). It establishes its visual involvement by having connections to both the medial occipital lobe (areas V2, V3, and V4) and lateral occipital lobe (areas PH, TPOJ1, FST, V4t, MT, and LO3). It furthers its visual functionality with connections to both the dorsal (areas V6, V6a, V3a, and V7) and the ventral visual streams.

Area 4 is critical for planning, initiating, and controlling voluntary movements by generating and modulating the neural signals responsible for muscle activation. Area also contributes to giving muscles their tone and producing forceful muscle contractions (Chouinard & Paus, 2006). Area 4 is involved in the execution of fine motor movements of the distal forearm and fingers, as well as it has also been suggested to play a role in the early stages of visual learning of motor-based skills (Coco et al., 2016).

Area 4 is structurally connected to the brainstem via the pyramidal tracts, area PFm in the parietal lobe, and parcellation 4, 6ma, and 6mp in the contralateral lobe. Its connections to pyramidal tracts descend through the posterior limb of the internal capsule and cerebral peduncle to eventually terminate in the brainstem. The parietal projections that connect area 4 to PFm are portions of the SLF. Area 4 is structurally connected to the HCP parcellations 4, 6ma, and 6mp in the contralateral hemisphere through tracts that travel through the body of corpus callosum. Additionally, local short association fibers connect area 4 with surrounding parcellations that include areas 3a, 3b, 2, 1, and 6v.

Brodmann area 5 (BA5)

Brodmann area 5 can be further divided into three other regions: 5m, 5l, and 5mv. These subunits are found in the posterior portion of the paracentral lobule. Area 5m is the medial portion, area 5l represents the lateral portion, and area 5mv makes up the medial-ventral section. We define these regions further below in the context of the speculated functional relevance (Glasser et al., 2016). For additional definitions and reasons for separating these subdivisions from other surrounding areas, see the supplementary material of Glasser et al. (2016).

Area 5m

Situated in the posterior superior region of the medial aspect of the paracentral lobule, area 5m (medial 5) shares its anterior boundary with area 4 and its posterior boundary with area 5l. Additionally, a minor anterior border is shared with area 24dd, while its inferior boundary is defined by area 5mv. On its lateral and superior surface, area 5m borders areas 3a and 3b.

Area 5m has functional connections both to the sensory strip (areas 1, 2, 3a, and 3b) and the motor strip (area 4). It has further integration with sensory processing as it is connected to Heschl's gyrus, in the primary auditory cortex, and the lower opercula through area A4. Area 5m strengthens its motor connection through having functional connections to area 6mp and 6d in the premotor regions. Within Brodmann area 5, it only has a connection to area 5l in the middle cingulate region.

Area 5m has been shown to play a role in the organization of both somatosensory and visuomotor data to facilitate movement. For example, it has been shown to participate in activities such as reaching or pointing when the movements are based on somatosensory information, more so than visual stimuli (Scheperjans et al., 2005). Another recent study also highlighted that area 5m is stimulated in the performance of complex tasks that require coordination between the right and left hand (Naito et al., 2008).

White matter tracts structurally connect area 5m to the contralateral hemisphere. Contralateral connections travel through the body of the corpus callosum to parcellations 4 and 5mv. There are no local short association fibers that have been visualized.

Area 5l

Area 5l (lateral 5) can be found on the posterior superior section of the postcentral gyrus, situated at the junction where the gyrus meets the interhemispheric surface. Inferiorly, it shares a border with area 2, while anteroinferiorly, it is adjacent to area 3b. Its posterior boundary is formed by area 7AM with 7AL making up the posteroinferiorly border. Area 5mv is positioned directly beneath area 5l, and area 5m lies anteriorly.

Like area 5mv, area 5l has functional connections to both the sensory strip (areas 1, 2, 3a, and 3b) and the motor strip through area 4. In addition to being connected to the motor strip, it is also connected to areas 6mp and 6d in the premotor regions. Within Brodmann area 5 it is connected to both area 5m and 5mv in the middle cingulate regions. It has connections to OP1 in the superior insula opercular region and to area A4 in the lower opercula and Heschl's gyrus regions. Lastly, it has connections to the parietal lobe by areas 7AL and 7PC.

The function of area 5l has been shown to be involved in the proprioceptive control of goal-directed movements (Scheperjans et al., 2005). Area 5l was found to not be activated if these movements were based on visual guidance (Grefkes et al., 2004). Like area 5m, area 5l is stimulated in the performance of complex task that require coordination between the right and left hand (Scheperjans et al., 2005).

Area 5l is connected to the brainstem through the pyramidal tracts. The pyramidal tracts descend to the posterior limb of the internal capsule and cerebral peduncle then finally to the brainstem. Contralateral white matter tracts go through the body of the corpus callosum to the parcellations 5m, 5l, and 5mv. Unlike areas 5m and 5mv, there are local short association fibers that connect with 5m and 5mv.

Area 5mv

Area 5mv (medial-ventral 5) is situated on the posterior inferior section of the paracentral lobule which constitutes the anterior bank of the cingulate sulcus's ascending ramus. Superiorly, it shares borders with areas 5m and 5l, while inferiorly, it is adjacent to area 23c across the ascending ramus of the cingulate sulcus. Area 24dd forms the anterior boundary of Area 5mv. Its posterior border is composed of Area 7am and PCV.

Area 5mv has a robust functional connectivity profile with strong connections to multiple areas of the brain. Area 5mv has connection to area 2 in the sensory strip and several connections in the premotor regions (areas SCEF, FEF, 6r, 6a, 6mp, and 6ma). In the middle and posterior cingulate regions, it is connected to areas 24dd, 24dv, a24prime, p24prime, p32prime, 23c, and 5L. Area 5mv has connection to the frontal lobe through areas 9-46d and 46 which are in the dorsolateral frontal lobe. Area 5mv is connected to the areas 43, OP4, PFcm, FOP1, FOP3, and FOP4 in the superior insula opercular regions. Like area 5m and 5l, it has connections to the lower opercula and Heschl's gyrus regions but through different areas that include areas 52, PoI1, PoI2, and MI. Area 5mv has connections to both the lateral parietal lobe (areas AIP, MIP, LIPv, LIPd, IP0, PGp, PFop, PF, PFt, 7AL, and 7PC) and the medial parietal lobe (7am, PCV, and DVT). It has connections to both the medial occipital lobe (areas V1, V2, and V3) and lateral occipital lobe (areas PHT, PH, TPOJ2, TPOJ3, FST, and LO3) with further connections to areas V6 and V6a in dorsal visual stream areas.

Area 5mv has been shown to be involved with somatosensation and motor response (Scheperjans et al., 2005). A recent study suggests that when a subject attempted to imitate upper body movements of someone else, their body movements were more accurate with increased stimulation of area 5mv (Kruger et al., 2014).

Area 5mv has structural connections through white matter tracts to both the contralateral hemisphere and the cingulate cortex. These contralateral connections travel through the body of the corpus callosum to reach parcellations 5mv, 4, 5l, and 5m. The fibers projecting to the cingulate cortex extend anteriorly from 5mv and terminate at 24sv and p24r. There are no local short association fibers that have been visualized.

Brodmann area 6 (BA6)

Area 6 can be further divided into 5 distinct subdivisions. Area 6ma and 6mp are in the supplementary motor regions. The remaining areas (6d, FEF, 55b, PEF, 6v, and 6r) make up most of the premotor area. We define these regions further below in the context of the speculated functional relevance (Glasser et al., 2016). For additional definitions and reasons for separating these subdivisions from other surrounding areas, see the supplementary material of Glasser et al. (2016).

Area 6ma

Area 6ma (medial anterior 6) comprises the lateral posterior section of the superior frontal gyrus, predominantly situated at the angle where the hemispheres intersect. This area is responsible for forming the lateral posterior portion of the superior frontal gyrus, primarily straddling the interhemispheric angle.

Area 6ma displays functional connectivity with various regions throughout the brain. In the premotor regions, area 6ma connects to areas SCEF, PEF, FEF, 6r, 6a, 6v, and 6mp. Within the middle and posterior cingulate regions, Area 6ma is linked to areas a24prime, p24prime, a32prime, p32prime, 23c, and 5mv. In the dorsolateral frontal lobe, connections are found within areas IFSa, 9-46d, a9-46v, p9-46v, and 46. Area 6ma is also connected to insula opercular regions, including PoI1, PoI2, AVI, MI, 43, PFcm, FOP1, FOP3, FOP4, and FOP5. Area 6ma connects to both the lateral parietal lobe (areas IP2, IP0, AIP, MIP, LIPd, PGp, PFop, PF, PFt, 7AL, 7PL, and 7PC) and to the medial parietal lobe (areas 7am, 7pm, PCV, and DVT).

Area 6ma displays distinct activation patterns in comparison to other adjacent Brodmann area 6 regions. For example, when individuals receive a visual instruction cue, 6ma demonstrates increased activation compared to areas 6mp, SFL, or s6-8. Conversely, when individuals are told a story, area 6ma exhibits greater deactivation than area SFL. Furthermore, area 6ma displays less functional activity compared to area s6-8 when individuals are required to match objects based on provided verbal categories (Glasser et al., 2016). Area 6ma was found to be one of four regions that comprise the supplementary motor area (Sheets et al., 2021).

Area 6ma features structural connections with the brainstem via the pyramidal tracts, the inferior frontal gyrus via the FAT, and the contralateral hemisphere via the corpus callosum. The connections to the pyramidal tracts descend through the posterior limb of the internal capsule and cerebral peduncle, ultimately reaching the brainstem. The FAT links Area 6ma to the inferior frontal gyrus and terminates at parcellations 44, FOP4, and AAIC. The contralateral connections of area 6ma travel through the body of the corpus callosum to reach 6ma. Additionally, local short association fibers establish structural connections between 6mp, SFL, 6a, i6-8, and s6-8.

Area 6mp

Area 6mp (6 medial posterior) constitutes the region where the superior frontal gyrus (SFG) merges with the precentral gyrus. It forms the medial bank of the SFG at this junction and extends onto the superior surface of the posterior-most SFG and the anterosuperior section of the precentral gyrus. Additionally, area 6mp comprises the most posterior superior part of the superior bank of the superior frontal sulcus. The boundaries of area 6mp include area 4 posteriorly, area 6d laterally, area 6a anterolaterally, areas 6ma and SCEF anteriorly, and area 24dd inferiorly on the medial surface.

Area 6mp exhibits functional connectivity to the sensory strip (areas 1, 2, 3a, 3b) and to the motor strip (area 4). Furthering its involvement with motor function it is connected to premotor regions (areas SCEF, 6a, 6ma, and 6d). In the middle cingulate regions, connections are found with areas 24dd, 24dv, p32prime, 5mv, and 5L. Area 6mp is also connected to superior insula opercular regions, including areas 43, OP1, OP4, PFcm, FOP1, and FOP2. In the lower opercula and Heschl's gyrus regions, it links to areas A4, RI, and PBelt. Additionally, area 6mp connects to parietal lobe (areas PFop, PFt, IPS1, 7AL, and 7PC) as well as the lateral occipital lobe through area FST.

Area 6mp was differentiated from neighboring parcellations based on variations in myelin thickness and functional activity. Area 6mp functional activity can be defined by direct comparison of its activity to other areas. For example, when individuals are given a visual instruction cue while moving their feet, area 6mp displays less activation relative to area 6ma. Area 6mp also exhibits greater deactivation than its lateral neighbor area 6d when individuals listen to a story or solve a math problem. Furthermore, area 6mp demonstrates less activation than area 6a in a setting of social interaction (Glasser et al., 2016). Area 6mp was found to be one of four regions that comprise the supplementary motor area (Sheets et al., 2021).

Area 6mp features structural connections with the brainstem via the pyramidal tract and the contralateral hemisphere via the corpus callosum. The connections to the pyramidal tracts descend through the posterior limb of the internal capsule and cerebral peduncle, ultimately reaching the brainstem. The contralateral connections of area 6mp travel through the body of the corpus callosum to reach areas 6ma, 6mp, FEF. Additionally, local short association fibers establish structural connections between areas 24dd, 6a, and 6d.

Area 6a

Area 6a (6 anterior) creates the posterior superior bank of the superior frontal sulcus and adjacent portions of the superior frontal gyrus. It forms the bank where the sulcus meets the precentral sulcus at a right angle. It is bordered by areas 6d and 6mp posteriorly and area 6ma medially. The anterior border comprises s6-8, area 8AD, and i6-8, with FEF as its inferior neighbor.

Area 6a has robust connections to various regions throughout the brain. In the sensory strip, it connects to area 2, while in the premotor regions, it links to areas SCEF, PEF, FEF, 6ma, 6mp, 6d, and 6v. Connections to the middle cingulate regions include areas a24prime, p32prime, 5mv, and 23c. In the lateral frontal lobe, area 6a is functionally associated with areas IFSa, IFJa, i6-8, 46, p9-46v, and 9-46d. Area 6a also connects to the superior insula opercular regions, which include areas OP4, PFcm, FOP4, and FOP2. In addition to the superior insula opercular, it also connects to the lower opercula and Heschl's gyrus regions through areas PoI1 and PoI2. In the temporal lobe, it is linked to areas TE2p, PHA3, and PHT. Area 6a has connections to both the lateral parietal lobe (areas AIP, MIP, VIP, LIPd, LIPv, PFop, PF, PFt, PGp, IP2, IP1, IP0, IPS1, 7AL, 7PL, and 7PC) and the medial parietal lobe (areas 7pm, 7am, DVT, and PCV). It also has connections to the lateral occipital lobe (areas PH, TPOJ2, TPOJ3, and FST) and the medial occipital lobe (area V2).

The exact function of area 6a is yet to be determined as it is a newly discovered region of the premotor cortex. Differences in myelin thickness and functional activity made it a unique area (Glasser et al., 2016). There was an observed functional difference with area 6a′s anterior neighbors. Area 6a was shown to have more involvement in solving math problems, performing object feature comparison tasks, and in the setting of social interactions than area s6-8. It also had increased activation with social interactions when compared to area i6-8, another anterior neighbor. Area 6a did have a relative decrease in activation when compared to area i6-8 in emotion identification and object feature comparison. Lastly, when compared to FEF, area 6a had a relative decrease with gambling and objection feature comparison (Glasser et al., 2016).

Area 6a features structural connections with the brainstem via the pyramidal tract and the parietal lobe. The connections to the pyramidal tracts descend through the posterior limb of the internal capsule and cerebral peduncle, ultimately reaching the brainstem. Parietal projections are portions of the SLF that connect with 3a, 3b, 7PC, and 7AL. Additionally, local short association fibers establish structural connections between areas FEF, i6-8, 55b, 8Av, 46, and 6r.

Area 6d

Area 6d (6 dorsal) occupies the anterosuperior section of the precentral gyrus, situated just below its intersection with the superior frontal gyrus (SFG). It constitutes the posterior bank of the neighboring precentral sulcus. Superiorly, area 6d shares a border with area 6mp, while FEF lies inferiorly. Its posterior border is formed by area 4, and area 6a creates its anterior border across the precentral sulcus.

Area 6d has functional connectivity to both the sensory strip (areas 1, 2, 3a, and 3b) and the motor strip (area 4). In addition to the motor strip, it is functionally connected to premotor regions through areas 6a, 6mp, and 6v. Area 6d is connected to the middle cingulate regions by areas 5L and 24dd. Area 6d relates to areas FOP2, OP4, OP1, A4, and PBelt in the insula opercula regions. It is also functionally associated with parietal lobe (areas 7PC, 7AL, and PFt) and the lateral occipital lobe (area FST).

Area 6d was differentiated from neighboring parcellations based on changes in myelin thickness and functional activity (Glasser et al., 2016). There is less stimulation of area 6d during gambling tasks, in the setting of social interactions, and during object feature comparison when compared to area FEF, which is the inferior border. Compared to the inferior border (area 6a), area 6d shows less activation in solving math problems and in the setting of social interaction settings (Glasser et al., 2016).

Area 6d features structural connections with the brainstem via the pyramidal tract and the areas 6mp, FEF, and 55b contralaterally. The connections to the pyramidal tracts descend through the posterior limb of the internal capsule and cerebral peduncle, ultimately reaching the brainstem. The contralateral connections of area 6d travel through the body of the corpus callosum to reach at 6mp, FEF, and 55b. Additionally, local short association fibers establish structural connections between areas 3a, 3b, 6a, FEF, i6-8, 8Av, 3a, 3b, 6ma, and 6d.

Area 6v

Area 6v (6 ventral) constitutes the anterior-inferior one-third of the precentral gyrus, with only a minor contribution to the posterior bank of the precentral sulcus, which is predominantly formed by area 6r. Superiorly, it shares a border with area 55b, while area 43 lies inferiorly. Its posterior boundary is marked by area 4, and areas 6r and PEF delineate its anterior border.

Area 6v is connected functionally both to the sensory strip (areas 1, 2, 3a, and 3b) and the motor strip (area 4). In addition to the motor strip, it is functionally connected to multiple premotor regions through areas SCEF, FEF PEF, 6ma, 6mp, 6r, 6d, and 6v. Area 6v is also connected to the middle cingulate regions through areas 24dd and p32prime. It has multiple functional connections to the superior insula opercular regions (43, OP4, OP2-3, OP1, PFcm, FOP1 FOP2, FOP3, and FOP4). The function connections to areas A4, PBelt, RI, and PoI2 create an association with lower opercula and Heschl's gyrus regions. Area 6v has connections to lateral parietal lobe through areas AIP, MIP, VIP, LIPv, PFop, PFt, 7AL, and 7PC. It has connections both to the dorsal visual steam (areas V3a, V3b, V6, V6a, and V7) and the ventral visual stream (areas FFC). It has further visual involvement by having connections to the lateral occipital lobe (areas PH, TPOJ2, MST, and FST).

Area 6v encompasses what is traditionally referred to as the premotor cortex, which serves various functions. The ventral premotor region is recognized for its involvement in managing hand movements during object manipulation, such as grasping and lifting (Chouinard & Paus, 2006). Area 6v has also been shown to have stronger coupling with somatosensory regions when compared to its anterior border area 6r (Neubert et al., 2014).

Area 6v features structural connection to the contralateral FEF via the corpus callosum and inferior parietal lobe parcellations via the superior longitudinal fasciculus. The contralateral connections of area 6v travel through the body of the corpus callosum to reach FEF. Through the superior longitudinal fasciculus, area 6v connects to areas PHT, FST, PH, and PF in the inferior parietal lobe. Additionally, local short association fibers establish structural connections between areas 4, 6r, PEF, 43, 3a, and 3b.

Area 6r

Area 6r (6 rostral area) is located in the inferior portion of the precentral sulcus. It encompasses the sulcus's floor and both banks, which means that area 6r spans both the anterior inferior part of the precentral gyrus and the posterior portion of the pars opercularis in the inferior frontal gyrus. The inferior end of the precentral sulcus forms a shovel-shaped cup near the opercular edge, which constitutes area 6r. Area 6r is bordered by area 44 anteriorly and area 43 posteriorly. Posterosuperiorly, area 6r shares a border with area 6v. Its main superior neighbor is PEF, while IFJp and IFJa form an anterior superior border. On its undersurface opercular surface, FOP1 and FOP4 create the inferior border of area 6r.

Area 6r exhibits functional connectivity to several regions throughout the brain. In the premotor regions, it connects to the areas SCEF, FEF, PEF, 6ma, 6a, and 6v. Within the middle cingulate regions, it has connections to areas 23c, 5mv, a24prime, p24prime, and p32prime. It connects to the lateral frontal lobe through areas 46, p9-46v, 9-46d, IFSa, IFJa, IFJp, and p47r. It connects to the superior insula opercular regions (areas 43, OP4, PFcm, FOP2, FOP3, FOP4, and FOP5) and the lower opercula and Heschl's gyrus (areas AVI, MI, PoI1, and PoI2). Connections to the temporal lobe include areas TE2p and PHT. In the lateral parietal lobe, area 6r connects to areas AIP, MIP, LIPd, LIPv, PFop, PF, PFt, IP0, 7PL, 7AL, and 7PC. Lastly, in the lateral occipital lobe, it connects to areas PH and FST.

Area 6r is not a well-studied parcellation but the available literature does provide clues to its function. Area 6r has strong function association with Broca's area, a known cortical area associated with language processing (Amunts et al., 2010; Amunts & Zilles, 2012). Compared to its posterosuperior neighbor 6v, area 6r shows a relative stronger connection to visuomotor areas (Neubert et al., 2014).

Area 6r is structurally connected to the posterior temporal parcellations via the superior longitudinal fasciculus and the superior frontal gyrus via the FAT. The structural connections with the superior longitudinal fasciculus link area 6r to posterior temporal parcellations TE1a and TE2a. The FAT connects area 6r with the superior frontal gyrus at parcellation SFL. Additionally, local short association fibers establish connections with areas 6v, 44, IFJa, IFJp, 8C, and IFSa.

Area FEF

Area FEF is situated on the anterior half of the precentral gyrus, roughly halfway down its length along the convexity, just beneath the point where the precentral and superior frontal sulci intersect. This area also constitutes the adjacent floor of the precentral sulci and slightly extends onto the posterior edge of the middle frontal gyrus. Area FEF shares borders with areas 6a and 6d superiorly and area 55b inferiorly. Area 4 is located posteriorly to the FEF, while area i6-8 forms its anterior border on the middle frontal gyrus.

Area FEF exhibits functional connectivity within the sensory strip (area 2) and premotor regions (areas SCEF, PEF, 6r, and 6v). It connects to the middle cingulate regions through areas a24prime, p32prime, 5mv, and 23c. In the lateral frontal lobe it connects to areas IFSa, IFJa, 46, and 9-46d. Area FEF connects to the superior insula opercular (areas 43, OP4, PFcm, FOP1, FOP3, FOP4, and FOP5) and the lower opercula and Heschl's gyrus regions (areas STV, LBelt, PBelt, A4, MI, 52, RI, PoI1, and PoI2). Furthermore, it connects with areas TE2p and PHT in the temporal lobe.

It establishes multiple functional connections with both the lateral parietal lobe (areas AIP, MIP, VIP, LIPd, LIPv, PFop, PF, PFt, PGp, IP0, IPS1, 7AL, 7PL, and 7PC) and medial parietal lobe (areas 7am, DVT, and PCV). It has strong visual involved exhibited by its functional connections to the medial occipital lobe (areas V1, V2, V3, and V4), the lateral occipital lobe (areas V3cd, LO1, LO2, LO3, PH, TPOJ1, TPOJ2, TPOJ3, V4t, MST, and FST), the dorsal visual stream (areas V3a, V3b, V6, V6a, and V7), and ventral visual stream (areas V8, PIT, FFC, VVC, VMV1, VMV2, and VMV3).

Area FEF is recognized for its involvement in controlling and coordinating rapid eye movements between fixed points, referred to as intentional saccadic movements. Additionally, it has been associated with smooth eye movements that enable the eyes to track a moving target, known as smooth pursuit eye movements. FEF also contributes to visual attention by creating a salience map that represents the importance of various visual stimuli in the environment. This helps to prioritize information processing and allocate attention to the most relevant stimuli for efficient perception and decision-making (Fecteau & Munoz, 2006; Paus, 1996; Petit et al., 1997; Pierrot-Deseilligny, 1994; Pierrot-Deseilligny et al., 1997).

Area FEF has structural connections to both the contralateral hemisphere and the superior longitudinal fasciculus. Contralateral connections travel through the body of the corpus callosum, linking FEF to i6-8 and SFL in the opposite hemisphere. The superior longitudinal fasciculus provides connections between FEF and the intraparietal sulcus as well as the inferior parietal lobe, terminating at areas IP1, IP2, and PGs. Additionally, local short association fibers establish connections between FEF and other nearby regions, including 6d, 55b, i6-8, 8Av, 6a, and PEF.

Area PEF

Area PEF is a compact region situated in the floor of the precentral sulcus, right at the intersection of the precentral and inferior frontal sulci. Unlike FEF and area 55b, PEF is primarily vertically oriented and extends slightly onto the adjacent precentral gyrus. It shares borders with area 55b superiorly and area 6r inferiorly. Area 6v creates its posterior border, while area 8C forms the anterior border on the middle frontal gyrus (MFG) and area IFJp creates the anterior border in inferior frontal sulcus.

Area PEF exhibits functional connectivity with areas SCEF, FEF, 6ma, 6r, 6a, and 6v in the premotor regions, areas a24 prime, p32prime, and 23c in the middle cingulate regions, and areas IFSa, IFJp, and 9-46d in the lateral frontal lobe. It connects to the superior insula opercular regions (areas 43, PFcm, FOP4, and FOP5) and the lower opercula and Heschl's gyrus regions (areas MI and PoI2). It established temporal lobe involvement with functional connections to TE2p and PHT. It has significant functional links to the lateral parietal lobe (areas AIP, MIP, LIPd, LIPv, PFop, PFt, PGp, IP0, 7PL, and

7PC) and the medial parietal lobe (area 7). Lastly, it connects to areas PH and FST in the lateral occipital lobe.

Although area PEF has been extensively studied there is still little known about the precise function. It has been shown to be involved in reflex saccades which are rapid, automatic eye movements that occur in response to sudden or unexpected changes in the visual field, such as the appearance of a new stimulus. Area PEF's function is supported by studies of patients with lesions in this area (Pierrot-Deseilligny, 1994; Pierrot-Deseilligny et al., 1997, 2005; Pouget, 2015; Schall, 2013).

Area PEF has an established structural connection to area PHT, TPOJ2, FST, and PFm in inferior parietal lobe via the superior longitudinal fasciculus. Connections to the contralateral hemisphere do exist but they have a high variation based on the individual. Additionally, local short association fibers connect with 6r, 8C, and IFJp.

Area 55b

Area 55b is situated on the anterior half of the precentral gyrus, approximately midway down its length along the convexity, just below FEF. It also occupies the adjacent floor of the precentral sulci and extends marginally onto the posterior edge of the middle frontal gyrus. Superiorly, area 55b shares a border with area FEF, while inferiorly, it borders PEF and area 6v. Area 4 acts as its posterior border, and areas 8AV and 8C form its anterior border across the precentral sulcus.

Area 55b demonstrates functional connectivity to the motor strip (area 4) and premotor regions (areas SCEF and SFL). It has connections to the lateral frontal lobe through areas IFSp, IFJa, 8AV, 44, 45, and 47L. It establishes connections in the temporal lobe through areas STSda and STSdp. Area 55b has connection to areas PSL and STV in the posterior opercular cortices and area TPOJ1 in the lateral occipital lobe.

Area 55b is a relatively previously uncharacterized region, but is of growing interest. A recent case series from 2021 suggests that area 55b could play a vital role as an integration cortical hub for both dorsal and ventral streams of language (Hazem et al., 2021). In fact, task-based fMRI functional data from the HCP suggests this region is activated in all tasks requiring language production. Recent work suggests area 55b may house a laryngeal motor cortex region, but ultimately we believe this newly discovered area 55b plays an important integrative role in the cortical organization of language.

Area 55b has structural connections with several parcellations in the contralateral hemisphere as well as the superior longitudinal fasciculus. The contralateral connections pass through the body of the corpus callosum to reach 6ma, 6a, and 6mp. The superior longitudinal fasciculus connects area 55b to parcellations PHT and PFm, ultimately terminating in the temporal lobe at TGd. Additionally, local short association fibers establish connections with 8Av, 8C, IFJp, 3a, 3b, and PEF.

Brodmann area 7 (BA7)

Brodmann area 7 can be further divided into 9 unique regions which include: 7Am, 7AL, 7Pm, 7PL, 7pc, 7m, VIP, AIP, MIP. We define these regions further below in the context of the speculated functional relevance. For additional definitions and reasons for separating these subdivisions from other surrounding areas, see the supplementary material of Glasser et al. (2016).

7AM

Area 7AM (7 anterior-medial) is situated on the anterosuperior surface of the medial face of the superior parietal lobule. It is bordered by area 7AL superior-laterally, VIP posterolaterally, and areas 7PL and 7PM posteriorly. Along its interhemispheric face, the anterior border is formed by areas 5L and 5MV. The inferior border comprises the PCV (precuneus visual area), while the posterior border consists of area 7PM.

Area 7AM exhibits functional connectivity with numerous brain regions. It connects to the premotor regions (areas SCEF, FEF, PEF, 6ma, 6a, and 6r). It has functional connections to both the lateral frontal lobe (areas IFSa, 46, and 9-46d) and medial frontal lobe (a24prime, a32prime, p32prime, 5mv, and 23c). Area 7AM also connects to insula opercular regions (areas MI, PoI1, PoI2, PFcm, and FOP4) and temporal lobe (areas PHA3, PHT, and TE2p). It connects both to the lateral parietal lobe (areas 7PC, 7AL, 7PL, AIP, VIP, MIP, LIPd, PFop, PFt, PF, PGp, IP2, and IP0) and the medial parietal lobe (7pm, PCV, POS2, and DVT). Additionally, it has functional associations with both the medial occipital lobe (area V1) and lateral occipital lobe (PH, TPOJ2, TPOJ3, and FST). There are also connections to both the dorsal visual stream (area V6) and ventral visual stream (FFC).

Area 7AM plays a crucial role in various types of information processing. These include spatial awareness, visual shape, motion perception, working memory, and execution (Wang et al., 2015). The anterior most portion has been linked to self-centered mental imagery and attentional processes (Scheperjans et al., 2008). A more recent study suggests that area 7AM supports neural networks that are associated with self-related processes, memory, and consciousness (Jitsuishi & Yamaguchi, 2023). Further description of area 7AM's function can be done by comparison to its neighbors during specific tasks. For instance, it shows less activation than area 7PL (its posterolateral neighbor) during working memory and auditory story tasks. Similarly, when compared to area 7PM (its posteromedial neighbor), Area 7AM demonstrates less activation during working memory and shape recognition tasks. These differences in activation patterns highlight the distinct roles and functional specializations of the various regions within the parietal cortex.

Area 7AM has structural connections to areas 7AM and 7PM in the contralateral hemisphere via the corpus callosum and to the thalamus. The contralateral connections of Area 7AM course through the corpus callosum and terminate in areas 7AM and 7PM. Thalamic connections project inferiorly through the posterolateral thalamus, extending to the brainstem and superior colliculus. In some individuals, connections with the IFOF have been observed, but these tracts are inconsistent and not present in all individuals. Additionally, local association bundles connect area 7AM with other nearby regions, such as VIP and 7PL. Some individuals have IFOF connections, but these tracts are inconsistent.

7AL

Area 7AL (7 anterior-lateral) is situated on the anterior superior part of the superior parietal lobule. Medially, it extends to the interhemispheric midline, and anteriorly, it reaches the postcentral sulcus. Area 7AL is bordered by area 2 anteriorly, area 7PC inferiorly, and VIP posteriorly. Its medial boundary is formed by areas 5L and 7AM, as both regions extend onto the interhemispheric medial surface. This positioning of Area 7AL within the brain allows it to interact with surrounding areas and contribute to various cognitive and sensory processes.

Area 7AL exhibits functional connectivity with various brain regions. It connects to the sensory regions (areas 1, 2, 3a, and 3b) and premotor regions (area SCEF, FEF, 6ma, 6a, 6d, and 6r). It has functional connections to both the lateral frontal lobe (areas 46 and 9-46d) and medial frontal lobe (areas 24dv, a24prime, p24prime, p32prime, 5L, 5mv, and 23c). Area 7AL also connects to insula opercular regions (areas 43, PFcm, FOP1, FOP2, FOP3, FOP4, FOP5, 52, MI, PoI1, PoI2, OP4, and OP1) and temporal lobe (areas PHT and TE2p). It connects both to the lateral parietal lobe (areas 7PC, 7PL, AIP, VIP, MIP, PFop, PF, PFt, and PGp) and the medial parietal lobe (areas 7AM, PCV, and DVT). Additionally, it has functional associations with lateral occipital lobe (areas TPOJ2 and FST). There also exist connections to the visual stream regions (area V6).

Area 7AL shares some functional similarities with area 7AM, as it is involved in various types of information processing, including spatial awareness, visual shape, motion processing, working memory, and execution (Wang et al., 2015). The most anterior portion of 7AL plays a role in self-centered mental imagery and attentional processes (Scheperjans et al., 2008). Additionally, area 7AL demonstrates functional connectivity with the premotor cortex, further highlighting its involvement in motor-related functions (Glasser et al., 2016). By comparing area 7AL to high bordering areas assists in better defining its function. When compared to its medial neighbor, area 7AM, area 7AL exhibits greater activity during the processing of an average compilation of motor functions. Furthermore, area 7AL is more deactivated when observing socially interacting geometric objects. These differences in activation patterns suggest that area 7AL may have distinct functional roles in the brain despite sharing some similarities with area 7AM (Glasser et al., 2016).

Area 7AL has structural connections with several brain regions, including areas 7AL and 7AM in the contralateral hemisphere, parcellations 9a, 9p, 6a, and 6ma via the IFOF, as well as the thalamus. IFOF connections travel through the posterior temporal lobe and the extreme/external capsule, extending to the frontal lobe and connecting to parcellations 9a, 9p, 6a, and 6ma. Thalamic connections project inferiorly through the posterior thalamus, reaching the brainstem and superior colliculus. Contralateral connections course through the corpus callosum and terminate at areas 7AL and 7AM in the opposite hemisphere. Most local association bundles project posteriorly to areas 7PL, IP0, IP1, IPS1, LIPd, and MIP. There are also structural connections with areas 2 and 5L. It is worth noting that the white matter connections from the right hemisphere's area 7AL do not consistently connect with the superior frontal gyrus via the IFOF between individuals. However, when these IFOF connections are present, the tract terminates at the lateral frontal lobe.

7PM

Area 7PM (7 posterior-medial) is in the posterior superior parietal lobule, at the point where the convexity surface of the SPL turns inferior to form its interhemispheric surface. Area 7PM occupies portions of both surfaces. Its borders include area 7AM anteriorly and area 7PL laterally on its superior surface. On its interhemispheric surface, area 7PM is bordered by PCV anteroinferiorly, area 7M inferiorly, and POS2 (parietooccipital sulcus 2) posteroinferiorly.

Area 7PM exhibits robust functional connectivity to the frontal lobe (areas i6-8, s6-8, 8AD, 8BM, 8C, a10p, p10p, 46, 9-46d, a9-46v, p9-46v, a32prime, 23c, and IFJp). It has functional associations with premotor regions (areas 6a and 6ma) and temporal lobe (areas PHA2, PreS, PHT, and TE1p). Area 7PM has connections to both the lateral parietal lobe (areas PGp, PGs, PFm, IP0, IP1, IP2, AIP, MIP, LIPd, and 7PL) and medial parietal lobe (areas 7m, 7AM, PCV, DVT, 31a, POS1, POS2, and RSC).

Area 7PM has lateralization of functionality with the left and right hemisphere having two distinct functions. Both the right and left sides share the functions of vision motion, space, vision shape, attention and working memory. Only in the right hemisphere there is additional functionality of motor learning and execution. Additionally, area 7PM plays a role in episodic memory retrieval and saccade-related activity (Scheperjans et al., 2008). When compared to its superolateral neighbor, area 7PL, area 7PM shows different activation patterns in certain tasks. Area 7PM is deactivated when viewing body images versus a compilation of tool, face, and place image, while 7PL is activated when viewing a compilation of tool, face, and place images. Area 7PM is less activated during emotional and social cue tasks compared to area 7PL (Glasser et al., 2016).

Area 7PM is structurally connected to parcellations 7PM, POS2, and PCV in the contralateral hemisphere and the thalamus. Contralateral connections run through the corpus callosum and terminate at the areas 7Pm, POS2, and PCV in the contralateral hemisphere. Thalamic connections travel inferior through the posterolateral thalamus to ultimately terminate at the brainstem and superior colliculus. Additionally, there are local short association bundles that directly connect to areas POS2, PCV, and 7AM.

7PL

Area 7PL (7 posterior-lateral) is located on the posterior superior surface of the superior parietal lobule. It shares borders with several neighboring regions, including MIP laterally, VIP anteriorly, and areas 7AM and 7PM medially. The posterior border of area 7PL is defined by small boundaries with areas IPS1, DVT, and POS2.

Area 7PL exhibits functional connectivity with multiple different brain regions. It connects to the premotor regions (areas PEF, 6ma, 6a, and 6r). It has functional connections to both the lateral frontal lobe (areas IFSa, IFJp, p9-46v, 46, and 9-46d) and medial frontal lobe (areas p32prime, 5mv, and 23c). Area 7PL also connects to insula opercular regions (areas FOP4) and temporal lobe (areas PHA3, PHT, and TE2p). It connects both to the lateral parietal lobe (areas 7PC, 7AL, AIP, VIP, MIP, LIPd, LIPv, PFop, PFt, PF, IPS1, IP2, and IP0) and the medial parietal lobe (areas 7AM, 7pm, PCV, and DVT). Additionally, it has functional associations with both the medial occipital lobe (area V1) and lateral occipital lobe (areas PH, TPOJ2, and FST). There also connections to both the dorsal visual stream (area V6) and ventral visual stream (areas FFC).

Area 7PL has lateralization of functionality with the left and right hemisphere having two distinct functions. Both the right and left sides share the functions of vision motion, space, vision shape, attention, and working memory. Only in the right hemisphere there is additional functionality of motor learning and execution (Wang et al., 2015). Additionally, area 7PL plays a role in episodic memory retrieval and saccade-related activity (Scheperjans et al., 2008). When compared to its inferomedial neighbor, area 7PM, area 7PL shows different activation patterns in certain tasks. It is activated when viewing body images versus a compilation of tool, face, and place images. While 7PM is deactivated, area 7PL is more activated during emotional and social cue tasks compared to area 7PM.

Area 7PL is structurally connected to several regions and fiber tracts, including the IFOF, thalamus, MdLF, and local parcellations. The IFOF connections pass through the posterior temporal lobe and the extreme/external capsule, extending to the superior frontal gyrus parcellations 9p, 8BL, and SFL. The MdLF terminates near the planum temporale at the parcellation MBelt. Thalamic connections project inferiorly through the posterolateral thalamus, reaching the brainstem and superior colliculus. Most of the local short association bundles connect posteriorly to V3A, V7, IP0, MIP, PGp, and IPS1, while also establishing connections with VIP, 7PL, and 7AM.

Area 7PC

Area 7PC is situated in the anterior and lower part of the superior parietal lobule. It extends into the adjacent posterior side of the postcentral sulcus. Area 7PC is bordered by area two to the anterior and AIP to the inferior. Its posterior boundary is made up of LIPv and VIP. Area 7AL is located to its medial (superior) side.

Area 7PC has extensive functional connections throughout the brain. It connects to the sensory strip (areas 1, 2, 3a, and 3b) and the motor strip (area 4), as well as the premotor regions (areas SCEF, FEF, PEF, 6ma, 6mp, 6a, 6d, 6r, and 6v). It also links to the medial frontal lobe (areas 24dd, 24dv, p32prime, 5L, 5mv, and 23c) and the insula opercular regions (areas A4, PBelt, PFcm, FOP2, OP4, and OP1). Area 7PC has connections with the temporal lobe (areas PHT and TE2p). It connect to both the lateral parietal lobe (areas 7PL, AIP, VIP, MIP, LIPv, LIPd, PFop, PFt, PGp, IP0, and IPS1) and to the medial parietal lobe (areas 7AM and DVT). It has functional connections to the medial occipital lobe (area V2) and lateral occipital lobe (areas V3CD, V4t, PH, LO3, TPOJ2, TPOJ3, MST, and FST) with further connections to the dorsal visual stream (areas V3b, V6a, and V6) and the ventral visual stream (area FFC).

Area 7PC has differing functions depending on the side. The left side is primarily associated with imagination while the right side is associated with vision shape, language comprehension, sexuality, and working memory (Mars et al., 2011). It has also shown to play a role in four cognitive domains: calculation, writing, finger gnosis, and left-right orientation (Shahab et al., 2022). Overall area 7PC has robust involvement with both visual and somatosensory stimulation.

Area 7PC is structurally connected to 8BL, 6a, 6ma, and SFL via the IFOF. The connections from the IFOF course through the posterior temporal gyrus and extreme/external capsule to the frontal lobe, ultimately ending at areas 8BL, 6a, 6ma, and SFL. Additional, local short association bundles are also abundant and establish connections with several areas, including LIPd, LIPv, MIP, 1, and 2. It should be noted that the presence of IFOF projections varies among individuals. However, majority of individuals have these projections. Additionally, white matter connections from 7PC in the right hemisphere have more consistent connections with the motor and somatosensory cortex.

VIP

Area VIP is located on the central part of the superior surface of the superior parietal lobule. While it does not extend to the intraparietal sulcus, it does come close to the interhemispheric fissure. The boundaries of area VIP include the areas LIPv and MIP laterally, areas 7AL and 7PC anteriorly, area 7AM medially, and area 7PL posteriorly.

Area VIP exhibits functional connectivity with a wide array of brain regions. It connects to the IT and has a function connection with both the sensory strip (areas 1 and 2) and the motor strip (area 4). It furthers its motor involvement by having functional connections to the premotor regions (areas SCEF, FEF, 6a, and 6v). Area VIP also connects to insula opercular regions (area FOP4) and temporal lobe (areas PHT). It connects both to the lateral parietal lobe (areas 7PC, 7PL, 7AL, AIP, VIP, MIP, LIPv,

LIPd, PGp, IP0, and IPS1) and the medial parietal lobe (area DVT). Additionally, it has functional associations with both the medial occipital lobe (area V2, V3, and V4) and lateral occipital lobe (areas V3CD, V4t, PH, LO1, LO2, LO3, MT, MST, and FST). There are also connections to both the dorsal visual stream (area V3a, V3b, V7, V6a, and V6) and ventral visual stream (areas V8, PIT, FFC, VVC, VMV1, VMV2, and VMV3).

Area VIP is known to be directionally selective, responsive to optic-flow, and transmits information about self-motion (Bzdok et al., 2015; Field et al., 2020; Galletti & Fattori, 2018). This allows the brain to process and interpret the information necessary for effective navigation, spatial awareness, and interaction with the environment (Grefkes & Fink, 2005).

Area VIP is structurally connected to various brain regions through multiple white matter tracts. The Inferior Frontal Occipital Fasciculus (IFOF) white matter tract connects area VIP with the parcellations 9p, 8BL, and SFL in superior frontal gyrus. The IFOF passes through the posterior temporal lobe and extreme/external capsule. The MdLF connects VIP with the planum temporale, terminating at the MBelt area. This tract runs deep to the parietal lobe and projects inferiorly. VIP is also connected to the thalamus. These connections pass through the posterolateral thalamus and extend inferiorly to the brainstem and superior colliculus. In addition to the long-range connections, area VIP has local connections with neighboring regions such as 7AL, 7AM, 7PL, MIP, and LIPv. These short association bundles facilitate the integration of information from various brain areas.

Area 7m

Area 7m is located in the posterior part of the precuneus (Dadario & Sughrue, 2023), just in front of the parietooccipital sulcus. It does not form the sides of the sulcus. Area 7m is bordered by several other brain regions, including POS2 to the posterior, area v23asb and POS1 to the inferior, area 31pd to the anterior, and area 7pm and PCV to the superior. These connections between different brain regions demonstrate the complex and interconnected nature of the brain.

Area 7m exhibits functional connectivity with multiple brain regions. It connects both to the lateral frontal lobe (areas 8AV, 8BL, 8AD, i6-8, 47s, 9a, 9p, 10d, 10v, and 10r) and the medial frontal lobe (areas 9m, a24, d32, and s32). Area 7M also connects to the temporal lobe (areas STSva, STSvp, TGd, TE1a, TE1m, TE1p, PreS, and the hippocampus). It connects to both the lateral parietal lobe (areas PFm, PGi, and PGs) and to the medial parietal lobe (areas 23d, v23ab, d23ab, POS2, POS1, PCV, RSC, 7pm, 31a, 31pv, and 31pd).

Area 7m's function is as a subunit of the precuneus which is implicated in several different functions such as visuospatial processing, episodic memory retrieval, self-processing, and consciousness. Recent task fMRI studies have shown that area 7m supports specific processes of the precuneus. For example, area 7m is activated with working memory processing of place, body, tool, and face images. Its function can be further determined by the differences in activation during two different stimuli which showed that it is more activated in: listening to stories over answering arithmetic questions; focusing on socially interacting objects over randomly moving geometric shapes; recognizing emotional faces over neutral objects; and comparing featural dimensions of objects versus matching objects based on verbal classifications (Bzdok et al., 2015; Cavanna & Trimble, 2006; Glasser et al., 2016).

Area 7m is structurally connected to multiple parcellations in the cingulum and the parcellations 7m and PCV in the contralateral hemisphere. The cingulum fibers project anteriorly from 7m and have connections along the midcingulate and anterior cingulate cortex to d32, a24, p24, a24pr, and p24pr. Additionally, connections through the splenium of the corpus callosum terminate at the contralateral areas 7M and PCV. In addition, short association bundles establish connections with several areas, including POS1, POS2, 7pm, and PCV.

AIP

Area AIP (anterior intraparietal area) is located on the superior bank of the intraparietal sulcus at its most anterior part, extending onto the superior surface of the adjacent superior parietal lobule. The anterior tip of area AIP is situated in the bank of the postcentral sulcus. Area AIP is bordered by area 2 anteriorly and PFt anteroinferiorly. Its anterosuperior border is formed by area 7PC, while its inferior border is shared with IP2 across the intraparietal sulcus. LIPv and LIPd make up the posterior boundaries of area AIP.

Area AIP exhibits functional connectivity with multiple brain regions. It connects to the sensory strip (area 2) and premotor regions (areas SCEF, FEF, PEF 6a, 6r, and 6ma). It has functional connections to both the lateral frontal lobe (areas IFSa, IFJp, 46, and p9-46v) and medial frontal lobe (areas 5mv and 23c). Area AIP also connects to insula opercular regions (areas PoI2, FOP2, FOP4, and OP4) and temporal lobe (areas PHA3, PHT, and TE2p). It connects both to the lateral parietal lobe (areas 7PC, 7PL, 7AL, VIP, MIP, LIPv, LIPd, PFop, PFt, PGp, IP2, IP1, IP0, and IPS1) and the medial parietal lobe (areas DVT, 7AM, and 7pm). Additionally, it has functional associations with both the medial occipital lobe (area V2) and lateral occipital lobe (areas V3CD, PH, TPOJ2, and FST) with connection to the dorsal visual stream (area FFC).

Area AIP is an essential part of the brain's parietal cortex involved in various aspects of grasping activity, object recognition, and sensory tactile processing (Fogassi et al., 2001; Galletti & Fattori, 2018; Grefkes & Fink, 2005). For grasping activity, area AIP ensures proper hand orientation and shape. This works in conjunction with the information provided from the inferior temporal cortex, ventral and dorsolateral visual stream (Fogassi et al., 2001; Galletti & Fattori, 2018). Area AIP plays a vital role not only in grasping actions but also in other aspects of object manipulation and perception. Area AIP has been shown to be involved in tactile shape processing and understanding orientation in space (Grefkes & Fink, 2005).

MIP

Area MIP (medial intraparietal area) is situated in the posterior section of the superior bank of the intraparietal sulcus. It extends to the superior surface of the adjacent part of the superior parietal lobule. Area MIP is bordered by several other brain regions, including IPS1 to the posterior, area 7PL and VIP to the medial, IP1 and IP0 to the inferior, and LIPv and LIPd to the anterior. These connections between different brain regions demonstrate the complex and interconnected nature of the brain.

Area MIP is functionally connected to a wide array of brain regions. It is linked to both the lateral frontal lobe (areas IFSa, IFJa, IFJp, 46, and p9-46v) and medial frontal lobe (areas 5mv and 23c). It also links to the premotor regions (areas SCEF, FEF, PEF 6a, 6v, 6r, and 6ma) and the temporal lobe (areas PHA3, PHT, and TE2p). It has connections with the lateral parietal lobe (areas 7PC, 7PL, 7AL, AIP, VIP, LIPv, LIPd, PF, PFt, PGp, IP2, IP1, IP0, and IPS1) and the medial parietal lobe (areas DVT, 7AM, and 7pm). Area MIP is also connected to the medial occipital lobe (areas V1 and V2), the ventral visual stream (area FFC), and the lateral occipital lobe (areas PH and FST).

Area MIP has been shown to be involved in the interpretation of visual stimuli and converting that information into precise motor movements. It has also been shown to play in correction of any errors associated with the movements (Grefkes & Fink, 2005). More specifically area MIP is involved in both components of prehension: arm-reaching movements and grasp (Galletti & Fattori, 2018). In addition, it has been shown to play a role in processing of horizontal oculomotor signal and modulated by visual and somatosensory stimulation (Huffman & Krubitzer, 2001; Ploner et al., 2000; Whitsel et al., 2019). All of these functions come together to perform actions like taking food and putting it into your mouth.

Area PCV

Area PCV (precuneus cuneal visual area) is located in the anterior precuneus, which is situated just behind the marginal ramus of the cingulate sulcus. Area PCV shares its anterior border with areas 5mv, 23c, and 31a, inferior border with area 31pd, posterior border with areas 7M and 7pm, and superior border with area 7am.

Area PCV exhibits functional connections with several regions across the cortex, including: the lateral frontal lobe (areas 46, 9-46d, 8AD); the medial frontal lobe (areas a24prime, 5mv, 23c, and s32); the premotor region (areas FEF, 6ma, and 6a); the insula opercular region (area STV); the temporal lobe (areas PHA2, PHA3, and PHT); the lateral parietal lobe (7AL, 7PL, IP0, LIPd, PF, and PGp); the medial parietal lobe (areas 23d, POS2, POS1, RSC, DVT, 7am, 7pm, 7m, 31a, and 31pd); the medial occipital lobe (areas V1 and V2); the dorsal visual stream (area V6); and the lateral occipital lobe (areas TPOJ2 and TPOJ3).

Area PCV also exhibits structural connections to local and contralateral areas. White fibers which extend from area PCV connect it to areas 5m, 7am, PCV, and 7AL on the contralateral hemisphere via projections that traverse the splenium of the corpus callosum. Other superior projecting short fibers connect area PCV to local areas 7am, 7pm, and 5m.

Area LIPv

Area LIPv (lateral intraparietal, ventral) is located on the inferior edge of the superior parietal sulcus, a prominent groove on the surface of the parietal lobe. Specifically, LIPv is contained within the superior and upper aspect of the superior parietal lobule. Area LIPv shares its borders anteriorly with areas AIP and 7PC, superiorly by area VIP, laterally by LIPv, and posteriorly by area MIP.

Area LIPv is functionally connected to several regions across the cortex, including: the sensory strip (areas 1 and 2); the motor strip (area 4); the premotor regions (areas SCEF, FEF, PEF 6a, 6r, and 6v); the medial frontal lobe (areas p32prime, 5mv, and 23c); the insula opercular regions (area FOP4); the temporal lobe (areas PHT and TE2p); the lateral parietal lobe (areas 7PC, 7PL, AIP, VIP, MIP, LIPv, PFop, PFt, PGp, IP0, and IPS1); the medial parietal lobe (area DVT); the medial occipital lobe (areas V1, V2, V3, and V4); the dorsal visual stream areas (areas V3a, V3b, V7, V6a, and V6); the ventral visual stream areas (area V8, PIT, FFC, VVC, VMV1, VMV2, and VMV3); and the lateral occipital lobe (areas V3CD, V4t, PH, LO1, LO2, LO3, TPOJ2, MT, MST, and FST).

Area LIPv has white fiber tracts that connect it to local areas, including areas AIP, 7PC, IP2, LIPd, LIPv, and MIP. Interestingly, area LIPv is found to have right hemispheric connections with the IFOF in some, but not all, individuals.

Area LIPd

Located centrally on the superior aspect of the intraparietal sulcus, Area LIPd (lateral intraparietal, dorsal) is positioned ventral to LIPv. Area LIPd is flanked by a number of areas, including AIP on its anterior aspect, MIP on its posterior aspect, areas LIPv and VIP on its superior aspect, and areas IP2 and IP1 on its inferior border that spans the intraparietal sulcus.

Area LIPd is functionally connected to several regions across the cortex, including: the premotor regions (areas SCEF, FEF, PEF 6a, and 6r), the lateral frontal lobe (areas IFSa, IFSp, IFJa, p47r, i6-8, 8C, 9-46d, 46, a9-46v, and p9-46v), the medial frontal lobe (areas 8BM, p32prime, 5mv, and 23c), the insula opercular regions (areas PoI2, FOP4, FOP5, MI, AVI, and PoI2), the temporal lobe (areas PHA3, PHT, and TE1p), the lateral parietal lobe (areas 7PC, 7PL, AIP, VIP, MIP, LIPv, PF, PGp, IP2, IP1, IP0, and IPS1), the medial parietal lobe (areas PCV, DVT, 7AM, and 7pm), the medial occipital lobe (areas V1, V2, and V3), and the lateral occipital lobe (areas PH and FST).

Area LIPd is intricately connected to both the premotor region and neighboring local areas. However, it's worth noting that some individuals exhibit a peculiar pattern of anterior projecting structural connections from LIPd to the premotor region that terminate within the motor cortex, rather than reaching the premotor regions. Nevertheless, connections to the premotor regions include areas 55b and PEF. In addition, area LIPd shows structural connections to several local regions, including areas PGs, AIP, IP1, IP2, and PGs.

Brodmann area 8 (BA8)

Brodmann area 8 can be divided into a number of HCP regions. In our definition, and in accordance with work by the HCP, BA8 has five area 8 subdivisions: areas 8BL and 8AD on the posterior half of the superior frontal gyrus, areas 8AV and 8C on the posterior half of the middle frontal gyrus, and area 8BM in the medial superior frontal gyrus (Baker et al., 2018b, 2018c). Two pre-SMA regions (SCEF and SFL) and two hybrid areas between areas 6 and 8 were also described by the HCP (s6-8 and i6-8), but are not described in detail in the current work (Glasser et al., 2016). The authors have recently described this anatomy in detail (Dadario et al., 2023b), and further summarize it in brief below.

Area 8C

Area 8C is found at the posterior aspect of the MFG and is lateral to area 8AV. As such, its borders are area 8Av medially and its lateral borders area IFSp, IFJa, and IFJp. Posterior borders include areas 55b and PEF, while its anterior borders include area p9-46v.

Area 8C demonstrates functional connectivity to the dorsolateral frontal lobe (s6-8, i6-8, a9-46v, p9-46v, a10p, 8BL, 8AD, and 8AV), the medial frontal lobe (8BM and d32), the inferior frontal lobe (IFSp, IFJp, a47r, p47r, and 44), the insula (AVI), the temporal lobe (TE1p, TE2a, STSva, and STSvp), the inferior parietal lobe (IP1, IP2, LIPd, PFm, PGi, and PGs), and the medial parietal lobe (7pm, 31pv, 31a, POS2, 23d, and d23ab).

Area 8C is primarily connected via the arcuate/SLF. These long-range fibers project posteriorly to ultimately reach the posterior temporal lobe and terminate at areas PH and PHT. Furthermore, contralateral connections travel through the corpus callosum to end at contralateral 8C.

Area 8AD

Area 8AD (8A dorsal) is located on the banks of the SFS and the lateral aspect of the posterior SFG. Its anterior border is a wedge that adjoins 9P, 9-46d, and 46, from medial to lateral. Other borders include 8BL medially, 8AV laterally, areas s6-8 and i6-8 posteriorly.

The functional connectivity of area 8AD can be seen extending to the dorsolateral frontal lobe (9p, s6-8, i6-8, 10d, p10p, 8C, and 8AV), the medial frontal lobe (10r, a24, p32, s32, and d32), the orbitofrontal regions (area 47m), the temporal lobe and hippocampus (STSva, PHA1, PHA2, PreS, EC, TE1p, TE1m, TE1a), and the inferior and medial parietal lobe (PFm, PGi, and PGs, 7m, 7pm, 31pd, 31pv, 31a, RSC, PCV, 23d, d23ab, and v23ab).

Structural connections from this region primarily link with local parcellations. These short bundles connect with areas 9a, 9p, s6-8, 8Av, 6a, and p10p.

Area 8Av

Area 8Av (8A ventral) is also found on the posterior aspect of the MFG, but medial to area 8C. Its borders include 46 anteriorly, 55b, FEF, and i6-8 posteriorly, and 8C laterally.

The functional connectivity of area 8Av includes the dorsolateral frontal lobe (9a, 9p, 9m, 8BL, 8AD, and 8C, i6-8 and s6-8), the medial frontal lobe (areas 8BM, SFL, 10d, and d32), the inferior frontal lobe (44, 45, A47r, 47l, and 47s), area 55b of the premotor region, the tempora lobe (TGd, TE1a, TE1m, TE1p, TE2a, STSva, and STSvp), the inferior parietal lobe (PFm, IP1, PGi, and PGs) and the medial parietal lobe (7m, 31pd, 31pv, 31a, 23d, d23ab, and v23ab).

Structural connections are similar to the previous area 8AD and include the arcuate/SLF and contralateral extending fibers. Contralateral hemisphere connections travel to the contralateral SFL through the corpus callosum body. Arcuate/SFL connections travel posteriorly to the parietal lobe and terminate at 6a, 7PC, MIP, PFm, and 2. Local fibers link 8Av with 8C, 8Ad, i6-8, and 46.

Area 8BL

Area 8BL (8B lateral) is on the superior aspect of the posterior SFG. Its borders include 8BM medially, 8Ad laterally, 9p and 9m anteriorly, and s6-8 and SFL posteriorly.

Functional connectivity is seen linking area 8BL with the dorsolateral frontal lobe (9a, 9m, 8BM, 8C, and 8AV), the medial frontal lobe (SFL, 10d, 10v, and d32), the inferior frontal lobe (a47r, 44, 45, 47l, and 47s), the temporal lobe (TGd, TE1a, TE1m, TE2a, STSdp, STSva, and STSvp), the inferior parietal lobe (PGi and PGs), and the medial parietal lobe (7m, 31pd, 31pv, 23d, d23ab, and v23ab).

The structural connectivity of area 8BL is relatively vast. It is connected through the IFOF, contralateral connections, the medial thalamus, and the FAT. These fibers connect with the following: contralateral connections through the genu of the corpus callosum terminate at 8BM and 9m with the FM; medial thalamus connections travel through the internal capsule (anterior limb); IFOF connections travel from 8BL through the external capsule and travel posteriorly to parietooccipital regions (V2, V3, 7PL, MIP, V6, and V6a); FAT links 8BL to the IFG to terminate at area 44. Local fibers are also found linking with areas SFL, 8BM, 9a, and 9p.

Area 8BM

Area 8BM (8B medial) is found on the medial aspect of the posterior SFG. Borders 8BL, SFL, and area 24 subdivisions superiorly, d32 and a32pr inferiorly, 9m anteriorly, and SCEF posteriorly.

Area 8BM demonstrates functional connectivity to all area 8 subdivisions and also the dorsolateral frontal lobe (i6-8, s6-8, a10p, a9-46v, and p9-46), temporal lobe (TE1p, TE1m, STSvp), and the lateral parietal (LIPv, IP1, IP2, PFm, PGi, and PGs) and medial parietal lobe (7pm, 31a, and d23ab).

Structural connections of area 8BM are primarily mediated through short association fibers and contralateral connections, the FAT, the IFOF, and thalamic connections. Contralateral connections extend to contralateral area 8 subdivision 8BM and also 9m. FAT fibers project inferiolaterally to area 44. Differently, IFOF fibers travel through the external capsule and temporal lobe to terminate at parietal-occipital 7PC and V1−V3 areas. Thalamic connections project to area 8BM and to the brainstem. Local fibers link 8BM with 9m, d32, and SCEF.

Brodmann area 9 (BA9)

BA9 can be subdivided into 4 HCP regions, including areas 9a and 9p in the anterior superior frontal gyrus (SFG), area 9m in the medial SFG, and area 9-46d within the superior frontal sulcus and middle frontal gyrus. A more detailed description of each area is provided below in the following section.

Area 9a

Area 9a is located on the cortical surface of the anterior SFG and is flanked by area 10d anteriorly and by area 9p posteriorly, with areas 9m and 9-46d covering its medial and lateral borders, respectively.

Area 9a exhibits many functional connections with regions across the frontal, temporal and parietal lobes. In the dorsal lateral frontal lobe, area 9a is functionally connected to areas 9p, 9m, 8BL, and 8AV, while in the medial frontal lobe, it shares functional connections with areas SFL and d32. Additionally, in the inferior frontal lobes, area 9a shows functional connections with areas 44, 45, 47l, and 47s. Area 9a also exhibits functional connectivity with various regions situated in the temporal and parietal lobes, including areas TGd, TE1a, SYSva, and STSvp in the temporal lobe, areas PGi and PFm in the inferior parietal lobe, and areas 7m, 31pd, 21pv, and 23d in the medial parietal lobe.

Importantly, BA9, in which area 9a is situated, has been found to be active in working memory tasks and to have a role in reinstating executive control in autonomic behaviors (Kubler et al., 2006; Roux et al., 2012).

One of the major fiber bundles connecting area 9a is the inferior frontooccipital fasciculus (IFOF), which is known as being one of the brain's longest white matter tracks (Schmahmann & Pandya, 2007). The IFOF bundle is formed by many layers and is responsible for visual perception and other cognitive functions (Roux et al., 2021). The IFOF connects area 9a via fibers which pass through the external capsule, through the posterior aspect of the temporal lobe, and terminate at occipital parcellations V1, V2, and V3. Area 9a is also connected to the contralateral hemisphere, via fibers which pass through the forceps minor of the corpus callosum and end at areas 9a, 9p, p10p, 10d, and 9m. Area 9a is also connected to adjacent area 9p via short association bundles and while some individuals have medial thalamic connections, not all do.

Area 9p

Area 9p is located on the superior surface of the anterior SFG, with its superior triangular apex border situated between areas 8BL and 8AD. Meanwhile its medial and lateral borders are composed of areas 9M and 9-46d, respectively.

Area 9a demonstrates functional connectivity to various regions throughout the frontal lobe, including the dorsolateral frontal lobe (9p, 9m, 8BL, and 8AV areas), the medial frontal lobe (SFL and d32 areas), and inferior frontal lobe (44, 45,47l, and 47s areas). As part of Brodmann area 9, it is not surprising that area 9p shares similar functions with area 9a, such as working memory and executive control.

Furthermore, area 9p has white matter connections to the IFOF, which traverse through the external capsule, pass through the posterior aspect of the temporal lobe, and finally terminate at occipital parcellations V1, V2, V3, V3A, and V6. Lastly, Area 9a is connected to nearby regions 9a, 8BL, and 9-46d, via local short association bundles.

Area 9m

Area 9m is found on the anterior aspect of the medial SFG and is surrounded by several neighboring areas, including areas 8BM and 8BL on its posterosuperior border, 9a and 9p on its superior border, and p32, a24, p24, and d32 running anterior to posterosuperior on its wedging inferior and posterior border.

Area 9m is found to have extensive functional connections dispersed throughout different areas of the frontal lobe, including the dorsolateral frontal lobe (areas 9a, 9p, 10d, 8BL, 8AD, and 8AV), the medial frontal lobe (areas SFL, a24, 10r, 10v, and d32), and the inferior frontal lobe (areas 44, 45, 47s, and 47l). Additionally, area 9m shares functional connections with the temporal lobe (TGd, TE1a, STSdp, STSva, and STSvp), the insula (area AVI), the lateral parietal lobe (areas PGi and PGs), and the medial parietal lobe (areas 7m, POS1, 31pv, 31pd, 23d, v23ab, and d23ab). Area 9m is part of the dorsolateral prefrontal cortex, which has been shown to be highly active in tasks which require the active monitoring of multiple pieces of spatial information.

Area 9m shares structural connections to the contralateral hemisphere through white matter fiber tracts which traverse the corpus callosum ending at parcel areas 9m, 8BL, and 8BM. Additional white matter connections have been identified that bridge area 9m to precuneus areas 31pv, v23ab, and POS1. These connections occur via posterior-projecting cingulum fibers which pass adjacent to the dorsal aspect of the corpus callosum. In addition, area 9m connects to local structures (p32, 10d, and a24) via short associate bundles as well.

Area 9-46d

Describing area 9-46d is challenging due to its unique topographic location which runs along the superior frontal sulcus (SFS) and contains a domain which extends into the middle frontal gyrus (MFG). While the majority of area 9-46d is situated along the lateral flank of the SFG, it also spans across the SFS at an oblique angle in the anterior to posterior direction. Area 9-46d′s central point is located approximately in the anterior portion of the SFS, with an additional lateral wedge in its anterior aspect protruding from the SFS into the MFG. Running anterior to posterior, area 9-46d is bordered by areas p10q, 9a, and 9p, with a small part of it posterior aspect bordering area 8AD. The lateral wedge of area 9-46d shares a border with area a9-46v and area 46, while its anterior wedge forms a pointed edge which borders areas p10p and a9-46v on the medial and lateral edges, respectively.

Area 9-46d demonstrates functional connectivity to neighboring areas 46, a9-46v, and p9-46v. Additionally, functional connectivity is seen with areas in the dorsolateral frontal lobe (FEF), the medial frontal lobe (SCEF, a32prime, p32prime, p24, and a24prime), the premotor regions (6a, 6r, PEF), the insula-opercular region (FOP areas 1–5, PSL, PFop, PFcm, AVI, 43, PoI1, MI), the temporal lobe (PHT), lateral and medial parietal regions (PF, 7AL, 7PL, LIPd, POS2, PCV, 7am, 7pm, 23c), and occipital regions (V1–4, V6). Interestingly, a metaanalysis conducted by Nitschke et al. investigated the functional properties of area 9-46d and found it to be highly active during the planning of behaviors (Nitschke et al., 2017). In addition, area9-46d is located within the dorsal lateral prefrontal cortex, which is known to be activated during tasks that require the control of premeditated behaviors (Petrides, 2005).

In terms of structural connectivity, area 9-46d exhibits a minority of its white matter fiber projections to the contralateral hemisphere and majority its white matter fibers to local parcellations. The contralateral projections traverse across the corpus callosum and terminate at areas 9m and 9p. On the other hand, short association bundles link area 9-46d to numerous local areas, including 8BL, 9P, 9-47D, 46, a9-46v, 8Ad, p47r, and a10p. Additionally, some individuals may have structural connections between area 9-46d and the IFOF, but this is not a consistent finding.

Area IFJa

Area IFJa is situated within the posterior cortical surface of the inferior frontal sulcus and has landscape that extends to the inferior aspect of the middle frontal gyrus. Furthermore, it can be found superior to the pars opercularis which resides in the inferior frontal gyrus. Area IFJa is bounded anteriorly by area IFSp, posteriorly by area IFJp, inferiorly by area 44, and superiorly by area 8c.

Area IFJa exhibits functional connectivity with several regions across the cortex, including: the dorsolateral frontal lobe (areas 44, IFSa, IFSp, IFJp, and p9-46v), the medial frontal lobe (area SCEF), the premotor areas (areas FEF, 55b, PEF, and 6r), the insular opercular regions (area FOP5 and PSL), the temporal lobe (areas PH, PHT, and TE2p); and the inferior parietal lobe (areas MIP, TPOJ1, and LIPd).

Area IFJa is connected to areas TE1a, TE1m, and TE2a of the middle and inferior temporal gyrus via superior longitudinal fasciculus connections that pass posteriorly and bend around the sylvian fissure. Additionally, area IFJa has white matter tracts that pass superiorly and terminate in area SFL. These superior projecting fibers are thought to be part of the frontal aslant tracts. Area IFJa also is structurally connected to areas 8c, IFJa, IFSp, 44, and 8Av via short association fibers.

Area IFJp

Area IFJp is situated on the posterior aspect of the inferior frontal sulcus, just superior to the pars opercularis within the inferior frontal gyrus. Area IFJp is bounded anteriorly by area IFJa, posteriorly by area PEF, medially by area 8C, and inferiorly by the superior borders of areas 6R and 6V.

Area IFJp exhibits functional connectivity with several regions across the cortex, including:the dorsolateral frontal lobe (areas p47r, IFSa, IFJa, IFJp, p9-46v, i6-8, and 8C), the medial frontal lobe (areas 8BM and 33prime), the premotor areas (areas 6r, 6a, and PEF), the temporal lobe (areas PH, PHT, TE1p, and TE2p), the inferior parietal lobe (areas 7PL, PFt, PF, IP0, IP1, IP2, AIP, MIP, and LIPd), and the medial parietal lobe (area 7PM).

Area IFJp also exhibits structural connections with the arcuate/superior longitudinal fasciculus and local areas. Area IFJp is structurally connected to areas PHT and FST within the temporal gyrus via connections to the superior longitudinal fasciculus extend posteriorly and bend around the sylvian fissure. Local structural connections to areas 8C, IFJa, IFJp, IFSp, 44, 6r, and PEF are also demonstrated.

Brodmann area 10 (BA10)

Area 10, consistent with HCP, is separated into 6 discrete regions: area 10pp, area 10d, area a10p, area p10p, area 10v, and 10r. We define these regions further below in the context of the speculated functional relevance (Glasser et al., 2016). For additional definitions and reasons for separating these subdivisions from other surrounding areas, see the supplementary material of Glasser et al. (2016).

Area 10pp

Area 10pp, located at the anterior tip of the frontal pole, is situated at the junction of the most anterior aspects of the superior and middle frontal gyri. This region has important connections with the orbitofrontal cortex, which is found along its inferoposterior boundary. Area 10pp is bordered by area 10d on its superior boundary and shares its lateral boundary with area a10p.

Area 10pp has limited evidence that shows strong functional connectivity with other brain areas when using a z-score cutoff of 20. There is some amount of evidence that suggests a functional connectivity with its lateral neighbor area a10p. It also exhibits some functional connectivity with areas 8Av and 8C in the posterior dorsolateral frontal lobe. Other possible functional connections include areas 31pd and 31pv in the posterior cingulate cortex, areas PFm and PGi in the inferior parietal lobule, and area STSvp in the superior temporal sulcus. It is important to note that these areas show only borderline functional connectivity.

Area 10pp is involved in episodic and working memory tasks. More broadly, BA10, which includes 10pp, is activated in response to increasing complexity of working memory tasks (Chahine et al., 2015). Additionally, area 10pp contributes to abstract cognitive functions, which may include reasoning, planning, and problem-solving (Bludau et al., 2014). Area 10pp can be differentiated from its lateral neighbor, area a10p, as it is less activated in working memory and more activated in language story versus math, theory of mind versus random, and faces versus shapes (Glasser et al., 2016).

Area 10pp is structurally connected to parcellations V1 and V1 in the occipital lobe via the IFOF and parcellations 10d and 10pp in the contralateral hemisphere. The IFOF connections originate from area 10pp and pass through the extreme/external capsule, continuing posteriorly until they reach the occipital lobe parcellations V1 and V2. Contralateral connections travel through the genu of the corpus callosum with the forceps minor and eventually terminate at areas 10d and 10pp in the contralateral hemisphere. Additionally, local short association bundles establish connections with area 10d.

Area 10d

Area 10d (10 dorsal) can be found in the anterior part of the superior frontal gyrus (SFG). It is situated within the interhemispheric fissure, on the medial side. Area 10d is bordered by area 10pp below it and area 9a to the rear in the SFG. It is also surrounded by area 10r on its medial side, and areas a10p and p10 on its lateral side.

Area 10d exhibits functional connectivity with multiple brain regions. It connects to the anterior frontal lobe (areas 10r, 10v, 9M, 9P, 8BL, and 8AV). It has connections to both the anterior cingulate cortex (areas s32, d32, and a24) and the posterior cingulate area (areas POS1, 7M, 31a, 31pv, 31pd, and v23ab). Area 10d also connects to the middle temporal gyrus region (areas STSva and TE1a), the inferior parietal lobule (areas PGs and PGi), and the hippocampus.

Similar to area 10pp, area 10d is involved in episodic and working memory tasks and is activated in response to increasing complexity of working memory tasks (Chahine et al., 2015). Additionally, area 10d contributes to abstract cognitive functions, which may include reasoning, planning, and problem-solving (Bludau et al., 2014). Area 10d can be further differentiated from its posterior neighbor 9A because it is more deactivated during gambling (Glasser et al., 2016).

Area 10d is structurally connected to the contralateral hemisphere's areas 9a and 10d. These white matter connections traverse through the corpus callosum genu via the forceps minor before reaching area 9a and 10d. Additionally, area 10d establishes connections with local short association bundles, including those in areas 10pp, a10p, 9a, 10r, 9m, and p32. These structural connections are also heavily myelinated than area 9A further differentiating them from neighboring parcellations (Glasser et al., 2016).

Area a10p

Area a10p (anterior 10 polar) is located at the junction of the anterior parts of the superior and middle frontal gyri. It is surrounded by several other brain regions, including area 11r on its inferior boundary, area p10p on its superior boundary, areas a47r and p47r as its lateral neighbors, and area 10pp as its medial neighbor.

Area a10p is functionally linked to various brain regions. It establishes connections with the dorsolateral frontal lobe (areas p10p, a9-46v, 8BM, and 8C), the anterior cingulate region (area d32), the lateral parietal lobe (area PFm), and the medial parietal lobe (areas POS2 and 7pm).

Similar to previous regions above, area a10p is activated in response to increasing complexity of working memory tasks (Chahine et al., 2015). Additionally, area a10p contributes to abstract cognitive functions, which may include reasoning, planning, and problem-solving (Bludau et al., 2014). Area a10p can be further differentiated by comparing it to its superoposterior neighbor area p10p. Area a10p is more activated in gambling. Additionally, in the left hemisphere relative to its posteriorly border area a9-46v, area a10p is less deactivated in the language story versus math contrast and more activated in the relational primary contrasts (Glasser et al., 2016).

Area a10p is structurally connected to the occipital lobe parcellations V1, V2, V3, V6, and V6a via the IFOF and to the contralateral hemisphere areas 9a and p10p. The contralateral connections travel through the corpus callosum genu via the forceps minor and ultimately end at areas 9a and p10p. Meanwhile, the IFOF connections originate from a10p and pass through the external capsule, continuing posteriorly until they reach the occipital lobe parcellations V1, V2, V3, V6, and V6a. In addition, local short association bundles establish connections with several areas, including 10d, 10pp, p10p, a9-46v, and 9-46d. Area a10p can also be differentiated from area p10p based on increased thickness (Glasser et al., 2016).

Area p10p

Area p10p (posterior 10 polar) is situated on the lateral side of the anterior part of the superior frontal gyrus (SFG), just behind area a10p. It is positioned at the end of the superior frontal sulcus (SFS). Area p10p is bordered by area a10p on its anterior boundary and area a9-46d, as well as a small part of area 9a, on its superior boundary. Area 9-46d is also its lateral neighbor. The shape of area p10p is slightly triangular, with its superior apex wedged between area 9a and area 9-46d.

Area p10p has functional connections with several brain regions. It links to the anterior frontal lobe (areas a10p and a9-46v), the anterior cingulate region (area d32), the posterior frontal lobe (area i6-8), the posterior cingulate region (areas 31a, 31pv, d23ab, POS2, retrosplenial cortex (RSC), and 7pm), and the inferior parietal lobule (area PFm).

Area p10p is involved in episodic and working memory tasks and facilitates increasing complexity of working memory tasks (Chahine et al., 2015). Additionally, area p10p contributes to abstract cognitive functions, which may include reasoning, planning, and problem-solving (Bludau et al., 2014). Compared to its lateral neighbor a9-46v, area p10p exhibits deactivation during gambling and emotion primary contrasts, and a weaker deactivation during the story versus math contrast. Conversely, when compared to its neighboring areas 9a and 10d, p10p shows higher activation in working memory tasks (Glasser et al., 2016).

Area p10p is structurally connected to the occipital lobe parcellations V1 and V2 via the IFOF, as well as to the contralateral hemisphere's area 9m. The contralateral connections travel through the corpus callosum genu via the forceps minor and ultimately end at area 9m. The IFOF connections originate from p10p and travel through the external capsule, extending posteriorly until they reach the occipital lobe parcellations V1 and V2. In addition, local short association bundles establish connections with several areas, including 9-46d, 10d, a10p, a9-46v, and p9-46v. Further structural differences make area p10p a unique parcellation within the local area. Area p10p is thinner than its posterior neighbor area 9-46d and has more myelin than its posteromedial neighbor area 9a (Glasser et al., 2016).

Area 10v

Area 10v is situated in the lower front part of the medial superior frontal gyrus (SFG). It has a long inferior border with the orbitofrontal cortex (OFC) and shares borders with several other brain regions. Areas 10pp and 10d form its anteroinferior borders, while area 10r is located above it as its superior neighbor. Superiorly, area 10v also has a border with area s32. It terminates in the OFC in the posterior region.

Area 10v is functionally connected to numerous brain regions. It establishes connections with both the dorsolateral frontal lobe (areas 10d, 47l, and 8BL) and the medial frontal lobe (areas 9m, 10r, and s32). It also has connections to the temporal lobe (areas TGd, TE1a, STSva, and STSvp). Lastly, it has functional connection to both the lateral parietal lobe (areas PGi and PGs), and the medial parietal lobe (areas 7m, POS1, 31pv, 31pd, and v23ab).

Area 10v represents a novel ventral subdivision of the preexisting area 10. Typically linked to the ventromedial prefrontal cortex, this region is believed to contribute to behavioral decision-making processes by combining value evaluations from the orbitofrontal cortex (OFC) and the anterior cingulate cortex (ACC) (Grabenhorst & Rolls, 2011).

Area 10v is structurally connected to the contralateral hemisphere's areas 10v and 10d and to parcellation TGd in the temporal lobe via uncinate fasciculus. The contralateral connections travel through the corpus callosum genu to end at areas 10v and 10d. The uncinate fasciculus fibers project through the limen insulae to temporal pole parcellation TGd. It should be noted that connections with the uncinate fasciculus are not consistent across individuals.

Area 10r

Area 10r (10 rostral) is situated in the front lower part of the medial superior frontal gyrus. It shares borders with several other brain regions, including area 10d and 10pp in the front and below, respectively. Area p32 is located on its superior boundary, while area s32 is on its posterior boundary. These connections between different brain regions demonstrate the complex and interconnected nature of the brain.

Area 10r displays functional connectivity with a range of brain regions. It connects to both the dorsolateral frontal lobe (areas 9p, 10d, and 8AD) and the medial frontal lobe (areas 9m, a24, 10v, s32, and p32). It also has functional connections to the temporal lobe (areas TGd, TE1a, STSva, PHA1, and the hippocampus) with additional connections to both the lateral parietal lobe (areas PGi and PGs) and the medial parietal lobe (areas 7m, POS1, 31pv, 31pd, v23ab, and d23ab).

The rostral portion of area 10 (area 10r) is a relatively newer parcellation of the preexisting area 10 that still has a lot to be discovered about its function. The research suggested that it facilitates either stimulus-oriented (SO) or stimulus-independent (SI) attending (Burgess et al., 2007). SO tasks require concentration on a current sensory input while SI attending is self-generated. The involvement of area 10r in these processes indicates that it is used in working memory and concentration.

Area 10r is structurally connected to areas 10r and 10v in the contralateral hemisphere and to the cingulum. The contralateral connections travel through the corpus callosum genu to end at areas 10r and 10v. The cingulum fibers project posteriorly from 10r to precuneus areas v23ab, POS1, and RSC. In addition, local short association bundles establish connections with several areas, including p32, 10d, and 10v. It should be noted that white matter tracts in the right hemisphere have less consistent connections with the cingulum.

Brodmann area 11l (BA11l)

Area 11l is a parcellation in the orbitofrontal region. This region has historically been poorly defined, only consisting of two regions: the gyrus rectus or the orbitofrontal gyri. Below we highlight and define relevant information about this parcellation. For additional definitions and reasons for separating these subdivisions from other surrounding areas, see the supplementary material of Glasser et al. (2016).

Area 11l

Area 11l is situated in the anterior portion of the orbitofrontal cortex (OFC) within the frontal lobe. Area 11l borders area a10p anteriorly, area a47r laterally, both areas 13l and 47m posteriorly, and OFC medially.

Area 11l demonstrates functional connectivity to both the frontal lobe (areas 46, a9-46v, p9-46v, and IFSa) and IP2 in the parietal lobe.

Investigations into the role of area 11l have unveiled its fundamental responsibility in the reception, integration, and regulation of olfactory cues. This region appears to be crucial for assessing food-related reinforcers, as well as scrutinizing satiety levels and the anticipation of forthcoming food rewards. By contributing to a more comprehensive understanding of these processes, the research on area 11l holds promise in unraveling the intricate mechanisms that govern appetite regulation and reward-based feeding behaviors (Gottfried, 2007; Ongur et al., 2003).

Area 11l exhibits structural connectivity with the occipital lobe via the inferior frontooccipital fasciculus, although its connection with the uncinate fasciculus is not always present in individuals. The inferior frontooccipital fasciculus fibers course through the extreme/external capsule to terminate at area V1 of the occipital lobe.

Brodmann area 13 (BA13)

Area 13 has been further specified as a single HCP parcellation—area 13L.

Area 13L

Area 13L is situated in the anterior portion of the orbitofrontal cortex (OFC) within the frontal lobe. Area 13L borders area 11L anteriorly, areas 47s and 47m make up the posterior and lateral border, and OFC medially.

Area 13L mainly exhibits functional connectivity to areas 47m and OFC.

Area 13L has been identified as a pivotal secondary nexus for the processing of olfactory, gustatory, visceral, and food texture information. This region adeptly integrates these diverse sensory inputs to evaluate satiety levels by juxtaposing food reward values against prevailing internal physiological states. Elucidating the role of area 13L in this sophisticated network can foster a more comprehensive understanding of the complex interplay between sensory inputs and internal cues that ultimately govern satiety and appetite regulation (Carmichael & Price, 1995; Ongur et al., 2003; Rolls, 2011). ‘

Area 13L's structural connections consist mostly of local parcellations. These parcellations primarily include 11L, 47m, and 47s.

Brodmann area 17 (BA17)

Area V1

Area V1, also referred to as the primary visual cortex, is situated on the banks of the calcarine sulcus and occupies both banks of the sulcus, extending onto the occipital pole. Area V1 is adjacent to area V2 which covers its entire length. Area V1 shares its anterior border with the prostriate cortex.

Area V1 is the first cortical region to receive information regarding visual stimuli. Within each hemisphere, area V1 is responsible for encoding an inverted, contralateral hemifield. This region exhibits a larger cortical surface area dedicated to processing visual stimuli perceived by the fovea, signifying enhanced processing capabilities for input situated closer to the center of the visual field (Bailey, 1951). Moreover, area V1 has been substantially implicated in the primary detection of motion, as well as in transmitting signals to the middle temporal area (MT) to facilitate the directional and spatial integration of global motion patterns. By understanding these complex interactions, researchers can gain insights into the underlying mechanisms that enable the brain to process and interpret dynamic visual information (Petrides et al., 2012).

Area V1 demonstrates functional connectivity with numerous brain regions. Area V1 is functionally connected to premotor regions (areas SCEF and FEF) and the frontal lobe (46, 9-46d, p9-46v, 23c, 5mv, a24prime, p32prime, a32prime, and p24). In the insula and opercular region, area V1 has functional connections with the parcellations FOP4, FOP5, AVI, PoI1, 43, LBelt, PBelt, and A1, with additional connection to area PHT in the temporal lobe. Area V1 demonstrates robust connectivity with the parietal lobe (areas 7am, 7PL, PGp, PFop, PF, MIP, LIPd, LIPv, IPS1, IP1, IP0, POS2, RSC, PCV, and DVT). Area V1 exhibits strong connections to multiple visual areas of the brain including functional connections to both the medial occipital lobe (ProS, V2, V3, and V4) and the lateral occipital lobe (V3cd, LO1, LO2, LO3, PH, and FST). It furthers its visual involvement by having several connections to both the dorsal visual stream (areas V3a, V3b, V7, V6, and V6a) and ventral visual stream (FFC, VVC, V8, PIT, VMV1, VMV2, and VMV3). These various connections emphasize the critical role of V1 in integrating visual information with various brain regions involved in processing and interpreting sensory input.

Area V1 is structurally connected to the parcellations 11I, a10p, and p10p in the frontal pole via the IFOF, parcellations STSda in the superior temporal gyrus via MdLF, the optic radiations (OR), and the contralateral area V1. Area V1 exhibits consistent connections with the forceps major which crosses the midline via the splenium of the corpus callosum to connect area V1 contralaterally. Additional short local association bundles are connected to other areas within the occipital lobe V2, V3, and V3B.

Area ProS

Area ProS is a cortical region that is located within the anterior portion of the calcarine fissure, more specifically within the prostriate cortex. This area is situated between the parietooccipital sulcus and the calcarine fissure, and is located posterior to the isthmus of the cingulate gyrus. Area ProS is flanked posteriorly by area V1, anteriorly by area PreS (presubiculum), superiorly by areas DVT, POS1, and RSC, and inferiorly by the inferior aspect of area V2.

Area ProS displays functional connectivity with several regions across the cortex, including: the temporal lobe (areas PHA1, PHA2, and PreS), the parietal lobe (areas IP0, DVT, POS1, and PGp), the medial occipital lobe (areas V1, V2, V3, and V4), the dorsal visual stream areas (areas V3a, V3b, V7, V6, and V6a), the ventral visual stream (VVC and V8), and the lateral occipital lobe (area V3cd).

Area ProS exhibits structural connectivity with the parahippocampal cingulum, terminating at PeEc. In addition, it has contralateral connections to area V1 through fibers that traverse the splenium of the corpus callosum via the forceps major. Local structural connections within the area also exist through fibers that project posteriorly to area V1 and superiorly to areas V6 and POS2.

Brodmann area 18 (BA18)

Area 18 is primarily divided into two visual regions, parcellations V2 and V3.

Area V2

Area V2 has a distinctive "C" shape that completely encompasses V1, with its apex slightly looping around the occipital pole onto the lateral surface. Area V2's upper limb is shorter than its lower limb but it constitutes a significant part of the cuneus. On the other hand, its lower limb comprises a substantial portion of the lingula. V2 is located interiorly to V1, while its external boundary is formed by V3. The anterior border of its upper limb is shared with V6 and DVT, and the anterior tip of its lower limb forms a wedge between ProS and VMV1.

Area V2 plays an essential role in processing visual information and demonstrates functional connectivity with a wide range of brain regions. Area V2 connects to the sensory strip (areas 1, 2, 3a, and 3b) and the motor strip (area 4) and has further motor connection to the premotor regions (areas SCEF, FEF, and 6a). It also has connections to the frontal lobe (areas 46, 9-46d, 23c, 5mv, 24dd, 24dv, a24prime, and p32prime), the insula and opercular region (area FOP3, FOP4, OP4, OP2-3, PoI1, PoI2, PFcm, 43, 52, RI, TA2, STV, PSL, MBelt, LBelt, PBelt, A1, and A4), and the temporal lobe (area PHT). In the parietal lobe, V2 connects with areas 7PC, PGp, PFop, AIP, VIP, MIP, LIPd, LIPv, IPS1, IP0, PCV, and DVT. It has strong functional connectivity with areas involved with processing of visual information having connections to both the medial occipital lobe (ProS, V1, V3, and V4) and lateral occipital lobe (areas TPOJ1, TPOJ2, TPOJ3, MST, MT, V3cd, V4t, LO1, LO2, LO3, PH, and FST). It has additional functional connection with both the dorsal visual stream (areas V3a, V3b, V7, V6, and V6a) and ventral visual stream (areas FFC, VVC, V8, PIT, VMV1, VMV2, and VMV3). These extensive functional connections emphasize the vital role of V2 in processing and integrating visual information with various brain regions involved in interpreting sensory input.

Area V2 delineates the visual field into two disjointed, inverted, contralateral quarterfield maps, which are separated by the horizontal meridian. This region is responsible for processing visual stimuli at wider angles from the foveal point compared to V1, suggesting that V2's perception of the visual field extends further from the fovea than that of V1 (Wandell et al., 2007). Furthermore, V2 has been associated with more intricate visual integration tasks following the reception of information from V1. Among its notable functions, V2 is thought to play a vital role in differentiating between a stimulus and its surrounding background. By examining the contributions of area V2 to visual processing, researchers can better comprehend the sophisticated neural mechanisms that facilitate the interpretation of complex visual scenes (Qiu & von der Heydt, 2005).

Area V2 is structurally connected to the parcellations 8BL, 9a, 9p, 10d, 10pp, a10p, and 10pp in the frontal pole via IFOF, parcellations STSda, TGd, and TE1a in the superior temporal gyrus via the MdLF, and to the contralateral hemisphere through connections to the forceps major (FM). The FM traverses through the splenium of the corpus callosum to the contralateral hemisphere to establish structural connections to the contralateral parcellations of V1, V2, V3, V4, V6, DVT, and V3A. Additionally, there are local short association bundles that are connected to areas V1, V2, V3, V4, V6, V3A, and DVT.

Area V3

Area V3 is a "C" shaped region that surrounds V2 in the occipital lobe. Its apex extends around the occipital pole onto the lateral surface, while its upper limb makes up a smaller portion of the cuneus and its inferior limb makes up the lower half of the lingula. Area V3 wraps around V2 on the interior and is bordered externally by V4. The superior boundary of its anterior limb is V3a, while the anterior boundary is V6. The anterior boundary of its inferior limb is VMV1.

Area V3 has similar important functional connections as area V2. Area V3 connects to the sensory strip (areas 1, 2, 3a, and 3b) and the motor strip (area 4) and has further motor connection to the premotor regions (areas SCEF, FEF, and 6a). It also has connections to the frontal lobe (areas 46, 9-46d, 23c, 5mv, a24prime, and p32prime), and the insula and opercular region (areas FOP3, FOP4, OP4, OP2-3, PoI1, PFcm, 43, 52, RI, TA2, STV, PSL, MBelt, LBelt, PBelt, A1, and A4). In the parietal lobe, V3 connects with areas PGp, PFop, VIP, LIPd, LIPv, IPS1, IP0, and DVT. It has strong functional connectivity with areas involved with processing of visual information having connections to both the medial occipital lobe (ProS, V1, V3, and V4) and lateral occipital lobe (areas TPOJ1, TPOJ2, MST, MT, V3cd, V4t, LO1, LO2, LO3, PH, and FST). It has additional functional connection with both the dorsal visual stream (areas V3a, V3b, V7, V6, and V6a) and ventral visual stream (areas FFC, VVC, V8, PIT, VMV1, VMV2, and VMV3). Both V3 and V2 have extensive functional connections but there are some differences that can be highlighted between them. Area V2 has functional connections in the parietal lobe (area 7PC, MIP, and PCV), the frontal lobe (areas 24dd and 24dv), and the temporal lobe (PHT) that do not exist in area V3.

Analogous to V2, area V3 processes visual input in the form of two distinct, inverted, contralateral quarterfield maps, which are separated by the horizontal meridian. Area V3 captures an even broader range of the visual field compared to V2, signifying its ability to perceive visual stimuli at more distant angles from the fovea than V2 (Wandell et al., 2007). Moreover, area V3, along with V2 to a lesser extent, has been demonstrated to be involved in the perception and integration of global motion. This is achieved by processing directionally specific, local visual motion within its expansive receptive fields and subsequently transmitting the information to area V6 and the dorsal visual stream for further processing. By examining the roles of areas V3 and V2 in motion perception, researchers can gain a deeper understanding of the complex neural pathways that enable the brain to interpret and integrate dynamic visual information (Furlan & Smith, 2016).

Area V3 is structurally connected to the parcellations 8BL, 9a, 9p, and 10s via the IFOF, area A5 in the superior temporal gyrus via MdLF, areas STGa and TGd via the ILF, and to the contralateral hemisphere through connections to the forceps major (FM). The IFOF projections start in the superior aspect of the area V2 before it bends inferiorly to the basal surface of the occipital lobe. The ILF projections start in the inferior portion of V3 along the basal surface of the occipital lobe. The FM traverses through the splenium of the corpus callosum to the contralateral hemisphere to establish structural connections to the contralateral parcellations of V1 and V2. Additional local short association bundles connection exist with areas V1, V2, V3A, and V4.

Brodmann area 19 (BA19)

Area 19 can be divided into 16 different unique HCP regions primarily on the surface of the occipital lobe. Area V4 is the only one found on the medial surface. While the superior surface has areas V6,

V6a, V7, V3a, and V3b, the lateral surface of the occipital lobe includes LO1, LO2, LO3, V4t, MT, MST, FST, and PH. Then on the basal surface area V8 is found. Lastly, the intraparietal sulcus has areas IPS1 and MIP.

Area V4

V4, which wraps around V3 in a semicircular shape, is mainly situated on the posterior aspect of the lateral occipital surface. Unlike V3 and V2, its upper limb is shorter and does not extend onto the medial surface, while its inferior limb extends onto the medial basal surface of the occipital lobe. V4 borders V3 internally, and its superior limb terminates in V3a, while its inferior limb ends in VMV2 and VMV3. Additionally, it shares a common external boundary with numerous visual processing areas, including V3b, V3cd, LO1, LO2, Pit, and V8, arranged from superior to inferior.

Area V4 exhibits functional connections with a wide array of brain regions. It links to both the sensory strip (areas 1, 3a, and 3b) and the motor strip (area 4) with additional connection to the premotor region (areas SCEF, FEF, and 6a). It also has functional connectivity to the frontal lobe (areas 9-46d, 23c, a24prime, and p32prime). Area V4 connects to the insula and opercular regions (areas FOP3, OP4, PFcm, 43, 52, RI, TA2, STV, PSL, MBelt, LBelt, PBelt, A1, and A4) and the parietal lobe (areas PFop, VIP, LIPv, IPS1, IP0, and DVT). It has both connections to the medial occipital lobe (areas ProS, V1, V3, and V4) and the lateral occipital lobe (areas TPOJ1, MST, MT, V3cd, V4t, LO1, LO2, LO3, PH, and FST). Area V4 has further connections with the dorsal visual stream (areas V3a, V3b, V7, V6, and V6a) and the ventral visual stream (areas FFC, VVC, V8, PIT, VMV1, VMV2, and VMV3).

Area V4 plays a critical role in object recognition through color processing, pattern identification, perception of form, and distinguishing patterns (Amunts & Zilles, 2015). Additionally, it also plays a role in the contralateral hemifield representations of the peripheral visual field (Bailey, 1951).

Area V4 has structural connections with the TF via ILF projections, IFOF, and the superior temporal gyrus (areas A4 and A5) via the MdLF. IFOF projections do not consistently terminate and so the parcellations connections are not known. Additionally, there are short association bundles that exhibit connections to V3, V3A, V3B, V3CD, LO1, LO2, and V6A.

Area V6

Area V6 is a vertically oriented visual area located in the anterosuperior region of the cuneus, just posterior to the superior parietooccipital sulcus. It shares a posterior border with the superior limbs of V2 and V3, and an anterior border with DVT. V6 is bounded inferiorly by V2 and superiorly by V6a.

Area V6 exhibits robust functional connections throughout the brain. It links to both the sensory strip (areas 1, 2, 3a, and 3b) and the motor strip (area 4) with additional connection to the premotor region (areas SCEF and FEF). It also has functional connectivity to the lateral frontal lobe (areas 9-46d and 46) and the medial frontal lobe (areas a24prime, p32prime, 5mv, and 23c). Area V6 connects to the insula opercular regions (areas FOP1, FOP3, FOP4, OP4, OP2-3, 43, PFcm, STV, PoI1, PoI2, MI, RI, TA2, 52, A4, MBelt, and PBelt) and the parietal lobe (areas 7PC, 7AL, 7am, VIP, LIPv, PGp, PFop, IPS1, IP0, PCV, and DVT). It has both connections to the medial occipital lobe (areas ProS, V1, V2, V3, and V4) and the lateral occipital lobe (areas TPOJ2, TPOJ3, V3cd, V4t, LO1, LO2, LO3, MT, MST, PH, and FST). Area V6 has further involvement functional connection with the dorsal visual stream (areas V3a, V3b, V7, and V6a) and the ventral visual stream (areas FFC, VVC, V8, VMV1, VMV2, and VMV3).

Area V6 acts as almost the motion detectors of the brain. Area V6 is very sensitive to detecting the direction of motion of visual stimuli. In patients that have damage to this area they are unable to actually detect visual motion (Petrides et al., 2012). Area V6 has shown to be mostly sensitive to the peripheral visual field (von Economo & Koskinas, 1925).

Area V6 is structurally connected to the frontal lobe (9a, 9p, 9m, and 8BL) via the IFOF, parcellations at MBelt, A5, and PI via the MdLF, and the areas at V6, V2, and V1 via the FM. The MdLF runs parallel to the IFOF then courses laterally to the superior temporal gyrus, as the IFOF courses medially between the lateral ventricle and insula. FM connections travel through the splenium of the corpus callosum to terminate at V6, V2, and V1. Additionally, there are short association bundles that exhibit connections to V7, V3, V3b, and V3a. The right hemisphere has consistent ILF projections.

Area V6A

Area V6A (visual area 6a) can be found along the angle of the interhemispheric cleft along the superior surface of the occipital lobe. It borders V3a posteriorly and DVT anteriorly with V7 and Ips1 making its lateral borders and V6 its inferior border.

Area V6A demonstrates vast functional connectivity similar to V6. Functional connections are seen with the sensory (areas 1,2,3a,3b), premotor (SCEF, FEF), and motor (area 4) strips. Furthermore, functional connections extend to the medial frontal lobe (a24prime, and p32prime, 5mv, 23c) the insula opercular cortices (OP4, 43, RI, A4, and PBelt), and the parietal lobe (7PC, VIP, LIPv, PGp, IPS1, IP0, and DVT). Occipital lobe functional connections are seen with medial occipital regions (ProS, V1, V2, V3, and V4), the dorsal visual stream (V3a, V3b, V7, and V6), the ventral visual stream (FFC, VVC, V8, VMV1, VMV2, and VMV3) and lateral occipital regions (TPOJ2, V3cd, V4t, LO1, LO2, LO3, MT, MST, PH, and FST).

Structural connectivity of this region is primarily through the IFOF and MdLF long-range connections. IFOF connections link area V6a to frontal regions 8BM and 8BL of BA8. MdLF fibers run alongside the IFOF and link this region with the superior temporal gyrus before ending at A4 and MBelt. Local connections link V6A with V3B and V7.

Area V7

Area V7 is located on the superior surface of the occipital lobe, close to the interhemispheric cleft situated lateral to the most superior point of the intraparietal sulcus. Area V7 shares a border with V3a at its posterior end and IPS1 at its anterior end. On its lateral side, V7 is bordered by V3b, while on its medial side, it is bordered by V6a.

Area V7 is functionally connected to both the sensory strip (areas 1, 2, 3a, and 3b) and the motor strip (area 4), with additional connections to the premotor region (areas SCEF and FEF). It also has functional connectivity to the medial frontal lobe (areas a24prime, p32prime, and 23c). Area V7 connects to the insula opercular regions (areas OP4, 43, MI, RI, A4, LBelt, and PBelt) and the parietal lobe (areas VIP, LIPv, IPS1, IP0, and DVT). It has both connections to the medial occipital lobe (areas ProS, V1, V2, V3, and V4) and the lateral occipital lobe (areas V3cd, V4t, LO1, LO2,

LO3, MT, MST, PH, and FST). Area V7 has further functional connection with the dorsal visual stream (areas V3a, V3b, V6, and V6a) and the ventral visual stream (areas FFC, VVC, V8, VMV1, VMV2, and VMV3).

Area V7 comprises a relatively focused retinotopic map of the central visual field, which encompasses the region surrounding the foveal point. Notably, V7 demonstrates increased activity when attention is directed toward a visual stimulus (Sarkissov et al., 1955). Area V7 incorporates the visual information it processes to construct spatial information within the central visual field (Vogt & Vogt, 1919).

Area V7 is structurally connected to the superior temporal gyrus (areas A5, STSda, STSdp, STSva, and TGd) via MdLF fibers, and areas TGv and TGd via the ILF, and the IFOF. The MdLF fibers course parallel to the IFOF then diverge laterally into the superior temporal gyrus to establish structural connections with areas A5, STSda, STSdp, STSva, and TGd. ILF white matter tracts pass through the temporal lobe to terminate at areas TGv and TGd. IFOF has variable patterns of termination across brains. Additionally, short structural connections are established to LO1, LO2, LO3, V3CD, V3A, V6A, and V6.

Area V3a

Area V3a (visual area 3a) is situated on the superior surface of the occipital lobe, close to the interhemispheric fissure. It is situated similarly to the superior limb of V4. V3a borders are as such: V3 on its inferior end, V4 on its posterior end, V3b on its lateral border, and V7 and V6a form its anterior border. Lastly, across the interhemispheric cleft, V6 serves as its anterior neighbor.

Area V3a displays robust functional connections with various brain regions. It links to both the sensory strip (areas 1, 3a, and 3b) and the motor strip (area 4) with additional connection to the premotor region (areas SCEF and FEF). It also has functional connectivity to the medial frontal lobe (areas a24prime, p32prime, 5mv, and 23c). Area V3a connects to the insula opercular regions (areas FOP1, FOP3, FOP4, OP4, OP2-3, 43, PFcm, PoI1, RI, 52, A4, LBelt, and PBelt) and the parietal lobe (areas VIP, LIPv, PGp, PFop, IPS1, IP0, and DVT). It has both connections to the medial occipital lobe (areas ProS, V1, V2, V3, and V4) and the lateral occipital lobe (areas TPOJ2, V3cd, V4t, LO1, LO2, LO3, MT, MST, PH, and FST). Area V3a has further functional connection with the dorsal visual stream (areas V3b, V7, V6, and V6a) and the ventral visual stream (areas FFC, VVC, V8, VMV1, VMV2, and VMV3).

Despite being a common focus of anatomical and imaging research, the functional role of area V3a in visual stimulus processing is not well established. Area V3a exhibits high sensitivity to motion and contrast in the central visual field that surrounds the foveal point (Zilles & Amunts, 2010). It also plays a role in integrating spatial data from both the contralateral and ipsilateral visual fields (Baker et al., 2018a). This functionality described makes area V3a a component of the motion-processing system of the visual cortex (Ploner et al., 2000).

Area V3a is structurally connected to the frontal lobe (areas 8BL, 9a, 9p, 9m, 10pp, and 10d) via the IFOF, the superior temporal gyrus (area STGa) via the MdLF, and to the contralateral parcellation V3a via the FM. FM connections course through the splenium of the corpus callosum to eventually end at contralateral V3a. Additionally, V3a has local structural connections through association bundles to the parcellations V6A, V3b, V6, V7, and V3. It also is important to note that the right hemisphere has inconsistent projections with the IFOF when compared to left hemisphere.

Area V3b

Area V3b (visual area 3b) is found on the superior occipital surface. Area V3b borders consist of IP0 anterolaterally, IPS1 anteromedially, V7 medially, and V3cd laterally. The posterior border is comprised of V3a and the superior limb of V4.

Area V3b has widespread functional connectivity throughout the brain. It connects with the premotor region (areas SCEF and FEF) and the medial frontal lobe (areas a24prime, p32prime, and 23c). It is also linked to the insula opercular regions (areas 43, RI, and PBelt) and the parietal lobe (areas 7PC, VIP, IPS1, IP0, and DVT). Area V3b has connections with both the medial occipital lobe (areas ProS, V1, V2, V3, and V4) and the lateral occipital lobe (areas V3cd, V4t, LO1, LO2, LO3, MT, MST, PH, and FST). It has further functional connection with both the dorsal visual stream (areas V3a, V7, V6, and V6a) and the ventral visual stream (areas FFC, VVC, V8, VMV1, VMV2, and VMV3).

Area V3b serves as a relay point, for collecting motion-sensitive information from the dorsal visual stream. Specifically, it assists in the processing of kinetic boundaries which are perceptual edges that arise when an object is in motion (Martuzzi et al., 2015). Kinetic boundaries play significant role in the ability to perceive and make sense of the dynamic world and help separate moving objects from their background.

Area V3b structural connections are primarily with the local parcellations MIP, IPO, IPS1, V4, and V3CD through short association bundles. However, structural connections do exist with IFOF and MdLF but do not exhibit consistent structural connections across brains.

Area V4t

Area V4t (visual area 4t) is a horizontal area that is situated in the central portion of the lateral occipital cortex. Area V4t borders include LO1 posteriorly, FST anteriorly, LO2 inferiorly, and its superior border is comprised of MT, MST, and LO3.

Area V4t exhibits functional connections with numerous brain regions. It is linked to both the sensory strip (areas 1, 2, 3a, and 3b) and the motor strip (area 4), as well as the premotor region (area FEF). Area V4t connects with the insula and opercular region (areas RI, PBelt, and A4) and the parietal lobe (areas VIP, LIPv, and IPS1). It also has functional connectivity with the medial occipital lobe (areas V2, V3, and V4) and the lateral occipital lobe (areas V3cd, MT, MST, LO3, PH, and FST). It has further connections with the dorsal visual stream (areas V3a, V3b, V7, V6, and V6a) and the ventral visual stream (areas VVC, V8, VMV1, VMV2, and VMV3).

Research has shown that area V4t plays a role in integrating information from both the ventral and dorsal streams. This region exhibits a high level of activity in response to both motion and shape-sensitive information, which suggests its importance in the integration of object processing and global-motion perception (Hyvarinen & Poranen, 1978).

Area V4t structural connections are primarily with the local parcellations V4t, LO3, MST, MT, LO1, and V3CD through short association bundles. However, structural connections do exist with SLF and ILF but do not exhibit consistent structural connections across individuals.

IPS1

Area IPS1 is located on the superior bank of the intraparietal sulcus, specifically on its posterior end. IPS1 represents the posterior most aspect of the superior bank of the intraparietal sulcus. Area IPS1 borders consist of IP0 inferiorly, MIP anteriorly, DVT medially, 7PL anteromedially, and the posterior border is a combination of V6a, V7, and V3b.

Area IPS1 displays functional connectivity with various brain regions. It is connected to the sensory strip (areas 1 and 2), the motor strip (area 4), and the premotor regions (areas SCEF and FEF). Additionally, area IPS1 is functionally connected to the temporal lobe (areas TE2p and PHT), lateral parietal lobe (areas AIP, MIP, VIP, LIPd, LIPv, PGp, PFt, IP0, 7PL, and 7PC), and the medial parietal lobe (area DVT). Area IPS1 is further functionally connected to the medial occipital lobe (areas V1, V2, V3, and V4), lateral occipital lobe (areas V3cd, V4t, LO1, LO2, LO3, PH, TPOJ2, MT, MST, and FST), the dorsal visual stream (areas V3a, V3b, V7, V6, and V6a), and the ventral visual stream (areas V8, PIT, FFC, VVC, VMV1, VMV2, and VMV3).

Area IPS1 contributes to different aspects of visuospatial processing and spatial attention (Choi et al., 2015; Vierck et al., 2013). It specifically processes visuospatial data that shows size and rotation-invariant representation, which is indicative in area IPS1's involvement of object recognition further promoting object constancy (Choi et al., 2015). Area IPS1 also plays a role in mental visualization of the action of grasping objects as well as the objects function (Choi et al., 2015). Area IPS1's role in top-down spatial attention signals is the transmission of signals to the early visual cortex (Vierck et al., 2013).

Area IPS1 is primarily connected through the IFOF, MdLF, and local short fibers. The IFOF pathway travels from area IPS1 through the posterior temporal lobe and extreme/external capsule to frontal lobe parcellations. The IFOF termination sites can vary among individuals, but typically end at the parcellations 8BL, 9a, and 9p in the superior temporal gyrus. The MdLF courses deep into the parietal lobe to make its way to the superior temporal gyrus and planum temporale to finally terminate at A4, PBelt, and MBelt. Additionally, local, short association fibers establish connection with areas MIP, V7, V6A, and V3B.

LO1

Area LO1 is a small vertically oriented area situated in the posterior occipital lobe and anterior to the occipital pole. Area LO1 borders consist of V4 posteriorly, LO3 anteriorly, V3cd superiorly, and both V4t and LO2 create the inferior boundary.

Area LO1 exhibits functional connections with various brain regions. It connects to the premotor region (area FEF) and the insula and opercular region (area A4). Additionally, area LO1 has functional connectivity with the parietal lobe (areas VIP, LIPv, IP0, DVT, and IPS1). It has functional connections to both the medial occipital lobe (areas V1, V2, V3, and V4) and the lateral occipital lobe (areas V3cd, V4t, MT, LO2, LO3, PH, and FST). It furthers its functional connections with the dorsal visual stream (areas V3a, V3b, V7, V6, and V6a) and the ventral visual stream (areas FFC, VVC, V8, PIT, VMV1, VMV2, and VMV3).

Area LO1 represents a higher-order visual region that integrates information about details, motion, and characteristics from both the dorsal and ventral streams of the brain. This region exhibits a selective activation pattern in response to orientation-specific and boundary information, which highlights its critical role in processing the form of objects. Through this specialized function, area LO1 plays a crucial role in complex visual processing, including object recognition and perception (Bailey, 1951; Huffman & Krubitzer, 2001; Whitsel et al., 2019).

Area LO1's structural connections are primarily with the local parcellations V3A, V3B, V3CD, LO2, and LO3 through short association bundles.

LO2

Area LO2 is a small horizontally oriented area that is situated anterior to the occipital pole and somewhat superior to the tentorium. Area LO2 borders consist of V4 posteriorly, PH anteriorly, V4t superiorly, and PIT inferiorly.

Area LO2 displays functional connections with multiple brain regions. It connects to the premotor region (area FEF) and the insula and opercular region (area PBelt). Area LO2 also has functional connectivity with the parietal lobe (areas VIP, LIPv, and IPS1). It also connects to both the medial occipital lobe (areas V1, V2, V3, and V4) and the lateral occipital lobe (areas V3cd, V4t, MT, LO1, LO3, PH, and FST). It has further structural connection with the dorsal visual stream (areas V3a, V3b, V7, V6, and V6a) and the ventral visual stream (areas FFC, VVC, V8, PIT, and VMV3).

Area LO2 exhibits preferential retinotopic activation in the peripheral visual field, allowing it to efficiently integrate inputs from the dorsal and ventral streams of the brain. This integration process encodes detailed information about the shapes of stimuli, which is crucial for object recognition and perception. Moreover, research has revealed that this region also demonstrates high activation in the processing and encoding of object-related information. Taken together, these findings underscore the essential role of area LO2 in the complex process of visual perception and object recognition (Bailey, 1951; Huffman & Krubitzer, 2001; Whitsel et al., 2019).

Area LO2 structural connections are primarily with the local parcellations V3A, V3B, V3CD, LO2, and LO3 through short association bundles.

LO3

Area LO3 is a vertically oriented area in the superior central portion of the lateral occipital lobe. It is located just inferior to the posterior angular gyrus. Area LO4 borders consist of V3cd and LO1 posteriorly, MT anteriorly, TPOJ3 and PGp superiorly, and LO1 and V4t inferiorly.

Area LO3 exhibits functional connectivity with multiple brain regions. It is functionally connected to both the sensory strip (areas 1, 2, 3a, and 3b) and the motor strip (area 4), as well as the premotor region (areas SCEF and FEF). Area LO3 is also linked to the cingulate regions (areas 5mv and 23c) and the insula and opercular region (areas OP4, PFcm, 43, RI, PBelt, and A4). Additionally, it demonstrates functional connections with the parietal lobe (areas 7PC, DVT, PFop, PGp, VIP, LIPv, IP0, and IPS1). It also has strong visual involvement as it is connected to both the medial occipital lobe (areas V1, V2, V3, and V4) and the lateral occipital lobe (areas TPOJ2, TPOJ3, V3cd, V4t, MT, LO1, LO2, PH, and FST). It has further visual involvement with functional connections with the dorsal visual stream (areas V3a, V3b, V7, V6, and V6a), the ventral visual stream (areas FFC, VVC, V8, PIT, VMV1, VMV2, and VMV3).

Although it is considered a recent discovery, the existing literature strongly suggests that LO3 plays a vital role as a central hub connecting the dorsal and ventral streams within the occipital lobe. Through its intricate neural connections, LO3 facilitates the integration, encoding, and processing of essential information related to details, motion, and shape. The culmination of these processes allows for effective object recognition and encoding, highlighting the significance of LO3 in the visual perception and cognitive processing of the brain (Bailey, 1951; Huffman & Krubitzer, 2001; Kuehn et al., 2018).

Area LO3 is structurally connected through the ILF and local connections. ILF white matter tracts pass through the temporal lobe to terminate at TGv. Area LO3 also has structural connections with the local parcellations V3A, V3B, V3CD, LO1, LO2, PH, and FST through short association bundles.

Area V8

Area V8 (visual area 8) is situated at the most posterior aspect of the fusiform gyrus. Area V8 borders consist of V4 medially, V4 posteriorly, VMV3 anteromedially, VVC anteromedially, FFC anterolaterally, VVC also creates part of the lateral border, and PIT posterolaterally.

Area V8 exhibits functional connectivity with various regions across the brain. It is connected to the premotor region (area FEF) and the insula and opercular region (areas 43, PBelt, and A4). Area V8 also demonstrates functional connections with the parietal lobe (areas VIP, LIPv, IPS1, IP0, and DVT). It has functional connections with both the medial occipital lobe (areas ProS, V1, V2, V3, and V4) and the lateral occipital lobe (areas V3cd, LO1, LO3, PH, and FST). It has further connections with both the dorsal visual stream (areas V3a, V3b, V7, V6, and V6a) and the ventral visual stream (areas VVC, V8, VMV2, and VMV3).

Area V8 is primarily responsible for color perception and processing within both the lower and upper visual fields. This region plays an essential role in the neural processing of chromatic information, contributing to the formation of color perception and recognition. Through its specialized function, area V8 provides foundational ability to perceive and distinguish colors in the world (Chouinard & Paus, 2006).

Area V8 is structurally connected with the ILF and VOF. It is important to note that the ILF connections vary across brains. The VOF tract travels mediodorsally originating from V8 to terminate at V3a, V3b, V3cd, and V7. Additional, short association bundles are connected to V4, PIT, FFC, and V3.

FST

Area FST is found in the anterior portion of the lateral occipital lobe that has an oblique to marginally vertically orientation. It is just posterior to the end of the MT gyrus. Area FST borders consist of MST and V4t posteriorly, PHT anteriorly, PH inferiorly, and TPOJ2 superiorly.

Area FST exhibits functional connectivity with a wide range of brain regions. It is connected to the sensory strip (areas 1, 2, 3a, and 3b), the motor strip (area 4), and the premotor region (areas SCEF, FEF, PEF, 6mp, 6r, 6v, 6a, and 6d). Area FST also demonstrates connections with the cingulate regions (areas 24dd, p32prime, 23c, and 5mv) and the insula and opercular region (areas FOP2, OP1, OP4,

PFcm, PoI1, PoI2, 43, RI, LBelt, PBelt, and A4). Area FST is functionally connected to the temporal lobe (areas TE2p and PHT) and the parietal lobe (areas 7PC, 7AL, 7PL, 7am, PFt, PFop, PFm, AIP, VIP, LIPv, DVT, IP0, and IPS1). It also has connections to both the medial occipital lobe (areas V2, V3, and V4) and the lateral occipital lobe (areas TPOJ2, TPOJ3, MT, MST, V3cd, V4t, MT, LO1, LO2, LO3, and PH). It has further connections to both the dorsal visual stream (areas V3a, V3b, V7, V6, and V6a) and the ventral visual stream (areas FFC, VVC, V8, PIT, VMV1, VMV2, and VMV3).

Area FST serves as a vital hub within the visual system, seamlessly integrating large amounts of visual information from both the dorsal and ventral streams of the brain (Dadario et al., 2023a). By combining essential detail, motion, and form-sensitive information, area FST serves a fundamental role in the perception of image content. Through this integration process, area FST contributes to the processing of spatial reference frames, leading to continuous global-motion perception and the formation of spatial maps. Additionally, this region exhibits heightened attention-based motion selectivity, which allows for effective stimulus filtering and enhances our perception of visual stimuli. Overall, the specialized functions of area FST makes it an essential part of the visual system in dynamic scenes (Coco et al., 2016; Glasser et al., 2016; Hyvarinen & Poranen, 1978).

Area FST has structural connections via SLF and to the ILF. The SLF consistently links this region with the premotor area 8C between individuals. ILF connections however have shown to be inconsistent across brains. Additional, short association bundles provide local connections to PHT, LO1, LO2, LO3, MST, MT PH, and V4. We expand on these connections with nonvisual cerebrum in a recent study by our team (Dadario et al., 2023a).

Area PH

Area PH is found in the anteroinferior region of the lateral occipital lobe and is a horizontally oriented structure. It is approximately positioned with the ITG and is primarily lateral to the occipitotemporal sulcus, which forms a small part of its lateral bank making it spill into part of the basal surface. Area PHs borders consist of TE1 and TE2p anteriorly, LO2 and PIT posteriorly, FFC medially on its basal surface, and a combination of TE1p, PHT, and FST create the superior border.

Area PH exhibits functional connectivity to a diverse range of brain regions. It is connected to the sensory strip (areas 1, 2, and 3a), the motor strip (area 4), and the premotor region (areas SCEF, FEF, PEF, 6r, 6v, and 6a). Area PH also demonstrates connections with the lateral frontal lobe (areas p9-46v, IFSa, IFSp, IFJa, and IFjp) and the cingulate regions (areas 23c and 5mv). Additionally, area PH is functionally connected to the insula and opercular region (areas PoI2, LBelt, PBelt, and A4), the temporal lobe (areas PeEc, PHA3, TE2p, and PHT), the parietal lobe (areas 7PC, 7PL, 7am, PGp, PFop, PF, AIP, MIP, VIP, LIPd, LIPv, DVT, IP2, IP0, and IPS1). It has functional connections to both the medial occipital lobe (areas V1, V2, V3, and V4) and the lateral occipital lobe (areas TPOJ2, TPOJ3, MT, MST, V3cd, V4t, MT, LO1, LO2, LO3, and FST). It has further functional connection to both the dorsal visual stream (areas V3a, V3b, V7, V6, and V6a) and the ventral visual stream (areas FFC, VVC, V8, PIT, VMV1, VMV2, and VMV3).

Area PH is a higher-level perception region in the visual system that acts as a processing center for ventral stream input. Both areas PH and FST area hubs of the visual system connect various visual and nonvisual cortices (Dadario et al., 2023a). It integrates "place-specific" information and thus assisting in spatial navigation. Through this unique capability, area PH encodes a comprehensive representation of the local scene, allowing for effective memory and recognition of specific places. This region plays a significant role in the formation of spatial maps and the encoding and recognition of place. In this way, area PH provides a vital foundation for our ability to navigate and perceive our surroundings, facilitating essential cognitive processing (Huffman & Krubitzer, 2001; Scheperjans et al., 2005).

Area PH is structurally connected via the ILF and SLF. The SLF consistently links PH with parcellations 44 and 45. ILF projections navigate through the temporal lobe to terminate at TGv and TGd. Additional short association bundles establish structural connections to FST, MST, MT, PHT, V4T, and TE1p. We expand on these connections with nonvisual cerebrum in a recent study by our team (Dadario et al., 2023a).

Area MST

Area MST is situated just below the angular gyrus and is vertically oriented as it runs parallel and lies slightly anterior to area MT. Area MST borders consist of MT posteriorly, FST anteriorly, TPOJ2 and TPOJ3 superiorly, and FST and V4t inferiorly.

Area MST exhibits functional connectivity with many different areas of the brain. It is connected to the sensory strip (areas 1, 2, 3a, and 3b), the motor strip (area 4), and the premotor region (areas SCEF, FEF, and 6v). Area MST also demonstrates connections with the cingulate regions (area 24dd) and the insula and opercular region (areas OP1, OP4, PFcm, 43, RI, LBelt, PBelt, and A4). Area MST is functionally connected to the parietal lobe (areas 7PC, VIP, LIPv, and IPS1), the medial occipital lobe (areas V2, V3, and V4), the lateral occipital lobe (areas TPOJ2, V3cd, V4t, MT, LO1, LO2, LO3, PH, and FST), the dorsal visual stream (areas V3a, V3b, V7, V6, and V6a), and the ventral visual stream (areas FFC, VVC, V8, PIT, and VMV3).

Area MST plays a role in the sophisticated task of perception of self-motion and processing global visual motion. Area MST responds to the motion and direction data it receives, which is largely from its posterior neighbor area MT. Additionally, this region is actively involved in the execution and continuation of smooth pursuit eye movements, in coordination with the frontal eye fields. Through these specialized functions, area MST contributes to our ability to perceive motion and movement within our environment and to effectively navigate our surroundings (Grefkes et al., 2004; Kruger et al., 2014; Naito et al., 2008).

Area MST is structurally connected via the ILF and has inconsistent connections with SLF. SFL connections have shown to vary across brains. The ILF projections course across the temporal lobe to end at TGv. There are also short association bundles that establish local structural connections to MT, PH, TE2p, and FST.

Area MT

Area MT is found in the superior aspect of the central lateral occipital lobe and position slightly inferior to the angular gyrus. Its borders consist of LO3 posterior, MST anteriorly, TPOJ3 superiorly, and V4t inferiorly.

Area MT has strong functional connections across many brain regions. It is connected to the sensory strip (areas 1, 2, 3a, and 3b), the motor strip (area 4), and the insula and opercular region (areas OP4, RI, PBelt, A4, and A5). Area MT is functionally connected to the parietal lobe (areas VIP, LIPv, and IPS1). It has functionally connections to both the medial occipital lobe (areas V2, V3, and V4) and the lateral occipital lobe (areas V3cd, V4t, MST, LO1, LO2, LO3, PH, and FST). It has further functional connection to both the dorsal visual stream (areas V3a, V3b, V7, V6, and V6a) and the ventral visual stream (areas FFC, VVC, V8, PIT, and VMV3).

Area MT demonstrates direction-sensitive responses to visual motion and is primarily responsible for integrating one-dimensional visual signals into a comprehensive two-dimensional visual motion pattern. Additionally, this region is involved in binocular disparity tuning, noise reduction, and segmentation of figure and background within complex and moving stimuli. Another critical role of area MT is the initiation of smooth-pursuit eye movements, which are necessary for effective visual tracking and perception of motion. Through these specialized capabilities, area MT contributes to our ability to perceive and effectively navigate our environment, enhancing our visual processing and perception (Grefkes et al., 2004; Kruger et al., 2014). A prominent difference of area MT when compared to area MST, is that area MT responds to less complex motion (Sheets et al., 2021). Another recent study showed that the area MT plays a role in the development of visually guided manual behaviors (Neubert et al., 2014).

Area MT has inconsistent structural connections to the ILF. The ILF structural connections vary across brains but when it does exist, the ILF projections cross through the temporal lobe to end at TF. Area MT has additional local short association fiber bundles that connect it to MST, LO1, LO2, LO3, TPOJ2, TPOJ3, FST, PH, V3b, and IPO.

Area POS2

Area POS2 is positioned on the superior portion of the anterior aspect of the parietooccipital sulcus's cortical surface. Area POS2 is bounded superiorly by areas 7PM and 7pl, inferiorly by area POS1, anteriorly by area 7M, and posteriorly by area DVT.

Area POS2 in the parietal lobe demonstrates functional connectivity to several regions across the cortex, including: the lateral frontal lobe (areas 46, 9-46d, i6-8, 8C, p10p, and a10p); the medial frontal lobe (areas 8BM, 9m, a24, p24, a32prime, p32, and d32); the lateral parietal lobe (areas IP1, IP2, PFm, PGp, and PGs); and the medial parietal lobe (areas 23d, d23ab, POS1, PCV, RSC, DVT, 7am, 7pm, 7m, 31a, 31pv, and 31pd).

Area POS2 also exhibits structural connectivity to local and contralateral parcellations, although most fiber projections are highly variable. The white matter pathways which connect area POS2 to area V1 on the contralateral hemisphere extend from area POS2 through the forceps major, although where the fiber ends is highly variable. Additionally, area POS2 is structurally connected to local areas DVT, POS1, V2, and 7pm through short association fibers.

Area PIT

Area PIT is situated on the posterior aspect of the occipitotemporal gyrus, specifically where it passes into the occipital pole. Area PIT is flanked medially by areas V8 and V4, anteriorly by areas FFC and PH, and superiorly by area LO2. Area V4 also forms a lateral border to area PIT.

Area PIT in the occipital lobe demonstrates functional connectivity to several regions across the cortex, including: the premotor region (area FEF); insula and opercular region (areas PBelt and A4); parietal lobe (areas VIP, LIPv, IPS1, and DVT); medial occipital lobe (areas V1, V2, V3, and V4); dorsal visual stream (areas V3a, V3b, V7, V6, and V6a); ventral visual stream (areas VVC, FFC, and V8); and lateral occipital lobe (areas V3cd, V4t, LO1, LO2, LO3, PH, MST, and FST).

Area PIT is primarily connected via the ventral occipital fasciculus (VOF). It exhibits structural connections to V2, V3, and V3a via the VOF. Additionally, area PIT is connected to local areas PH, V8, V4, and V1 via short association fibers.

Area POS1

Area POS1 is located within the anterior aspect of the parietooccipital sulcus, and forms the inferior half of it. Area POS1 borders a number of regions, including areas 7M and d2ab anteriorly, area DVT posteriorly, area POS2 superiorly, and areas ProS and RSC inferiorly.

Area POS1 in the parietal lobe demonstrates functional connectivity to several regions across the cortex, including: the lateral frontal lobe (areas 8AD, i6-8, 47m, 10d, 10v, and 10r); medial frontal lobe (areas 9m, a24, s32, p32, and d32); temporal lobe (areas STSva, TE1a, PHA1, PHA2, PHA3, PreS, and the hippocampus); lateral parietal lobe (areas PGi, PGp, and PGs); and medial parietal lobe (areas ProS, v23ab, d23ab, POS2, PCV, RSC, DVT, 7pm, 7m, 31a, 31pv, and 31pd).

Area POS1 exhibits structural connectivity to the anterior and parahippocampal cingulum fibers, as well as the contralateral hemisphere. White matter tracts which connect area POS1 to areas a24, a24pr, and p24 within the anterior cingulate cortex traverse the anterior cingulum fibers, while connections to area EC of the parahippocampal gyrus pass through cingulum fibers as they bend around the splenium of the corpus callosum. Furthermore, contralateral connections from area POS1 to area V1 exist via FM fibers. Area POS1 also exhibits structural connectivity to local regions, including area V1, V2, POS2, and V6 via short associations fibers.

Area TPOJ3

Area temporal-parietal-occipital junction 3 (TPOJ3) is located on the inferior parietal lobule, specifically on its posterior inferior aspect. Area TPOJ3 is flanked by a number of regions, including area PGI in its superior aspect, TPOJ2 in its anterior aspect, areas MST, MT, and LO3 in its inferior aspect, and area PGp in its posterior aspect.

Area TPOJ3 in the parietal lobe demonstrates functional connectivity to several regions across the cortex, including: the sensory strip (area 2); premotor areas (areas 6a and FEF); the cingulate regions (areas 5mv and 23c); the insula opercular region (areas OP4, PFcm, and STV); the temporal lobe (areas PHA2, PHA3, and TE2p); the lateral parietal lobe (areas PGp, IP0, and 7PC); the medial parietal lobe (areas 7AM, PCV, and DVT); the medial occipital lobe (area V2); the dorsal visual stream (area V6); and the lateral occipital lobe (areas FST, PH, LO3, MST, TPOJ1, and TPOJ2).

Area TPOJ3 also exhibits structural connectivity to the inferior longitudinal fasciculus (ILF) and to adjacent regions. Area TPOJ3 exhibits white matter fiber connections to areas TF, TGv, and PeEc, via ILF connections. Other ILF fibers connect area TPOJ3 to the premotor cortex, although this not a consistent finding. Local structural connections exist to areas MT, MST, LO3, FST, TPOJ3, TE2p, PH, and PGi via local short association fibers.

Area DVT

The Dorsal Visual Transitional Area (DVT) is a lengthy subdivision that encompasses the entire parietooccipital sulcus. Its anterior border is contiguous with areas POS2 and PO1, while its posterior border abuts the anterior aspect of area V2. DVT's inferior boundary is adjacent to the ProS area, and its superior pointed border connects with V6 and POS2.

Functional connectivity analysis reveals that area DVT exhibits connections to several regions across the cortex, including: the premotor region (areas SCEF, FEF, 6ma, and 6a), the lateral frontal lobes (areas 9-46d and 46), the cingulate regions (areas p32prime, a24prime, 5mv, and 23c), the insula and opercular region (areas FOP4, PFcm, 43, and 52), the temporal lobe (areas PHA1, PHA2, PHA3, and PHT), the lateral parietal lobe (areas 7PC, 7AL, 7PL, PGp PF, PFop, AIP, VIP, LIPd, LIPv, IP0, and IPS1), the medial occipital lobe (areas 7am, 7pm, RSC, POS2, POS1, and PCV), the medial occipital lobe (areas ProS, V1, V2, V3, and V4), the dorsal visual stream (areas V3a, V3b, V7, V6, and V6a), the ventral visual stream (areas FFC, VVC, V8, VMV1, VMV2, and VMV3), and the lateral occipital lobe (areas TPOJ2, TPOJ3, V3cd, LO1, LO3, PH, and FST).

Area DVT is a region of the brain that exhibits extensive connectivity to areas throughout the cortex. It has structural connections to various brain regions via different white matter tracts, including the inferior frontooccipital fasciculus (IFOF), middle longitudinal fasciculus (MdLF), and the forceps major of the corpus callosum. Specifically, Area DVT is structurally connected to areas 8BL and a10p within the anterior pole of the frontal lobe via fibers that traverse the IFOF and pass posteriorly to the insula. Additionally, Area DVT has structural connections to area STGa within the superior temporal gyrus via fibers that pass through the MdLF. Furthermore, fibers from area DVT connect to areas V1, V2, V3, DVT, POS1, and POS2 in the contralateral hemisphere via the forceps major and the splenium of the corpus callosum. Finally, Area DVT also has local connections to areas V6, DVT, POS1, and POS2 via extensions of short association fibers.

Area V6A

Area V6A (visual area 6a) is a small region located in the superior occipital lobe of the brain, specifically within the angle of the interhemispheric cleft. Area V6A is flanked posteriorly by area V3a, anteriorly by DVT, laterally by V7 and IPS1, and inferiorly by V6.

Area V6A exhibits functional connectivity with several regions across the cortex, including: the sensory strip (areas 1, 2, 3a, and 3b, the motor strip area 4); the premotor cortex (areas SCEF and FEF), the medial frontal lobe (areas a24prime, p32prime, 5mv, and 23c), the insula opercular (areas OP4, 43, RI, A4, and PBelt), parietal lobe (areas 7PC, VIP, LIPv, PGp, IPS1, IP0, and DVT), the medial occipital lobe regions (areas ProS, V1, V2, V3, and V4), the dorsal visual stream (areas V3a, V3b, V7, and V6), the ventral visual stream (areas FFC, VVC, V8, VMV1, VMV2, and VMV3), and the lateral occipital lobe (areas TPOJ2, V3cd, V4t, LO1, LO2, LO3, MT, MST, PH, and FST).

Area V6A is structurally connected via short range fibers to the IFOF and MdLF. IFOF fibers link V6A with areas 8BM and 8BL in the frontal lobe. In addition, V6A is connected to areas A4 and MBelt via the MdLF, running parallel to the IFOF. Short association fibers also exist, linking V6A to nearby areas V3B and V7. Although not consistently observed in all individuals, some have shown structural connectivity between V6A and the FM.

Area V3CD

Area V3CD is a vertical stave located on the topographical surface of the posterior-superior aspect of the lateral occipital lobe, on the posterior aspect of the angular gyrus. Area V3cd is bounded anteriorly by area PGp and LO3, inferiorly by area LO1, posteriorly by area V4, and superiorly by area V3b.

Area V3CD exhibits functional connectivity with several regions across the cortex, including: the premotor cortex (area FEF), the insula and opercular region (area PBelt), the parietal lobe (areas 7PC, AIP, VIP, LIPv, PGp, IPS1, IP0, and DVT), the medial occipital lobe (areas ProS, V1, V2, V3, and V4), the dorsal visual stream (areas V3a, V3b, V7, V6, and V6a), the ventral visual stream (areas VVC, V8, VMV1, VMV2, and VMV3), and the lateral occipital lobe (areas MT, MST, V4t, LO1, LO2, LO3, PH, and FST).

Area V3CD exhibits structural connectivity to mostly local regions, including areas V3B, V7, V4t, V3CD, LO1, LO2, LO3, MST, and MT via short association bundles.

Brodmann area 20 (BA20)

BA20 is primarily divided into four different HCP parcellations in the temporal lobe. These parcellations consist of areas TE2a, TGV, TE2p, and TF.

Area TE2a

Area TE2a is located on the anterior portion of the ITG, occupying the anterior half of the inferior sulcus and the lateral bank of the occipitotemporal sulcus. It is adjacent to TF along its basal-medial edge and shares a superior border with areas TE1a and TE1m. At its anterior end, TE2a borders areas TGd and TGv, while its posterior end is wedged between areas TE1p and TE2p.

Area TE2a exhibits functional connectivity to the frontal lobe (areas 8AV, 8BL, 8C, and a47r), the temporal lobe (areas STSvp, TE1m, and TE1p), and the parietal lobe (areas PGs, PGi, and PFm).

The primary function of area TE2a seems to be predominantly associated with visual pathways. TE2a exhibits a functional profile like that of TE1m, which is situated superiorly adjacent to the region. Both areas display activation during visual working memory secondary contrast tasks and deactivation in response to language tasks. However, when compared to TE1m, TE2a demonstrates reduced activation in visual working memory tasks and diminished deactivation during language tasks.

Area TE2a is structurally connected to the arcuate/SLF and ILF. The arcuate/SLF tracts wrap around the Sylvian fissure and project toward the frontal lobe before turning medially to terminate at various areas, including 6r, 6v, 8C, p9-46v, IFJa, IFJp, and IFSp. In addition, there are posterior projections from the arcuate/SLF that terminate at the inferior parietal lobule at PF and PFm. There are additional local short association fibers that establish structural connections TE2a to TE1p and TGd.

Area TGv

Area TGv (TG ventral) is situated in the inferior temporal polar region, just anterior to the ITG and fusiform gyrus. It is bordered posteriorly by TE2a on its lateral surface, TF on its basal surface, and TGd superiorly. On its medial basal surface, PeEc makes up its border.

Area TGv primarily demonstrates functional connection to area 45 and TGd.

Areas TGv and TGd together constitute the temporal polar cortex. Similar to TGd, area TGv is activated during language-related task contrasts, implying a potential role in ventral stream language processing. In comparison to the superiorly located TGd, TGv exhibits activation in motor tasks in response to visual cues and relational primary contrasts. Area TGv can be further distinguished from surrounding areas with comparison to its posterior neighbor, area TE2a. Area TGv is more activated with motor cue-average, language story, theory of mind versus random contrasts (Amunts & Zilles, 2015).

Area TGv is structurally connected to the parcellations of V1, V2, V3, and V4 via the ILF. The ILF traverses through the inferior temporal lobe to terminate at these parcellations. Further distinction between area TGv and area TE2a could be made because area TGv is thicker. Also, relative to its inferoposterior neighbor TF, area TGv has more myelination (Amunts & Zilles, 2015).

Area TE2p

Area TE2P can be found on the posterior aspect of the occipital temporal sulcus and the medial aspect of the posterolateral fusiform gyrus. Area TE2p is flanked posteriorly by area PH, medially by area FFC, anteriorly by area TF, and laterally by areas TE2a and TE1p.

Area TE2p is functionally connected to several regions across the cortex, including the frontal lobe (areas FEF, PEF, IFSa, IFJa, IFJp, p9-46v, 6a, and 6r), the insula (area PoI2), the temporal lobe (area PHT), the parietal lobe (areas PGp, AIP, MIP, LIPv, LIPd, IPS1, IP0, PFop, 7PC, 7PL, and 7AL), and the occipital lobe (areas PH, FFC, FST, TPOJ2, and TPOJ3).

Area TE2P exhibits significant structural connectivity with both local areas and the superior longitudinal fasciculus. The white matter connections linking TE2P to the SLF extend and terminate within the frontal lobe, but their precise termination points within this area are not yet fully understood. It is interesting to note that white matter pathways originating from area TE2P's right hemispheric parcellation connect it to regions within the occipital cortex. Additionally, TE2P is also connected to FFC, PH, TE2p, FFC, TE1m, TF, and TE2a through local structural connections via short association fibers.

Area TF

Area TF is a distinct cortical region situated at the intersection of the anterior aspect of the fusiform gyrus and the occipitotemporal sulcus. It also occupies a portion of the lateral bank of the collateral sulcus. Along its anterior border it is bounded by area TGv, while its lateral border is bounded by areas TE2a and TE2p laterally. Furthermore, it shares its medial border with area PeEC, and posterior border with areas with multiple regions, including, areas FFC, ventral visual complex (VVC), parahippocampal area 2 (PHA2), and PHA3.

Area TF exhibits limited functional connections to only areas PeEC and TE2p.

In terms of structural connectivity, Area TF exhibits connections with the superior longitudinal fasciculus (SLF), inferior longitudinal fasciculus (ILF), and local parcellations. The structural connections with the SLF extend around the Sylvian fissure, projecting toward the frontal lobe, and terminate at areas IFSa and 46.

Meanwhile, the structural connections with the ILF extend toward the inferior temporal lobe and end at areas V2, V3, V4, V3A, and V3b. Additionally, local connections from area TF to areas PeEc, VVC, TE2p, and Te2a are also present through short association bundle fibers.

Brodmann area 21 (BA21)

Area 21 can be further divided into 6 unique regions. The middle temporal gyrus includes four of the areas: TE1a, TE1m, TE1p, and PHT, while the other two regions, which are STSda, STSva, are centered around superior temporal sulcus (STS).

TE1a

Area TE1a is located on the lateral face of the MTG, covering its anterior third, extending up to the edge of the inferior temporal sulcus. It borders STSva superiorly, TGd anteriorly, TE2a inferiorly, and TE1m posteriorly.

Area TE1a exhibits functional connectivity to the frontal lobe (areas 8AV, 8AD, 8BL, 9a, 9p, 9m, a24, 45, 47s, 47L, 10d, 10v, 10r, and SFL), the temporal lobe (areas STSva, STSvp, STSda, TGv, entorhinal cortex (EC), the hippocampus, and TE1m), and the parietal lobe (areas PGs, PGi, 7m, d23ab, 31pv, and 31pd).

In general, the inferior TE is postulated to represent the final stage of the ventral visual processing pathway, and as such, is likely responsible for processing and representing information pertaining to complex visual objects. TE plays a crucial role in the short-term maintenance of visual object information as part of working memory (Ranganath, 2006). Both TE1 and TE2 are generally considered to be primarily unimodal visual areas, with minimal involvement in processing auditory or somatosensory inputs (Rolls, 2007).

Area TE1a is structurally connected to areas 44, FOP4, and FOP3 via the arcuate/SLF tracts, areas PFm, PF, PSL, and STV in the inferior parietal lobe via the arcuate/SLF tracts, and areas V1 and V2 via the ILF. The arcuate/SLF tracts project around the Sylvian fissure and turn medially to finally terminate at the frontal lobe (areas 44, FOP4, and FOP3). Additionally, there are posterior projections from the arcuate/SLF that wraps around the Sylvian fissure, terminating at the inferior parietal lobule (areas PFm, PF, PSL, and STV). The ILF travels through the inferior temporal lobe to end at V1 and V2. The local short association fibers of TE1a, which include "u" fibers of the occipitotemporal system, connect to STSva, STSvp, and TE1m. It is important to note that in the right hemisphere, the white matter tracts of TE1a have more consistent connections with the MdLF.

Area TE1m

Area TE1m is located on the lateral surface of the middle portion of the MTG and extends to the corresponding aspect of the inferior temporal sulcus, as well as occupying part of the superior portion of the corresponding ITG. It borders TE1a anteriorly and TE1p posteriorly. STSvp forms its superior border, and TE2a is its inferior border.

Area TE1m exhibits functional connectivity to the frontal lobe (a47r, 8AV 8BL, 8AD, 8C, 9p, 47L, i6-8, and s6-8), the temporal lobe (areas STSvp, TE2a, TE1p, and TE1a), and the parietal lobe (areas PGs, PFm, IP1, 7m, d23ab, 31pv, and 31pd).

The primary function of area TE1m seems predominantly associated with visual pathways. Similar to TE1p, TE1m exhibits greater activation in the visual working memory secondary contrast compared to area TE1a. When contrasted with TE1p, TE1m demonstrates reduced deactivation during language tasks, but increased deactivation in theory of mind tasks. Remarkably, TE1m displays more deactivation in theory of mind tasks than any of its neighboring regions.

Area TE1m is structurally connected to areas 44, 45, IFJa, IFJp, and 8C via the arcuate/SLF tracts and areas PGi and PFm in the inferior parietal lobe via the arcuate/SLF tracts. The arcuate/SLF tracts project around the Sylvian fissure and turn medially to finally terminate at the frontal lobe (areas 44, 45, IFJa, IFJp, and 8C). Additionally, there are posterior projections from the arcuate/SLF that wrap around the Sylvian fissure, terminating at the inferior parietal lobule (areas PGi and PFm). The local short association fibers of TE1m, which include "u" fibers of the occipitotemporal system, connect to TE1p and TE1a. Area TE1m has structural differences than area TE1m as it has shown to be thinner and less myelinated (Glasser et al., 2016).

Area TE1p

Area TE1p is situated in the posterior portions of the middle and inferior temporal gyri. It extends to the inferior temporal sulcus. It also leaks onto the basal face of the temporal lobe and up to the occipitotemporal sulcus. TE1p borders TE1m and TE2a on its anterior surface, while PHT forms its posterior border on the lateral surface and PH on the basal surface. Its inferobasal edge is formed by TE2p, and its superior edge is formed by STSvp.

Area TE1p exhibits functional connectivity to the frontal lobe (33prime, 8AV, 8AD, 8BM, 8C, IFSa, IFSp, IFJp, a47r, p47r, 47m, a9-46v, p9-46v, i6-8, and s6-8), the temporal lobe (STSvp, PHT, TE1m, and TE2), and the parietal lobe (areas PGs, PGi, PFm, IP2, IP1, IP0, 7pm, 7m, d23ab, and 31a).

The primary function of area TE1p seems predominantly associated with visual pathways. Like TE1m, TE1p exhibits greater activation in the visual working memory compared to area TE1a. In contrast to TE1m, TE1p demonstrates increased deactivation during language tasks and heightened activation during facial recognition tasks.

Area TE1p is structurally connected to area 45 via the arcuate/SLF tracts and areas STV, PFm, PSL, PGi, TPOJ1, TPOJ2, and STV in the inferior parietal lobe via the arcuate/SLF tracts. The arcuate/SLF tracts project around the Sylvian fissure and turn medially to finally terminate at the frontal lobe (area 45). Additionally, there are posterior projections from the arcuate/SLF that wraps around the Sylvian fissure, terminating at the inferior parietal lobule (STV, PFm, PSL, PGi, TPOJ1, TPOJ2, and STV). The local short association fibers of TE1p, which include "u" fibers of the occipitotemporal system, connect to TE2a and perirhinal ectorhinal cortex. Area TE1p is structurally different when compared to its poster-superior neighbor area TE2a by having more myelin (Glasser et al., 2016).

Area PHT

Area PHT is found on the anterior portion of subcentral gyrus, where the postcentral gyrus connects to the Sylvian fissure. It includes the lateral surface of the operculum as well as the inferior aspect that faces the Sylvian fissure. Area PHT borders area 6r anteriorly and OP4 posteriorly. Its superior border includes area 6v as well as areas 4 and 3a. Its inferior borders include FOP1 and FOP2.

Area PHT exhibits functional connectivity to the frontal lobe (IFSa, IFJa, IFJp, 6a, 6ma, 6r, 46, 9-46d, p9-46v, p47r, FEF, PEF, SCEF, a24prime, p24prime, p32prime, 33prime, 23c, and 5mv), the temporal lobe (TE1p, TE2p and PHA3), the insula opercular area (FOP1, FOP3, FOP4, FOP5, 43, PFcm, 52, MI, PoI1, and PoI2), the parietal lobe (AIP, MIP, VIP, LIPv, LIPd, IPS1, IP0, IP1, IP2, PF, PFop, PFt, PGp, 7PC, 7pm, 7AL, 7PL, PCV, and DVT), and the occipital lobe (V1, V2, FST, PH, and TPOJ2).

In contrast to other parcellations of the lateral temporal cortex and temporal pole (TE1p, TE1m, TE1a, TE2p, TE2a, TGv, TGd, and TF), which are all strongly associated with the task-negative network, PHT is notably linked to the task-positive network. Moreover, PHT, similar to TE1p, demonstrates deactivation during language recognition tasks (Glasser et al., 2016).

Area PHT is structurally connected to areas 44, IFJa, IFJp, and IFSp via the arcuate/SLF tracts and areas PGs, STV, PFm, PGi, TPOJ1, and TPOJ2 in the inferior parietal lobe via the arcuate/SLF tracts. The arcuate/SLF tracts project around the Sylvian fissure and turn medially to finally terminate at the frontal lobe (areas 44, IFJa, IFJp, and IFSp). Additionally, there are posterior projections from the arcuate/SLF that wraps around the Sylvian fissure, terminating at the inferior parietal lobule (PGs, STV, PFm, PGi, TPOJ1, and TPOJ2).

Area STSda

Area STSda is located on the anterior half of the lateral surface of the superior temporal gyrus (STG) and the anterior half of the superior bank of the superior temporal sulcus (STS). It borders area STSdp on its posterior surface, STSva and TGd on its inferior and anteroinferior surface, STGa on its anterior surface, and TA2 and A5 on its superior surface.

Area STSda exhibits functional connectivity to the frontal lobe (areas 9m, 45, 47L, SFL, and 55b), the insula opercular area (areas STV, PSL, A5, and STGa), the temporal lobe (areas STSva, STSvp, STSdp, TGd, and TE1a), and the parietal lobe (areas PGI and 31pd).

The STS has generally been implicated in theory of mind, alongside the inferior temporal sulcus, medial prefrontal cortex, and temporal parietal junction (Glasser et al., 2016). Additionally, the STS is involved in motion processing, speech processing, and facial processing (Hein & Knight, 2008). Beyond its activation in response to unimodal visual and auditory inputs, the STS exhibits even stronger activation when presented with combined audiovisual stimuli, indicating a role in audiovisual integration (Glasser et al., 2016). Area STSda is principally associated with speech processing (Hein & Knight, 2008). It is also activated during language-related task contrasts. In comparison to the inferiorly situated STSva, STSda demonstrates increased activation in motor tasks and reduced deactivation in tasks involving reward processing such as gambling (Glasser et al., 2016).

Area STSda is structurally connected to areas FOP1, FOP3, and FOP4 via the arcuate/SLF tracts and areas V1, V2, V3, and V4 via the MdLF. The arcuate/SLF tracts project around the Sylvian fissure and turn medially to finally terminate at areas FOP1, FOP3, and FOP4. MdLF tracts traverse through the temporal lobe to eventually terminate at V1, V2, V3, and V4. The local short association fibers of area STSda, which include "u" fibers of the occipitotemporal system, connect to areas TGd, STSva, and A5. It is important to note that it has been observed that in some individuals, the STSda is connected to the superior temporal terminations of the inferior longitudinal fasciculus (ILF). However, this connection is inconsistent across individuals and can be difficult to differentiate from the superior tracts of the MdLF.

Area STSva

Located on the anterior half of the inferior bank of the superior temporal sulcus, area STSva shares borders with TGd on its anterior side, area STSvp on its posterior side, TE1a on its inferior side, and area STSda on its superior side.

Area STSva exhibits functional connectivity to the frontal lobe (areas 8AV, 8BL, 8AD, 9a, 9p, 9m, 45, 47s, 47L, 10d, 10r, 10v, and SFL), the temporal lobe (areas STSda, STSvp, STSdp, TGd, the hippocampus, and TE1a), and the parietal lobe (areas PGi, 7m, d23ab, 23d, 31pv, and 31pd).

The specific function of area STSva can be determined by the differences in activation relative to its neighbors. As stated previously, it demonstrates decreased activation in motor tasks and increased deactivation in tasks involving reward processing such as gambling relative to its superior neighbor area STSda. Further comparison to its inferior neighbor, area TE1a, area STSva is more activated in story telling versus math and more deactivation during tasks that include gambling and emotion (Glasser et al., 2016).

Area STSva is structurally connected to areas 44, IFJa, 6r, and 43via the arcuate/SLF tracts and areas V1, V2, V3, V4, and V3a via the MdLF. The arcuate/SLF tracts project around the Sylvian fissure and turn medially to finally terminate at areas 44, IFJa, 6r, and 43. MdLF tracts traverse through the temporal lobe to eventually terminate at V1, V2, V3, V4, and V3a. The MdLF tract potentially has involvement of the superior portion of the ILF. Additionally, the local short association fibers of area STSva, which include "u" fibers of the occipitotemporal system, connect to areas STSda, STSdp, STSvp, PSL, PFm, and PF.

Brodmann area 22 (BA22)

Area 22 can be further divided into 7 regions—areas 4, 5, STGa in the temporal "hypotenuse" regions; areas STSvp and STSdp are centered around the superior temporal sulcus (STG); lastly, areas PSL and STV are in the parietal apex regions.

Area A4

Area A4 (auditory area 4) is located on the superior surface of the posterior half of the superior temporal gyrus, occupying the portion of the gyrus posterior to its junction with Heschl's gyrus. It is bordered laterally by A5 and medially by PBelt, while its posterior border is with PSL and STV. Its anterior border is with MBelt and TA2.

Area A4 exhibits functional connectivity to various areas of the brain. It has functional connectivity to both the sensory strip (areas 1, 2, 3a, and 3b) and the motor strip (area A4). It furthers its motor involvement by being functionally associated with several premotor regions (areas SCEF, FEF, 6mp, and 6v). Area A4 also has functional connections with medial frontal lobe (areas 5m, 5l, and 24dd), superior insula opercular regions (areas 43, PFcm, OP4, OP2-3, and OP1), the lower opercula and Heschl's gyrus regions (areas STV, 52, RI, TA2, PI, MI, PBelt, MBelt, LBelt, A1, A5, PoI1, and PoI2), and lateral parietal lobe (areas 7PC and PFop). It has strong visual involvement by being functionally connected to both the medial occipital lobe (areas V2, V3, and V4) and the lateral occipital lobe (areas TPOJ1, TPOJ2, LO1, LO3, MT, MST, PH, and FST). Strengthening its visual involvement it is functionally connected to both the dorsal visual stream (areas V6, V6a, V7, and V3a) and the ventral visual stream (areas V8, FFC, VVC, and PIT).

Area A4 is a recently identified region within the auditory association cortex. Research suggests that this area of the brain is involved in the processing of both perceptual and conceptual acoustic sounds (Amunts & Zilles, 2015). Area A4 was distinguished from area TA2 based on differences in functional activity observed during arithmetic and auditory story tasks (Bailey, 1951).

Area A4 demonstrates structural connections parcellations 45 and FOP5 in the inferior frontal gyrus via the arcuate/SLF tract and to the parcellations V2, V3, V6, V7, MIP, LIPd, LIPv, and IP1 via middle longitudinal fasciculus. The arcuate/SLF tracts project around the Sylvian fissure to terminate at parcellations 45 and FOP5. The middle fasciculus travels posterior and runs lateral from Area A4 to finally terminate at the parcellations V2, V3, V6, V7, MIP, LIPd, LIPv, and IP1 within the occipital and parietal lobes. Additional, local short association bundles establish structural connections with areas A5, PFop, RI, STV, LBelt, MBelt, PBelt, and PFcm.

Area A5

Area A5 (auditory area 5) is a relatively thin ribbon that runs anterior-posteriorly on the superior lateral surface of the posterior portions of the STG. It shares its main medial border with A4 and is primarily bordered by TPOJ1 posteriorly, STSdp and STSda laterally, and TA2 anteriorly.

Area A5 exhibits functional connectivity to various areas of the brain. It has functional connectivity to both the sensory strip (areas 1, 2, 3a, and 3b) and the motor strip (area 4). Area A5 also has functional connections with the superior insula opercular region (area OP4), the lower opercula and Heschl's gyrus regions (areas STV, RI, STGa, TA2, PBelt, LBelt, A4, and PSL), and the temporal lobe (areas STSda and STSdp). It has visual involvement by being functionally connected to the lateral occipital lobe (areas TPOJ1 and MT) and the ventral visual stream (area FFC).

Area A5 is a recently identified region within the auditory association cortex. Research suggests that this area of the brain is involved in the processing of both perceptual and conceptual acoustic sounds (Amunts & Zilles, 2015). Area A4 was distinguished from area TA2 based on differences in functional activity observed during arithmetic, auditory story, and social interaction tasks. Area A5 also showed more activation in language and math than its two laterally positioned neighbor areas STSda and STSdp (Bailey, 1951).

Area A5 demonstrated structural connections with the parcellation 44 in the inferior frontal gyrus via the arcuate/SLF tract and to the parcellations V6 and V6a via middle longitudinal fasciculus. The arcuate/SLF tracts project around the Sylvian fissure to terminate at parcellation 44. The middle longitudinal fasciculus travels posterior and runs lateral to the ventricle from area A5 to finally terminate at parcellations V6 and V6a within the intraparietal sulcus. Additional, local short association bundles establish structural connections with areas A4, STSda, and STSdp. Further distinction between areas A5 and A4 can be made as A5 is less myelinated than area A4 (Bailey, 1951).

Area STGa

Area STGa (superior temporal gyrus region a) is situated on the anterior superior surface of the superior temporal gyrus and does not extend onto the lateral surface. It is bordered medially by PI, anteromedially by TA2, posterolaterally by STSda, and anteriorly by TGd.

Area STGa exhibits limited functional connectivity. It has functional connections to the insula opercular regions (areas A5 and Pi) and the temporal lobe (TGd, STSda, STSdp).

Area STGa is a recently identified region within the auditory association cortex. Research suggests that this area of the brain is involved in the processing of both perceptual and conceptual acoustic sounds (Amunts & Zilles, 2015). Area STGa was distinguished from its neighboring areas TGd, PI, TA2, and STSda based on differences in fMRI activity observed during arithmetic, auditory story, and social interaction tasks (Bailey, 1951).

Area STGa has a structural connection to the parcellations V1, V2, V3, V3A, and V3CD in the occipital lobe via the inferior longitudinal fasciculus. The inferior longitudinal fasciculus runs posterior and travels through the temporal lobe to eventually terminate at the occipital lobe. Additionally, local short association tracts connect area STGa to parcellations TA2 and STSda. Area STGa can also be further differentiated from its medially neighbor area PI because it is substantially thicker (Bailey, 1951).

Area STSvp

Area STSvp is located on the posterior half of the inferior bank of the STS. It shares its anterior border with STSva, its inferior border with TE1m and temporal area 1 posterior (TE1p), its posterior border with TPOJ1 and PHT, and its superior border with STSdp.

Area STSvp exhibits functional connectivity robust connection to the frontal lobe (areas 8AV, 8BL, 8BM, 8C, 9a, 9p, 9m, 44, 45, 47s, 47L, a47r, IFSp, d32, 10v, SFL, and 55b) and the temporal lobe (areas STSva, STSda, STSdp, TGd, TE1a, TE1m, TE1p, and TE2a). It also has functional connections to the parietal lobe (areas PGs, PGi, 7m, POS1, d23ab, 31pv, and 31pd) and the insula opercular region (area PSL).

The posterior half of STSvp, like the posterior half of STSdp, exhibits strong activation during the story-math secondary contrast, suggesting a role in language comprehension. STSvp does not respond as robustly as STSva to primary language tasks and demonstrates less activity in social cognition and motor tasks (Bailey, 1951).

Area STSvp is structurally connected to 6r, IFJp, IFJa, FOP2, FOP3, FOP4, and 44 via the arcuate/SLF tracts and areas PF, PFm, PSL, PGi, and STV in the inferior parietal lobe via the arcuate/SLF tracts. The arcuate/SLF tracts project around the Sylvian fissure and turn medially to finally terminate at areas 6r, IFJp, IFJa, FOP2, FOP3, FOP4, and 44. Additionally, there are posterior projections from the arcuate/SLF that wraps around the Sylvian fissure, terminating at the inferior parietal lobule (PF, PFm, PSL, PGi, and STV). The local short association fibers of STSvp, which include "u" fibers of the occipitotemporal system, connect to STSdp, STSva, TE1p, TE1m, PF, PFm, PSL, PGi, and STV.

Area STSdp

Area STSdp is situated on the posterior half of the lateral face of the STG and the posterior half of the superior bank of the STS. It borders area STSda anteriorly, STSvp inferiorly, TPOJ1 posteriorly, and A5 superiorly.

Area STSdp exhibits functional connections to the frontal lobe (9m, 8BL, 44 45, 47L, 47s, IFSp, SFL, and 55b), the insula opercular (areas STV, PSL, A5, and STGa), the temporal lobe (areas STSva, STSvp, STSda, and TGd), lateral occipital lobe (TPOJ1), and the parietal lobe (PGi and 31pd).

Area STSdp is structurally connected to parcellations 44, FOP4, IFJa, IFJp, and IFSp4 via the arcuate/SLF tracts. The arcuate/SLF tracts project around the Sylvian fissure and turn medially to finally terminate at areas 44, FOP4, IFJa, IFJp, and IFSp. Additional structural connections exist with the local short association fibers of STSdp, which include "u" fibers of the occipitotemporal system, connect to STSda, STSva, STSvp, PSL, and PF.

Area PSL

The perisylvian language area (PSL) is situated at the crown of the posterior Sylvian fissure, in the lower region of the supramarginal gyrus. PSL shares borders with STV inferiorly, PFm posteriorly, PF superiorly, and PFcm anteriorly. Area PSL also has a border with RI on its internal surface.

Area PSL exhibits functional connectivity to paracingulate area through area SCEF, the premotor area through area 55bm, and the lateral frontal lobe (areas IFJa, 9–46d, 44, and 45). It also has connections to the inferior insula opercular region (areas STV and A5) and the temporal lobe (areas STSda, STSdp, and STSvp). Area PSL establishes visual involvement by having functional connections to both the medial occipital lobe (areas V2, V3, and V4) and the lateral occipital lobe (area TPOJ1). It furthers it visual association with connections to the dorsal visual stream (areas V6, V67a, V7, V3a, and V3b) and the ventral visual stream (areas V8, FFC, Pit, and VVC). Area PSL has been shown to have stronger functional connection with area 55b than area A4, PFM, PF, and STV (Bailey, 1951).

Area PSL is a newly defined area of the brain parcellated from the temporo-parieto-occipital junction (TPOJ). The TPOJ in general is believed to be associated with complex cognitive functions such as core information processing, working memory, emotional regulation, and cognitive control (Petrides et al., 2012). Area PSL was differentiated from surrounding areas through differences in fMRI activity. Area PSL can be distinguished from areas PFcm and RI based on differences in fMRI activity during arithmetic and auditory story tasks. Area PSL has been shown to be more activated in motor tasks than its anteroinferior neighbor area A4. There also exists a laterality to area PSL. The left PSL is more activated than the right in the relational versus match contrast (Bailey, 1951).

Area PSL has structural connections to 6r via the arcuate/SLF as well as area TE1a, STSdp, STSva, and STSvp via the arcuate/SLF. The arcuate/SLF white matter tracts course anteriorly and then run superior to the insula to terminate at parcellation 6r. The arcuate/SLF fibers also have another path that travels inferiorly through the temporal lobe to end at area TE1a, STSdp, STSva, and STSvp. Additional structural connections are created by local short association bundles that connect area PSL with the parcellations PBelt, STV, RI, A5, and PF.

Area STV

Area STV (superior temporal visual area) is located on the inferior posterior portion of the superior temporal gyrus and extends across the posterior part of the superior temporal sulcus with its posterior aspect on the anterior angular gyrus. It borders PSL superiorly and TPOJ1 inferiorly, with its posterior boundaries consisting of PGi and a smaller portion of PFm. The primary anterior border of STV is with A4.

Area STV exhibits functional connectivity to multiple areas of the brain. It has functional connectivity to both the sensory strip (areas 1, 2, 3a, and 3b) and the motor strip (area 4). It furthers its motor involvement by being functionally associated with several premotor regions (areas FEF and 55b). Area STV also has functional connections with superior insula opercular regions (areas 43, PFcm, and OP4), the inferior opercular regions (areas PSL, A4, A5, PBelt, LBelt, PoI1, PI, RI, and 52), parietal lobe (area PCV), and the temporal lobe (area STSda and STSdp). It has strong visual involvement by being functionally connected to both the medial occipital lobe (areas V2, V3, and V4) and the lateral occipital lobe (areas TPOJ3 and TPOJ2). Strengthening its visual involvement, it is functionally connected to both the dorsal visual stream (area V6) and the ventral visual stream (area FFC).

Area STV is a newly defined area of the brain parcellated from the temporo-parieto-occipital junction (TPOJ). The TPOJ in general is believed to be associated with complex cognitive functions such as core information processing, working memory, emotional regulation, and cognitive control (Petrides et al., 2012). Area STV can be differentiated from surrounding parcellations through activity levels during various tasks. For example, area STV is more deactivated in working memory, motor tasks, and language math tasks than area PSL (Bailey, 1951).

Area STV demonstrated structural connections parcellations TE1a, STSdp, STSda, STSva, and STSvp in the middle frontal gyrus via the arcuate/SLF tract. Additional local short association fibers establish connections with parcellations PSL, STV, PGi, PFm, and PFcm.

Area TA2

Area TA2 comprises a region within the medial planum polare and is found just anterior to the anterolateral aspect of Heschl's gyrus.

Area TA2 is bounded medially by area PI, laterally by areas STGa and A5, anteriorly by TGd, and posteriorly with the anterior aspects of areas A4 and MBelt.

Area TA2 exhibits functional connectivity with various regions across the cortex, including: the sensory strip (1, 2, 3a, and 3b); the motor strip (4); the insula opercular regions (OP4, 43, PFcm, A4, A5, MBelt, PBelt, LBelt, RI, and PI); the medial occipital lobe (V2, V3, and V4); and the dorsal visual stream (V6).

Area TA2 is structurally connected to both the inferior longitudinal fasciculus and local parcellations. Fibers from the ILF project posteriorly from TA2 through the temporal lobe, ultimately terminating at occipital lobe parcellations V1, V2, V3, and V6A. Additionally, TA2 is linked to nearby brain regions via short association bundle fibers, including areas 52, MBelt, A5, PI, and STGa.

Brodmann area 23 (BA23)

Area 23d

Area 23d is found immediately superior to the posterior aspect of the corpus callosum along the posterior-superior aspect of the cingulate gyrus. Surrounding area 23d are areas 23c and RSC in the superior and inferior borders, respectively. Additionally, areas p24prime and d23ab are adjacent to area 23d on its anterior and posterior borders, respectively.

Area 23d exhibits numerous functional connections to a variety of HCP parcellations across the frontal, temporal, and parietal lobes. Specifically, it shares functional connectivity to HCP areas within the lateral frontal lobe (a47r, p10p, s6-8, 8AV, and 9p), medial frontal lobe (areas

8BM, 8C 9m, a24, p24, p24prime, p32, and d32), temporal lobe (area TE1m), lateral parietal lobe (areas IP1, PGi, PFm, and PGs), and medial parietal lobe (areas d23ab, POS2, RSC, 7m, 31a, 31pv, and 31pd).

Most of the area 23d's structural white matter connections traverse the cingulum. Area 23d's cingulum white matter fibers extend in the anterior direction and connect to the anterior cingulate cortex and cingulate sulcus. From there, the white matter fibers continue and bend around the genu of the corpus callosum, with some fibers ending at areas a24, p24, and a32pr, while others continue anteriorly to area 9 or pass through the rostrum of the corpus callosum to end at area 25. In addition to having structural connections with the cingulum, area 23d also has local structural connections which extend in the posterior direction, connecting it to areas d23ab and v23ab.

Area 23c

Area 23c is an elongated and thin parcellation of BA23 that is located on the lower edge of the posterior cingulate sulcus and forms the posterior edge of the ascending marginal ramus. Area 23c borders PCV on its posterior aspect with areas 24dv and p23prime setting on its anterior aspect. Area 23c shares its superior border with two areas, including area 5mv posteriorly and area 24dd anterosuperiorly. Furthermore, area 23c has two adjacent neighbors on its elongated inferior edge, including area 31a posteriorly, and area 23d anteroinferiorly.

Area 23c is functionally connected to various regions across the cortex, including: the premotor regions (areas SCEF, FEF, PEF, 6r, 6a, 6ma); the middle and posterior cingulate regions (areas 24dv, a24prime, p24prime, a32prime, p32prime, and 5mv); the dorsolateral frontal lobe (areas IFSa, 9-46d, and 46); the superior insula opercular regions (43, OP4, PFcm, FOP1, FOP3, FOP4, and FOP5); lower opercula and Heschl's gyrus regions (areas 52, PoI1, PoI2, and MI); the temporal lobe (area PHA3); the lateral parietal lobe (areas AIP, MIP, LIPv, LIPd, IP0, PGp, PFop, PF, PFt, 7AL, and 7PC); the medial parietal lobe (areas 31a, POS2, RSC, 7am, 7pm, PCV, and DVT); the medial occipital lobe (areas V1, V2, V3, and V4); dorsal visual stream areas (areas V3b, V7, V6, and V6a); and the lateral occipital lobe (areas PHT, PH, TPOJ2, TPOJ3, FST, and LO3).

Area 23c has white matter fiber tracts that project both contralaterally and to local areas. Fiber tracts which connect area 23c to the contralateral hemisphere traverse the corpus callosum and terminate at areas 5mv, 5m, 23c, and 5l, while short association bundles extend to superior structures, terminating at areas 24dd, 24dv, 5l, 5m, and 5mv.

Area d23ab

Area d23ab is situated on the posterior aspect of the cingulate gyrus and superior to the splenium of the corpus callosum. Area d23ab is surrounded by a number of HCP subdivisions, including area 23d anteriorly, area v23ab posteriorly, areas 31a and 31pv superiorly, and area RSC inferiorly.

Area d23ab demonstrates functional connectivity to various parcellations throughout the cortex, including: the lateral frontal lobe (areas a47r, p10p, i6-8, s6-8, 10d, 8AD, 8AV, 8BL, 8C, and 9p); the medial frontal lobe (areas 8BM, 9m, 10r, a24, and d32); the temporal lobe (areas STSva, STSvp, TE1a, TE1m, TE1p, PreS, and the hippocampus); the lateral parietal lobe (areas IP1, PGi, PGs, and PFm); and the medial parietal lobe (areas 23d, v23ab, POS2, POS1, RSC, 7m, 31a, 31pv, and 31pd).

Area d23ab is connected to the cingulum, with fibers extending in the posterior direction that wrap around the genu of the corpus callosum, head toward cingulate gyrus and sulcus, and finally end at areas a32pr and p24. In addition, area d23ab is linked to its neighboring region v23ab through short association bundle fibers.

Area v23ab

Area v23ab is situated on the cortical surface of the posterior aspect of the posterior cingulate area, just superior to the cingulate isthmus. Area v23ab is bounded anteriorly by areas d23ab and 21pv, posteriorly by area POS1, superiorly by RSC, and anterior-inferiorly by RSC.

Area v23ab is functionally connected with a number of cortical regions, including: the lateral frontal lobe (areas a47r, p10p, s6-8, 10d, 8AD, 8AV, 8BL, and 9p); the medial frontal lobe (areas 8BM, 9m, 10r, 10v, a24, s32, and d32); the temporal lobe (areas STSva, TGd, TE1a, TE1m, PreS, and the hippocampus); the lateral parietal lobe (areas IP1, PGi, and PGs); and the medial parietal lobe (areas d23ab, POS1, RSC, 7m, 31a, 31pv, and 31pd).

In addition to extensive functional connections, area v23ab is also structurally connected to a number of areas identified by the HCP via extensions that traverse the cingulum. A subset of fibers extends anteriorly and terminates at areas a24, p24, and 32pr within the cingulate cortex. Other fibers continue along cingulum and wrap around the genu of the corpus callosum and split into an anterior branch, terminating at areas p32 and 10r in the anterior aspect of the frontal lobe. The remaining fibers continue along the rostrum of the corpus callosum and terminate at area 25. Additionally, there are fibers which project locally from area v23ab to nearby regions POS1 and 7m.

Brodmann area 24 (BA24)

Area 24 can be divided further into 6 subunits. Despite appearing confusing at first, the organization of the anterior cingulate regions is actually quite simple to understand. There are three parallel, C-shaped rows of areas that follow the curvature of the anterior cingulate from posterior to anterior, and bend downward to reach the subcallosal region. The middle of the three parallel regions is the one that contains the subdivision of area 24. The area going posterior to anterior are areas 24dd and 24dv, followed by the prime regions, area posterior 24prime and area anterior 24prime, followed by the standard pairs, p24 and a24.

Area a24

Area a24 (anterior 24) is situated slightly anterior to the genu of the corpus callosum and is in the anterior cingulate gyrus. Area a24 borders area p24 superiorly and area 25 posteriorly. It is adjacent to the genu of the corpus callosum and has s32 as its inferior border, while its anterior border is composed of both p32 and 9m.

Area a24 exhibits functional connectivity to both the medial frontal lobe (areas p24, d32, p32, s32, 10d, 10r, 9p, and 9m) and the lateral frontal lobe (area 8ad). It has further functional connections with temporal lobe (area TE1a), lateral parietal lobe (area PGs), and the posterior cingulate areas (RSC, 31pv, 31pd, 23d, d23ab, v23ab, 7m, and POS1).

Area a24 is associated with the "affect division" of the anterior cingulate cortex (ACC) and has been connected to the evaluation of both internal and external states. This region is believed to contribute to emotional expression and motivation, highlighting its importance in understanding the neural mechanisms underlying affective processing and goal-directed behavior (Devinsky et al., 1995; Drevets et al., 2008).

Area a24 is structurally connected to area 25 in the cingulum. The cingulum fibers project anteriorly to terminate near the rostrum of the corpus callosum at area 25. Further structural connections exist from the posterior cingulum fibers that end at the precuneus near the splenium of the corpus callosum establishing connection with areas 31pv and 23d. Additionally, there are local short association bundles that connect area a24 with p32 and 9m.

Area p24

Area p24 (posterior 24) is situated within the anterior cingulate gyrus, spanning the entire gyrus and positioned just anterosuperior to the genu of the corpus callosum. It is neighbored by areas d32 and a32pr superiorly, while area a24 lies anteroinferiorly. Area p24 shares a posterior boundary with a24pr and 33pr, and its inferior border is demarcated by the callosal sulcus. Understanding the anatomical connections and positioning of area p24 within the anterior cingulate gyrus can provide valuable insights into its functional contributions and interactions with neighboring regions in the brain.

Area p24 exhibits functional connectivity to both the medial frontal lobe (areas a24, d32, 23c, a24prime) and the lateral frontal lobe (area 9-46d). It has additional functional connections with the medial parietal lobe (areas RSC and POS2) and the medical occipital lobe (area V1).

Area p24 is functionally distinct from its anterior counterpart in that it plays a more prominent role in selective attention, coordination of conscious eye movements with complicated finger movement sequences, and stimulus/response selection. Differences between area p24 and area a24pr can be observed by activity levels in fMRI. It was shown that area p24 is more activated than a24pr in language story tasks.

Area p24 is functionally distinct from its anterior counterpart, as it plays a more significant role in selective attention, the coordination of conscious eye movements with complex finger movement sequences, and stimulus/response selection (Beckmann et al., 2009; Glasser et al., 2016). Functional distinctions between area p24 and area a24pr can be observed through differences in activity levels in functional MRI (fMRI) studies. For instance, area p24 demonstrates greater activation than a24pr during language story tasks (Glasser et al., 2016).

Area p24 is anatomically linked to the cingulum, a major white matter tract in the brain. Anterior cingulum fibers project toward frontal lobe parcellations p32 and 10r, and they also curve around the rostrum of the corpus callosum, terminating at area 25. Meanwhile, posterior cingulum fibers extend to the precuneus, near the splenium of the corpus callosum, connecting to areas POS1, v23ab, and RSC. Additionally, local short association bundles establish connections with areas a24, d32, and a32pr.

Area a24pr

Area a24pr (anterior 24prime) is situated within the middle cingulate gyrus, predominantly occupying the superior half of the gyrus and extending into the inferior bank of the cingulate sulcus. It is bordered by area 33prime inferiorly, p24pr posteriorly, p24 anteriorly, and shares its portion of the cingulate with p32pr superiorly.

Area a24pr exhibits functional connections to the cingulate areas (areas 33prime, 5mv, 23d, 24dv, p24, p24prime, a32prime, and p32prime), the premotor regions (areas SCEF, FEF, PEF, 6a, and 6r), the lateral frontal lobe (area 9-46d and 46), and the temporal lobe (area PHT). Area a24pr has robust functional connections to the insula opercular area (areas FOP1, FOP3, FOP4, FOP5, OP4, PFcm, MI, 43, 52, PoI2, and PoI1). It has connections to both the lateral parietal lobe (areas PGp, PF, PFop, and 7AL) and the medial parietal lobe (areas DVT, 7am, and PCV). Area a24pr establishes visual involvement by having functional connections to the medial occipital lobe (areas V1, V2, V3, and V4) and the dorsal visual stream (areas V6, V6a, V3a, V3b, and V7).

Area a24pr exhibits little response to affective processes. However, it is implicated in cognitive response selection and has been associated with word and sentence selection during language-based tasks (Devinsky et al., 1995).

Structurally, area a24pr is connected to the cingulum. Anterior cingulum fibers project above the corpus callosum, terminating at p32, and continue to curve around the rostrum of the corpus callosum to end at area 25. The posterior cingulum fibers reach the precuneus, ending at areas POS1 and 31pv, and curve around the splenium of the corpus callosum to terminate at the RSC. There are some individuals that exhibit contralateral connections via the body of the corpus callosum, but these structural connections are not consistent across all individuals. Area a24pr can be further differentiated from its surrounding neighbors because it has significantly less myelin than p32p and a32pr.

Area p24pr

Area p24pr (posterior 24 prime) is situated within the middle cingulate gyrus. It shares borders with area 24dv superiorly, a24pr anteriorly, 33prime inferiorly, and 23d posteriorly.

Area p24pr exhibits functional connections to the cingulate (areas 33prime, 24dd, 24dv, 5mv, 23d, a24prime, and p32prime), premotor region (areas 6ma and 6r), lateral frontal lobe (area 46), the insula opercular region (areas FOP4, FOP5, PFcm, MI, 43, and PoI1), the parietal lobe (areas PF, PFop, and 7AL), and the lateral occipital lobe (area TPOJ2).

Area p24pr has been implicated as part of the "cognitive division" of the anterior cingulate cortex (ACC) and is involved in stimulus and response selection during cognitively demanding tasks that may require movement execution (Devinsky et al., 1995; Gasquoine, 2013).

Anatomically, area p24pr is connected to the marginal branch of the cingulate sulcus and the precuneus. Fibers from p24pr project posteriorly to the marginal branch of the cingulate sulcus and precuneus, terminating at parcellations 23c, 31a, 31pd, and 7m.

Area 24dd

Area 24dd (24 dorsal-dorsal) is situated in the anterior inferior paracentral lobule, extending into the upper bank of the cingulate sulcus. From anterior to posterior, area 24dd is bordered superiorly by the supplementary and cingulate eye field (SCEF), area 6mp, area 4, and area 5m. Posteriorly and inferiorly, it adjoins the marginal ramus of the cingulate sulcus and area 5mv, while its inferior border comprises area 24dv and area 23c.

Area 24dd establishes functional connections to both the sensory strip (areas 1, 2, 3a, and 3b) and the motor strip (area 4). It has further motor involvement by exhibiting functional connections to premotor regions (areas 6mp, 6v, and 6d). It also has functional connection to the middle cingulate regions (areas 5mv, SCEF, and 24dv), the superior opercular areas (areas OP4, OP1, and 43), Heschl's gyrus regions (areas PBelt and A4), and the parietal lobe (area 7PC). Area 24dd interacts with visual areas of the brain by having functional connections with V2 in the primary visual areas and areas FST and MST in the lateral occipital lobe.

Area 24dd plays contributes to complex motor planning and regulation of muscles in the lower limb and lower trunk. This is achieved through coordination with the supplementary motor area and connections to the spinal cord, allowing for the precise control and execution of movements in the lower body (Vogt & Vogt, 2003).

Area 24dd is structurally connected to the contralateral hemisphere. Fibers originating from 24dd traverse the body of the corpus callosum, terminating at 24dd, 4, 5mv, 23c, SCEF, 6mp, and 24dv in the opposite hemisphere. Additional, local short association bundles establish connections with areas 4, 5mv, 23c, SCEF, and 6mp.

Area 24dv

Area 24dv (24 dorsal-ventral) is situated in the anterior inferior paracentral lobule, extending into the upper bank of the cingulate sulcus. Area 24dv shares borders with 24dd superiorly, p32pr and a small portion of SCEF anteriorly, p24pr inferiorly, and 24dd posteriorly. Notably, this area causes the 24 and 32 regions to be slightly misaligned in their anterior-to-posterior arrangement.

Area 24dv establishes functional connections to both the sensory strip (areas 1, 2, 3a, and 3b) and the motor strip (area 4). It has further motor involvement by exhibiting functional connections to premotor regions (areas 6mp). It also has functional connection to the middle cingulate regions (areas 5mv, SCEF, a24prime, p24prime, p32prime, 23c, and 24dd), the superior opercular regions (areas FOP1, FOP3, FOP4, PFcm, OP4, OP1, and 43), the lower opercula and Heschl's gyrus regions (areas PoI1 and 52), and the parietal lobe (areas 7AL, 7PC, and PFop). Area 24dv has one functional connection with a visual area and that is V2 in the primary visual area.

Area 24dv is involved in complex motor planning and regulation of muscles in the upper limb and upper trunk. This is achieved through coordination with the supplementary motor area and connections to the spinal cord, allowing for precise control and execution of movements in the upper body (Vogt & Vogt, 2003).

Area 24dv is structurally connected to the contralateral hemisphere, marginal branch of the cingulate sulcus, and precuneus. Contralateral connections pass through the body of the corpus callosum, linking to 24dv, SCEF, and p32pr in the opposite hemisphere. Fibers originating from 24dv project posteriorly to the marginal branch of the cingulate sulcus and precuneus areas, connecting to areas 23c and 31a. Local short association fibers establish connections with SCEF, p32pr, and 24dv.

Brodmann area 25 (BA25)

BA25 is primarily made up into a single HCP parcellation—area 25.

Area 25

Area 25 is situated in the most posterior portion of the subcallosal area. It is bordered by area a24 and s32 anteriorly, while its inferior and posterior borders include the orbitofrontal cortex (OFC) and posterior orbitofrontal cortex (pOFC) areas.

Area 25 exhibits functional connectivity only to area s32. Area 25 is considered part of the "affect division" of the anterior cingulate cortex (ACC) and has been linked to various emotional processes. These include conditioned emotional learning, emotional expression, assessment of motivational content, assignment of emotional valence to internal and external stimuli, and maternal-infant interactions. By playing a role in these processes, area 25 contributes to the regulation and interpretation of emotions and social interactions (Devinsky et al., 1995; Drevets et al., 2008).

Area 25 is structurally connected to the cingulum. Fibers from area 25 project posteriorly above the corpus callosum, connecting to areas v23ab and 31pv. White matter tracts from this region are variable with some individuals having connections with anterior parcellations, but this is inconsistent across individuals.

Brodmann areas 26, 29, and 30 (BA26, BA29, BA30)

BA26, BA29, and BA30 in our working definition together constitute the retrosplenial cortex, or HCP parcellation RSC.

Area RSC

Area RSC is a slim region of the posterior cingulate cortex located adjacent to the callosal sulcus, wrapping around the splenium. It extends from just above the midsection of the corpus callosum body, following the cingulate sulcus to the bottom of the cingulum's isthmus, where the parahippocampal gyrus starts. The corpus callosum borders the entire inferior boundary of area RSC. Anteriorly, it borders area 33prime, while its posteroinferior border is with the presubiculum. Area RSC has a long superior border that includes multiple parcellations. Moving anterior to posterior, the superior border is comprised of area 23d, area d23ab, area v23ab, area POS1, and area ProS.

Area RSC exhibits functional connectivity to both the lateral frontal lobe (areas 9-46d, p10p, and 8AD) and the medial frontal lobe (areas 8BM, 9m, a24, p24, a32prime, p32, and d32). It has further connections to both the lateral parietal lobe (areas IP1, IP2, PFm, and PGs) and medial parietal lobe (areas 23d, v23ab, d23ab, POS2, POS1, PCV, DVT, 7pm, 7m, 31a, 31pv, and 31pd).

Area RSC is primarily involved in transitioning between allocentric (view-independent) and egocentric (view-dependent) spatial perspectives. This shifting and relating perspective for spatial cognition assists in spatial navigation, episodic memory, future planning, and imagination. The RSC is thought to be involved in retrieving recent autobiographical information from memory. Area RSC has also been shown to predict and correct errors for current sensory states with internal representations of the environment (Aggleton et al., 2014; Alexander et al., 2023; Bzdok et al., 2015; Glasser et al., 2016; Vann et al., 2009).

Area RSC is structurally connected to the cingulum. The cingulum fibers project both anteriorly and posteriorly and travel from area RSC along the entire length of the cingulate cortex. The posterior projections continue around the splenium of the corpus callosum, terminating at the area EC in the parahippocampal gyrus. The anterior cingulum projections divide near the genu of the corpus callosum, extending superiorly to end at areas 10d and 9m. There is also an inferior portion which establishes connections to area 25. White matter tracts in the right hemisphere demonstrate more consistent connections with the precuneus compared to the left hemisphere.

Brodmann area 27 (BA27)

BA27 consists of one HCP region—area PreS.

Area PreS

Area PreS (presubiculum) is located on the posterior superior surface of the parahippocampal gyrus. It is bordered medially by the hippocampus and anteriorly by the entorhinal cortex. The retrosplenial cortex (RSC) and the prostriate region (ProS) form its posterior border, while PHA1 creates the inferior border.

Area exhibits functional connectivity to the frontal lobe (area 8AD and i6-8), the temporal lobe (areas PHA1, PHA2, and the hippocampus), the parietal lobe (areas RSC, Pros, d23ab, v23ab, 31a, 31pv, 7m, 7pm, POS1, POS2, IP1, and PGs), and the occipital lobe (area V1).

Area PreS is structurally connected to the cingulum, precuneus, and occipital lobe. Cingulum projections extend superiorly to the corpus callosum, terminating at anterior cingulate cortex and frontal lobe parcellations such as a24, 9m, 10d, and p32. PreS fibers also project posteriorly to occipital and precuneus areas, including V1, V2, V6, POS1, POS2, and 7m. Local short association fibers connect with the entorhinal cortex (EC) and perirhinal cortex (PeEc). Structurally, area PreS can be distinguished from its neighboring regions because it is significantly more myelinated. Only the RSC is more myelinated than area PreS (Glasser et al., 2016).

The presubiculum is situated medial to the subiculum, a region within the hippocampus thought to be involved in spatial information processing based on primate studies (Aggleton & Christiansen, 2015). Compared to its inferior neighbor PHA1, area PreS exhibits less activity during tasks related to working memory, language processing, and theory of mind, although area PreS demonstrates greater activity during motor tasks when compared to PHA1.

Brodmann area 28 (BA28)

BA28 consists of one HCP region—area PeEC.

Area PeEC

Area PeEC (perirhinal ectorhinal cortex) is situated on the anterior sections and inferior surface of the uncus, extending to the collateral sulcus. It is bordered by areas TGv and TGd anteriorly, TF laterally, and PHA2 and PHA3 posteriorly.

Area PeEC exhibits functional connectivity to the area IFSa in the frontal lobe, multiple areas in the temporal lobe (areas EC, TF, PHA2, and PHA3), area IP0 in the parietal lobe, and area PH in the occipital lobe.

The authors of HCP could not consistently differentiate between the ectorhinal cortex and perirhinal cortex. This led the authors to combine the two regions into a single complex (Glasser et al., 2016). The perirhinal cortex alone has been shown to facilitate the transmission of declarative memories between cortical areas and the hippocampus. By adding semantic knowledge, the perirhinal cortex contributes to tasks such as item identification. It also integrates item information with spatiotemporal information, which eventually is communicated to the hippocampus by the entorhinal cortex (Naya, 2016). Area PeEC is distinguishable from adjacent regions due to heightened activation during primary working memory tasks and selective facial recognition. The area's predilection for facial recognition tasks, in contrast to its neighboring regions, prompted the HCP authors to postulate that it might represent the location of the anterior temporal face patch (Glasser et al., 2016; Rajimehr et al., 2009; Tsao et al., 2008).

Area PeEC has structural connections to the parcellations PH, TPOJ3, and MT via the ILF. The ILF projections course through the inferior temporal lobe to finally terminate at areas PH, TPOJ3, and MT. It is important that additional structural connections are seen that run parallel to the ILF that connect area PeEC to area V1 in the medial occipital lobe. This connection is inconsistent across individuals.

Brodmann area 31 (BA31)

BA31 can be seen splitting into 3 different HCP subdivisions. These unique regions include areas 31a, 31pd, and 31pv.

Area 31a

Area 31a is located on the subparietal gyrus, specifically on its anterior aspect just posterior to the marginal sulcus. Area 31a is flanked by areas 31pd, 31pv, and PCV on its posterior boundary, and areas d23ab and 23d on its inferior aspect. Additionally, its elongated superior boundary is flanked by area 23c.

Area 31a exhibits functional connections to various regions across the cortex, including: the lateral frontal lobe (areas a9-46v, p10p, 10d, 8AD, 8AV, 8C, s6-8, and i6-8); the medial frontal lobe (areas 8BM, p32, and d32); the temporal lobe (areas PreS and TE1p); the lateral parietal lobe (areas PGi, PGs, IP2, and IP1); the medial parietal lobe (areas 23d, v23ab, d23ab, parietooccipital sulcus 2 (POS2); and the parietooccipital sulcus 1 (POS1), PCV, RSC, 7pm, 7m, 31pv, and 31pd).

Area 31a has structural connections to the cingulum and to local areas within the precuneus. Some individuals exhibit structural connections from area 31a to the contralateral hemisphere via tracts that traverse the corpus callosum; however, these findings are not found in all individuals. Area 31a′s connections to the cingulum project anteriorly and terminate at areas p24, d32, and a24pr within the cingulate sulcus and the superior frontal gyrus. White fiber bundles projecting posteriorly from area 31a extend to area 7m via short association bundles.

Area 31pd

Area 31pd is located on the cortical surface of the posterior superior aspect of the subparietal gyrus and is flanked inferiorly by area 31pv, anteriorly by area 31a, superiorly by area PCV, and posteriorly by area 7M.

Area 31pd is functionally connected to a number of regions across the cortex, including: the lateral frontal lobe (a9-46v, 45, 47l, 47s, 10d, 8AD, 8AV, 8BL, 9a, and 9p); medial frontal lobe (SFL, 9m, 10r, 10v, and d32); the temporal lobe (TGd, STSva, STSvp, STSda, STSdp, TE1a, and the hippocampus); the lateral parietal lobe (PGi, PGs, and IP2); and the medial parietal lobe (23d, v23ab, d23ab, PCV, POS2, POS1, RSC, 7m, 31pv, and 31a).

Structural connectivity of area 31pd includes the cingulum, contralateral hemisphere, and local parcellations of the precuneus. Connections to the cingulum extend anteriorly and exhibit inconsistent connections to the cingulate sulcus and the superior frontal gyrus. Contralateral projections extend through the corpus callosum, and terminate at areas 31a, 7m, and 31pd within the contralateral precuneus. Local projecting fibers connect area 31pd to areas 7m and PCV via short association bundles.

Area 31pv

Area 31pv is mostly situated on the posterior inferior subparietal gyrus and extends over the cingulate sulcus onto the posterior cingulate gyrus. Area 31pv is bounded anteriorly by area 31a, superiorly by area 31pd, posteriorly by area 7m and area v23ab, and inferiorly by area d23ab.

Area 31pv is functionally connected to a number of regions across the cortex, including: the lateral frontal lobe (areas 47l, 47s, p10p, 10d, 8AD, 8AV, 8BL, 8C, 9a, and 9p); the medial frontal lobe (areas SFL, 9m, 10r, 10v, a24, p32, and d32); the temporal lobe (areas TGd, STSva, STSvp, TE1a, TE1m, PreS, and the hippocampus); the lateral parietal lobe (areas PGi, PGs, and PFm); and the medial parietal lobe (areas 23d, v23ab, d23ab, POS2, POS1, RSC, 7m, 31a, and 31pd).

Area 31pv, like area 31pd, exhibits structural connectivity to the cingulum, contralateral hemisphere, and areas near to the precuneus. Connections to the cingulum extend in the anterior direction and exhibit inconsistent connections within the cingulate sulcus and superior frontal gyrus. Structural connections extend from area 31pv to the contralateral hemisphere which pass through the corpus callosum and terminate at areas 31pv, 31a, and 31pd within the contralateral precuneus. Local projecting fibers terminate at areas 23c23d, 31a, 31pd, 31pv, and 7m via short association fiber bundles.

Brodmann area 32 (BA32)

BA32 can be seen dividing up into 5 unique HCP regions. These divisions include areas p32, s32, d32, a32pr, and p32pr.

Area p32

Area p32 covers the central region of the medial superior frontal gyrus and is adjacent to the anterior curve of the callosal sulcus. Its superior and anterior borders are formed by areas 9m and 10d, respectively, while areas 10r and s32 mark its inferior border. Additionally, area p32 is mainly bounded inferiorly by area a24, with some contribution from area s32.

Area p32 exhibits functional connections to regions across the cortex, including the medial frontal lobe (areas d32, a24, p24, a32prime, and 10r), the lateral frontal lobe (area 8AD), and the posterior cingulate areas (areas RSC, 31a, 31pv, POS2, and POS1).

Apart from its functional connections, area p32 also has structural connections to the cingulum, contralateral hemisphere, and local parcellations. Connections to the cingulum extend in the posterior directions and terminate at areas POS1, 31pv, RSC, and v23ab within the precuneus, while connections to the contralateral hemisphere traverse the genu of the corpus callosum and end at areas p32, 9m, 10d, and 10v. Additional local projecting fibers connect area p32 structurally to areas 9m, 10d, 10r, and p24.

Area s32

Area s32 is a region located in the subcallosal gyri. Its anterior and superior broders include areas a24, p32, and 10r. Inferior boundaries and posterior boundaries are areas 10v and OFC (inferior) and area 25 (posterior).

Area s32 is functionally connected to several regions across the human cerebrum, including the medial frontal lobe (areas a24, 25, 10d, and 10r), the lateral frontal lobe (area 8AD), the lateral parietal lobe (area PGs), and the posterior cingulate areas (areas v23ab and POS1).

Area s32 exhibits structural connections mostly to local parcellations. Additional structural connectivity extends posteriorly to area 25 and anteriorly to area p32.

Area d32

Area d32 is a vertically elongated parcel situated within the superior frontal gyrus. It is flanked by several regions, including 8bm at its superior aspect, area 9m at its anterior aspect, area 32pr at its posterior aspect, and area p24 at its inferior aspect.

Area d32 is functionally connected to several regions across the cortex, including the medial frontal lobe (areas p32, a24, p24, a32prime, p32, 9m, 10d, 8BM, and SFL), the lateral frontal lobe (areas 8AV, 8AD, 8BL, 8C, 9a, 9p, a10p, p10p, i6-8, s6-8, and 47s), the insula (area AVI), the temporal lobe (area STSvp), the lateral parietal lobe (areas PFm, PGs, and PGi), and the posterior cingulate areas (areas RSC, 31a, 31pv, 31pd, d23ab, 7m, POS2, and POS1).

Area d32 exhibits structural connections to the cingulum, contralateral hemisphere, and local areas. Fiber bundles which extend to the cingulum extend in the posterior direction to areas 31pv, POS1, v23ab, and RSC within the precuneus. White matter pathways bridging area d32 to the contralateral hemisphere cross the corpus callosum and terminate at areas d32, 8BM, and 9m. Local projections structurally unite area d32 to areas 8BM, 9m, a32pr, and 10d.

Area a32pr

Area a32pr resides within the posterior inferior surface of the superior frontal gyrus as it curves into the superior aspect of the cingulate sulcus. Area a32pr is abutted superiorly by area 8BM, posteriorly by area p32pr, inferiorly by a24pr and p24, and d32 anteriorly.

Functional links between area a32pr and several cortical regions have been observed. These include functional connections with the medial frontal lobe (areas 8BM, SCEF, p24, d32, 23c, p24prime, and p32prime), the lateral frontal lobe (areas a9-46v, 9—46d, and 46), the insula opercular areas (areas FOP4, FOP5, AVI, and MI), the medial parietal lobe (areas 7pm, RSC, and POS2), and the medial occipital lobe (area V1).

In addition to functional connections, area a32pr is also structurally connected to the cingulum and the contralateral hemisphere. Connections from area a32pr to the contralateral hemisphere pass through the corpus callosum and terminate at areas a32pr, p32pr, 8BM, and 9m. Fibers connecting to the cingulum proceed posteriorly and terminate at areas 7m, 31a, 31pd, 31pv, RSC, and v23ab within the precuneus. Local structural connections have also been demonstrated between area a32pr and areas p24, 8BM, SCEF, and d32.

Area p32pr

p32pr is located in the inferior posterior region of the superior frontal gyrus, and it follows a curved path that leads into the upper part of the cingulate sulcus. Its superior border is demarcated by SCEF, its posterior border by 24dv, its inferior border by a24pr, and its anterior border by a32pr.

Area p32pr exhibits functional connectivity to various regions throughout the cortex, including: the sensory strip (areas 2); the cingulate (5mv, 23d, 24dv, p24, p24prime, and a32prime); the premotor region (areas SCEF, PEF, 6a, 6v, 6ma, and 6mp); the lateral frontal lobe (area 9-46d and 46); the insula opercular region (areas FOP1, FOP3, FOP4, FOP5, OP4, PFcm, MI, 43, 52, PoI2, and PoI1); the temporal lobe (area PHT); the lateral parietal lobe (areas PGp, PFt, PF, PFop, LIPd, 7PC, 7PL, and 7AL); the medial parietal lobe (areas DVT and 7am); the medial occipital lobe (areas V1, V2, V3, and V4); the dorsal visual stream (areas V6, V6a, V3a, V3b, and V7); and the lateral occipital lobe (area FST).

Area p32pr exhibits structural connectivity with projections locally, contralaterally, and to the cingulum. Connections from area p32pr to the contralateral hemisphere pass the corpus callosum and terminate at parcellations p32pr, SCEF, and a32pr. Interestingly, local connections correspond exactly to those which project contralaterally, as short association bundles connect area p32pr to areas SCEF and a32pr in the ipsilateral hemisphere as well. Furthermore, white matter pathways which reach the cingulum continue posteriorly and terminate are areas 31a, 31pd, 31pv, PCV, and v23ab within the precuneus.

Brodmann area 33 (BA33)

BA33 is split into 6 unique HCP regions. These HCP parcel divisions include areas PFm, PGs, PGi, IP1, IP0, and 33pr.

Area PFm

Area PFm is situated along the anterior aspect of the superior cortical surface of the angular gyrus and contains a portion of the adjacent posterior superior aspect of the supramarginal gyrus and the sulcus in-between. Area PFm is flanked superiorly by intraparietal 1 (IP1) and intraparietal 2, anteriorly by PF, inferiorly by PSL and STV, and posteriorly by PGI and PGS.

Functional connectivity of area PFm extends to various regions across the cortex, including: the lateral frontal lobe (areas 8AV, 8AD, 8BL, 8C, a47r, p47r, a10p, p10p, 9a, a9-46v); medial frontal lobe (area d32); insula (area AVI); temporal lobe (areas STSvp, TE1m, TE1p, and TE2a); lateral parietal lobe (areas PGs, PGi, IP2, and IP1); and medial parietal lobe (areas 7m, 7pm, POS2, 31a, 31pv, d23ab, 23d, and RSC).

Area PFm exhibits structural connectivity to the superior longitudinal fasciculus and to local parcellations. Connections bridging area PFm to the superior longitudinal fasciculus extend anteriorly to areas 8C and 8BM or posteriorly to areas TE1a, TE1m, TE1p, STSva, STSvp, and PHT of the middle temporal gyrus. Local projecting white matter fiber tracts connect area PFm to HCP parcellations AIP, 7PC, IP1, IP2, LIPd, LIPv, PGi, PGs, 2, and 1 via short association bundle fibers.

Area PGs

Area PGs is topographically located on the superior portion of the angular gyrus. Its boundaries are demarcated by areas PFm and PGi on its anterior and inferior boundary, respectively. Additionally, PGs is flanked by area PGp posteriorly, and areas IP1 and IP0 superiorly.

Functional connectivity of area PGs encompasses various regions across the cortex, including: the lateral frontal lobe (areas 8AD, 8AV, 8BL, 8C, a47r, 47m, 10d, and 9p); medial frontal lobe (areas 9m, 10v, 10r, 8BM, a24, d32, and s32); temporal lobe (areas PHA1, PHA2, PreS, EC, the hippocampus, STSva, STSvp, TGd, TE1a, TE1m, TE1p, and TE2a); lateral parietal lobe (areas PFm, PGi, and IP1); and medial parietal lobe (areas 7m, 7pm, POS1, POS2, 31a, 31pd, 31pv, d23ab, v23ab, 23d, and RSC).

Area PGs exhibits white matter connections to various brain regions, including the superior longitudinal fasciculus and local parcellations. Specifically, connections to the superior longitudinal fasciculus extend anteriorly to areas IFJa, IFJp, and IFSp of the inferior frontal sulcus, as well as inferiorly to areas PHT, FST, and TPOJ2 closest to the occipitotemporal junction. Additionally, area PGs is structurally connected to nearby regions such as areas PFm, PGi, PGp, IP0, IP1, IP2, and MIP.

Area PGi

Area PGi is situated on the cortical surface of the inferior angular gyrus and is bounded superiorly by area PGs and anteriorly by PFm and STV. Its inferior border is defined by TPOJ1 at its anterior-most inferior boundary, TPOJ3 in the posterior-most inferior boundary, and TPOJ2 in between.

Area PGi has structural connections to areas within the occipitotemporal junction, and some, but not all, individuals display additional connections to the superior longitudinal fasciculus that extend to the premotor cortex. Connections from area PGi to the occipitotemporal junction terminate at areas PHT, TE1p, STSvp, STSdp, TPOJ1, and TE1m via short fiber bundles.

Area IP1

Area IP1 is located in the central portion of the inferior aspect of the intraparietal sulcus. Area IP1 is flanked anteriorly by area IP2, posteriorly by IP0, inferiorly by areas PFm and PGs, and superiorly by MIP and LIPv.

Area IP1 demonstrates functional connectivity to a number of regions across the cortex, including: premotor region (area 6a); middle cingulate regions (areas 33prime and 8BM); lateral frontal lobe (areas IFSa, IFSp, IFJp, a9-46v, p9-46v, 8AV, 8C, i6-8, a47r, and p47r); temporal lobe (areas TE1m, TE1p, PHT, and PreS); lateral parietal lobe (areas PFm, PGs, IP0, IP2, AIP, MIP, and LIPd); medial parietal lobe (areas 7pm, 31a, d23ab, POS2, and RSC); and occipital lobe (area V1).

Area IP1 exhibits structural connections to the superior longitudinal fasciculus. Fibers which traverse the superior longitudinal fasciculus continue anteriorly and terminate at areas 55b, FEF, and PEF within the premotor cortex. Addition fibers are observed bridging area IP1to local areas PFm, LIPd, IP0, IPS1, and PGs. Interestingly, fiber bundles that extend inferiorly and originate in area IP1 specifically in the right hemisphere exhibit more pronounced connectivity with the inferior frontal gyrus.

Area IP0

Area IP0 is situated on the posterior extreme of the inferior intraparietal sulcus. Its anterior boundary is adjacent to areas IPO and IP1, while its posterior limit borders areas V3b and V3cd. The anterior-inferior aspect is delimited by two regions, PGp and PGs, with PGp comprising the larger portion. Additionally, area IPS1 forms its superior border.

Area IP0 demonstrates functional connectivity to several regions across the cortex, including: the premotor regions (SCEF, FEF, PEF, 6a, 6r, and 6ma); the middle cingulate regions (5mv and 23c); the lateral frontal lobe (IFSa, IFSp, IFJa, IFJp, i6-8, p9-46v, and 46); the temporal lobe (PeEc, PHA2, PHA3, TE1p, TE2p, and PHT); the lateral parietal lobe (AIP, MIP, VIP, LIPd, LIPv, PFop, PF, PFt, PGp, IP2, IP1, IPS1, 7PL, and 7PC); the medial parietal lobe (7AM, 7pm, DVT, and PCV); the medial occipital lobe (ProS, V1, V2, V3, and V4); the dorsal visual stream (V3a, V3b, V7, V6, and V6a); the ventral visual stream (FFC, VVC, V8, VMV2, and VMV3); and the lateral occipital lobe (V3cd, LO1, LO3, PH, TPOJ2, TPOJ3, and FST).

In addition to its functional connectivity, area IP0 is also structurally connected to both local and distant brain regions. Its connections to the middle longitudinal fasciculus span across the inferior parietal lobe and superior temporal gyrus and terminate at area A4. Furthermore, area IP0 is extensively connected to nearby regions such as areas IP1, IPS1, V3B, V6A, DVT, and V6.

Area 33prime

Area 33prime is found on the cortical surface of the anterior callosal sulcus, and is bounded posteriorly by the retrosplenial cortex and superiorly by areas a24pr, p24pr, and p24.

Area 33prime is functionally connected to several regions across the cortex, including: the lateral frontal lobe (areas IFSa, IFJa, and p9-46v), the medial frontal lobe (areas a24prime and p24prime), the temporal lobe (areas TE1p and PHT), and the parietal lobe (areas IP1 and IP2).

Area 33prime exhibits structural connectivity to the cingulum. Anterior projecting fibers terminate at areas a24, p32, and 10r via fibers that extend anteriorly and over the corpus callosum. Other fibers also extend anteriorly but terminate at area 25 via fibers that bend around the rostrum of the corpus callosum. Like its anterior projections, area 33prime′s posterior projecting fibers have two different courses. Some posterior fibers have extensions which terminate at areas 7m and v23ab within the precuneus, while others terminate at the RSC via fibers that curve around the splenium of the corpus callosum.

Brodmann area 34 (BA34)

BA34 is primarily divided into the HCP parcel area EC.

Area EC

Area EC (entorhinal cortex) is situated on the medial aspect of the posterior uncus, with its boundaries delineated by the presubiculum and area PHA1 at its posterior limit, and its inferior and anterior sides flanked by area PeEC.

Area EC demonstrates functional connections to various regions across the cortex, including the frontal lobe (specifically area 8ad), the temporal lobe (areas TE1a, PeEC, and the hippocampus), and the parietal lobe (area PGs).

In addition to its functional connections, area EC has been found to be structurally connected to the cingulum. These connections course through the cingulate cortex, with some fibers extending anteriorly and terminating at areas a24pr and p24, while others project inferiorly to areas DVT and POS2 located within the parietooccipital sulcus.

Brodmann area 35 (BA35)

BA35 primarily includes the parahippocampal (PHA) regions. As such, BA35 can be seen including PHA1, PHA2, PHA3.

Area PHA1

Area PHA1 is localized on the cortical surface of the medical aspect of the parahippocampal gyrus and shares its superior border with area PreS, inferior border with area PHA2, posterior border with area VMV1, and anterior border with the entorhinal cortex (EC).

Area PH1 demonstrates functional connectivity across the cerebral cortex, including connections to the frontal lobe (area 10r and 8ad), the temporal lobe (areas PHA2, PHA3, VMV1, PreS, and the hippocampus), and the parietal lobe (areas ProS, POS1, PGp, and PGs).

Additionally, area PHA1 is structurally connected to local regions via local fibers. White matter tracts project from area PHA1 anteriorly and posteriorly connecting to areas PeEC and PH, respectively. It is important to note that while these structural connections have been identified, the specific areas where they terminate is not shared by all connections.

Area PHA2

Area PHA2 is situated on the topographic inferior face of the parahippocampal gyrus, specifically alongside the collateral sulcus. Area PHA2 is flanked by PHA1 superiorly and PHA3 inferiorly, with a minimal shared border with VMV2 posteriorly.

Area PHA2 has been found to have functional connections to various regions across the cortex, including: the frontal lobe (areas 8AD, 47m, and i6-8); the temporal lobe (areas PHA1, PHA3, PreS, and the hippocampus); the parietal lobe (areas ProS, 7pm, PCV, DVT, POS1, IP0, PGp, and PGs); and the occipital lobe (area TPOJ3).

Area PHA2 is structurally connected to primarily the inferior longitudinal fasciculus (ILF). These fibers then extend in two directions with some fibers terminating anteriorly at areas V1, V2, and V3, while others terminate posteriorly at area PeEc. In addition to these long-range connections, there are also short, local structural connections to nearby areas. However, it should be noted that not all patients exhibit the same structural connections.

According to the HCP authors, like both PHA1 and PHA3, PHA2 is activated in the PLACE-AVG contrast and deactivated in the FACE-AVG contrast, suggesting a role in place/scene recognition rather than face recognition. Area PHA2 is more deactivated than PHA1 in facial recognition tasks.

Area PHA3

Area PHA3 is situated within the collateral sulcus of the parahippocampal gyrus and shares its superior boundary with PHA2, inferior boundary with VVC and TF, its posterior boundary with VMV2 and VMV3, and anterior boundary with PeEC.

Area PHA3 demonstrates functional connectivity to various regions across the cortex, including: the frontal lobe (areas 6a and IFSa); the temporal lobe (areas PHT, PHA1, PHA2, VMV2, and PeEc); the parietal lobe (areas 23c, 7pl, 7am, 7pm, PCV, DVT, POS1, MIP, LIPd, AIP, PGp, and PFt); and the occipital lobe (areas TPOJ3 and VVC).

Area PH3, like area PHA 2, has white matter pathways which project through the ILF. There are anterior connections which link area PH3 through the ILF to terminate at area TGd, while those which extend posteriorly terminate at areas VVC and FFC. Additional white fiber tracts connect area PH3 to the occipital lobe via tracts which run parallel to the ILF. Besides its long-range connections, there also exist short, nearby structural connections, although it is worth mentioning that these local projecting structural connections vary among patients.

Brodmann area 37 (BA37)

BA37 is split into 8 unique HCP regions. These HCP parcels include TPOJ1, TPOJ1, TE2a, FFC, VVC, VMV1, VMV2, and VMV3.

Area TPOJ1

Area temporal-parietal-occipital junction 1 (TPOJ1) is topographically located within the posterior aspect of the superior temporal sulcus. Area TPOJ1's posterior most portion slopes upward protruding into the angular gyrus. Area TPOJ is bounded superiorly by area STV, inferiorly by area PHT, anteriorly by STSdp, STSvp, A4, and A5, posterosuperiorly by PGi, and posteriorly by TPOJ2.

Area TPOJ1 demonstrates functional connectivity to various regions across the cortex, including: the sensory strip (areas 1, 2, 3a, and 3b); the motor strip (area 4); premotor areas (areas 55b and FEF); the lateral frontal lobe (areas IFJa and IFSp); the insula opercular region (areas OP4, PFcm, RI, STV, PSL, A4, A5, PBelt, and LBelt); the temporal lobe (area STSdp); the medial occipital lobe (areas V2, V3, and V4); the ventral visual stream (area FFC); and the lateral occipital lobe (areas TPOJ2 and TPOJ3).

Area TPOJ1 is connected to the superior longitudinal fasciculus through white matter fibers, which run closely to the sylvian fissure, projecting anteriorly to the frontal lobe, specifically to regions within the premotor cortex, such as areas IFJa, IFJp, and 6r. Furthermore, some fibers extend from the superior longitudinal fasciculus to areas PFm, PGi, and PGs within the inferior parietal lobule. There are also additional local structural connections from area TPOJ1 to nearby areas TPOJ2, STSvp, and STSdp.

Area TPOJ2

Area TPOJ2 (temporal-parietal-occipital junction 2) is situated on the anterosuperior aspect of the lateral occipital cortex. Specifically, it courses the posteroinferior portion of the angular gyrus, and runs along its sulcus. Furthermore, area TPOJ2 is bounded anteriorly by area TPOJ1, superiorly by area PGi, inferiorly by areas PHT, FST, and MST, and posteriorly by area TPOJ3.

Area TPOJ2's functional connections are broadly distributed throughout the cortex and include the following locations: the sensory strip (area 2); the premotor regions (areas 6a, 6v, SCEF, and FEF); the cingulate regions (areas 5mv, 23c, and p24prime); the insula opercular region (areas OP4, 43, PFcm, 52, RI, STV, A4, PoI1, PoI2, and PBelt); the temporal lobe (areas PHT and TE2p); the lateral parietal lobe (areas PFop, PGp, IPS1, IP0, AIP, LIPv, 7AL, 7PL, and 7PC); the medial parietal lobe (areas 7AM, PCV, and DVT); the medial occipital lobe (areas V2 and V3); the dorsal visual stream (areas V3a, V6, and V6a); the ventral visual stream (area FFC); and the lateral occipital lobe (areas FST, PH, LO3, MST, TPOJ1, and TPOJ3).

Like TPOJ1, area TPOJ2 is characterized by its structural connections that traverse the intermediary superior longitudinal fasciculus to the frontal lobe and inferior parietal lobule. Fibers extending from the superior longitudinal fasciculus to the frontal lobe terminate at premotor areas, such as IFJa, IFJp, 6v, and 6r, while those that project to the inferior parietal lobule end at PFm, PGi, and PGs. Additionally, TPOJ2 is connected to its anterior neighbor, area TPOJ1, via short association bundles.

Area TE2a

Area TE2a is situated in the anterior portion of the inferior temporal gyrus, with some of its parts extending into the anterior aspect of the inferior sulcus and others occupying the lateral aspect of the occipitotemporal sulcus. Area TE2a borders several HCP parcellations, including area TF along its basal-medial edge, areas TE1a and TE1m on its superior boundary, areas TGd and TGv anteriorly, and areas TE1p and TE2p along its posterior pointed end.

Area TE2a is functionally connected to other parcellations throughout the cortex, including: the frontal lobe (areas 8AV, 8BL, 8C, and a47r); the temporal lobe (areas STSvp, TE1m, and TE1p); and the parietal lobe (areas PGs, PGi, and PFm).

Area TE2A has been demonstrated to have structural connections to two major white matter bundles, the superior longitudinal fasciculus and the inferior longitudinal fasciculus. However, projections from area TE2A to the inferior longitudinal fasciculus are not a consistent finding in all individuals. White matter pathways that extend to the superior longitudinal fasciculus bend around the Sylvian fissure in the frontal lobe's direction, and then turn inwards, finally terminating at areas 6r, 6v, 8C, p9-46v, IFJa, IFJp, and IFSp. Furthermore, there are fibers that extend in the posterior direction from the superior longitudinal fasciculus to the inferior parietal lobule and end at areas PF and PFm. Alongside these long-range connections, TE2a has structural connections to local parcellations TE1p and TGd.

Area FFC

Area FFC (fusiform face complex) is situated on the cortical surface of the posterior fusiform gyrus. Specifically, it is located within the lateral aspect of the posterior FFC, and it is immediately adjacent to the medial aspect of the occipitotemporal sulcus. Area FFC is flanked by a number of HCP parcellations, including area TF in its anterior aspect, areas TE2p and PH in its lateral aspect, areas V8 and PIT in its posterior boundary, and area VVC in its medial boundary.

Area FFC demonstrates functional connectivity to various regions across the cortex, including: the sensory strip (areas 1, 2, 3a, and 3b); the motor strip (area 4); the premotor region (areas SCEF, FEF, and 6v); the insula and opercular region (areas OP4, STV, PFcm, 43, RI, LBelt, PBelt, A4, and A5); the temporal lobe (area TE2p); the parietal lobe (areas 7PC, PGp, PFt, AIP, MIP, VIP, LIPv, DVT, IP0, and IPS1); the medial occipital lobe (areas V1, V2, V3, and V4); the dorsal visual stream (areas V3a, V3b, V7, V6, and V6a); the ventral visual stream (areas VVC, V8, PIT, VMV1, VMV2, and VMV3); and the lateral occipital lobe (areas TPOJ1, TPOJ2, TPOJ3, V3cd, V4t, MT, LO1, LO2, LO3, MT, MST, PH, and FST).

Area FFC is structurally connected to the ILF and to local parcellations. Structural connections to the ILF further extend toward the temporal pole and terminate at TGv, while connections to adjacent areas exist via white matter tracts that initially project in a medial-superior direction and then curve back toward their origin. In some brains, fiber bundles connect with the SLF, but this is not always the case. Furthermore, area FFC exhibits connections to areas V4, PIT, LO3, FST, MST, MT, PH, and TPOJ2 via short associated bundles.

Area VVC

Area VVC (ventral visual complex) is located on the cortical surface of the posterior aspect of the fusiform gyrus. Specifically, area VVC resides within the medical half of the posterior fusiform gyrus, and extends into the lateral aspect of the collateral sulcus. Area VVC is bounded by area TF and area FFC, anteriorly and laterally. It also is flanked by V8 posteriorly, and areas PHA3 and VMV3 along its medical aspect.

In terms of function connectivity, Area VVC is connected to a number of cortical regions. These functional connections include the premotor region (area FEF), the insula and opercular region (areas PBelt and A4), the parietal lobe (areas VIP, LIPv, PGp, IPS1, IP0, and DVT), the medial occipital lobe (areas ProS, V1, V2, V3, and V4), the dorsal visual stream (areas V3a, V3b, V7, V6, and V6a), the ventral visual stream (areas VVC, V8, VMV2, and VMV3), and the lateral occipital lobe (areas V3cd, MT, MST, V4t, LO1, LO2, LO3, PH, and FST).

Area VVC exhibits structural connections to the inferior longitudinal fasciculus and the ventral occipital fasciculus (VOF). Connections to the inferior longitudinal fasciculus further extend to the temporal pole, terminating at area TGv, while connections to the VOF further project in the dorsal-medial direction, terminating at area V3b. Additionally, local projecting white fiber tracts terminate at areas FFC, PIT, V3B, V3CD, V8, VMV1, VMV2, and VMV3 via short association bundles.

Area VMV1

Area VMV1 is localized on the topographic surface of the anteromedial aspect of the lingula, specifically within the anterior aspect of the lingula, at the point where the lingula courses anteriorly into the lower portion of the parahippocampal gyrus. Area VMV1 is flanked in its superior and posterior borders by the superior aspect of area V2 and the inferior aspect of V3, respectively. Additionally, area VMV1 is inferiorly bounded by area VMV2, anteriorly by PHA1, and in some individuals, is bounded by ProS in its superior aspect. Considering functional connectivity studies, it is evident that area VMV1 is associated with significant connections to various cortical regions. The functional connections from area VMV1 include the premotor region (area FEF), the temporal lobe (area PHA1), and the parietal lobe (VIP, LIPv, IPS1, and DVT). Additionally, it shows functional connectivity to the medial occipital lobe (areas V1, V2, V3, and V4), the dorsal visual stream (V3a, V3b, V7, V6, and V6a), the ventral visual stream (areas FFC, VVC, V8, VMV2, and VMV3), and the lateral occipital lobe (areas V3cd, V4t, LO1, LO3, PH, and FST).

Area VMV1, like area VVC, is structurally connected with the inferior longitudinal fasciculus, and the VOF. White pattern pathways which connected VMV1 to the inferior longitudinal fasciculus continue laterally toward the temporal pole and end at area TGd. Connections to the VOF further extend dorsal-medially and terminate within area V3a. Additionally, local fiber projections to areas V3A, V3, V6, V1, V8, and VMV3 via short association bundles have also been observed.

Area VMV2

Area VMV2 is situated on the cortical surface of the posterior portion of the superior aspect of the collateral sulcus, and like VMV1, VMV2 also contains a region where the lingula courses anteriorly into the lower portion of the parahippocampal gyrus. Area VMV2 is bounded posteriorly and anteriorly by V4 and PHA3, respectively. It is also flanked by area VMV1 on its superior boundary, and VMV3 on its inferior boundary.

Area VMV2 is functionally connected to various regions across the cortex, including: the premotor region (area FEF); the temporal lobe (area PHA3); the parietal lobe (areas VIP, LIPv, PGp, IPS1, IP0, and DVT); the medial occipital lobe (areas ProS, V1, V2, V3, and V4); the dorsal visual stream (areas V3a, V3b, V7, V6, and V6a); the ventral visual stream (areas FFC, VVC, V8, VMV1, and VMV3); and the lateral occipital lobe (areas V3cd, V4t, LO1, LO2, LO3, PH, and FST).

Area VMV2 has structural connections to three major white matter bundle fibers; the inferior longitudinal fasciculus (ILF), the inferior frontooccipital fasciculus (IFOF), and the vertical occipital fasciculus (VOF). Structural connections from area VMV2 to the ILF terminate at area TGv within the temporal lobe, while those which join the IFOF extend to the pole of the frontal lobe. Connections from area VMV2 to area V3a are established via the VOF, a white matter pathway which runs almost perpendicular to the ILF. In terms of local connections, area VMV2 is structurally linked to V1, V2, V3, V3A, V6A, FFC, and VVC via short association bundles.

Area VMV3

Area VMV3 is situated in the extreme-posterior aspect of collateral sulcus, specifically along its lateral side. Area VMV3's inferior aspect contains an anterior border bounded by PHA3, and a posterior border bounded by areas V4 and V8. Additionally, area VMV3's superior and inferior boundaries are adjacent to areas VMV2 and VVC, respectively.

Area VMV3 exhibits functional connections to various regions across the cortex, including: the premotor region (area FEF); the temporal lobe (area PHA3); the parietal lobe (areas VIP, LIPv, PGp, IPS1, IP0, and DVT); the medial occipital lobe (areas V1, V2, V3, and V4); the dorsal visual stream (areas V3a, V3b, V7, V6, and V6a); the ventral visual stream (areas FFC, PIT, VVC, V8, VMV2, and VMV3); and the lateral occipital lobe (areas MT, MST, V3cd, V4t, LO1, LO2, LO3, PH, and FST).

Area VMV3 has structural connections to the ILF and the VOF. Connections from area VMV3 to temporal lobe area PeEc traverse the ILF, while connections to V3a and V3b course the VOF. Additional local connections to areas V4, V3, VMV1, VMV2, PIT, V8, and PH also exist via short association bundles.

Brodmann area 38 (BA38)

BA38 primarily contains a single HCP parcellation—area TGd.

Area TGd

Area TGd (TG dorsal) is situated on the cortical surface of the temporopolar region. Specifically, it is located in the most rostral aspect of the superior temporal and middle temporal gyri and covers the surface of the temporal planum polare. Area TGd's boundaries are demarcated by areas STGa, STSda, and TE1a posteriorly and TE2a and TGv inferiorly. Areas Pir and PI are found adjacent to its anteromedial aspect, while PeEC on its posterior mesial boundary.

Area TGd has extensive functional connections throughout the cortex, including the frontal lobe (areas 8AV, 8BL, 9a, 9p, 9m, 44, 45, 47s, 47L, 10v, 10r, and SFL), the temporal lobe (areas STSva, STSvp, STSda, STSdp, TGv, STGa, PeEc, and the hippocampus), and the parietal lobe (areas PGs, PGi, 7m, d23ab, 31pv).

Apart from its functional connections, area TGd is structurally connected to two major white matter fiber bundles, the uncinate fasciculus and the inferior longitudinal fasciculus (ILF). Of note, structural connections from area TGd to the external capsule that arrive and terminate within the occipital lobe have been shown, although this finding is not consistent across all individuals. Connections to the uncinate fasciculus extend posteriorly through the external capsule and pass the posterior aspects of the temporal lobe. These projections then continue into the occipital lobe, and finally terminate at areas DVT, V1, V3, V2, V6, and 7PL. Additional fibers pass through the uncinate fasciculus, continue through the insula, and reach areas FOP4, FOP5, 44, and 45 within the frontal lobe. Lastly, connections to the inferior longitudinal fasciculus, which has white matter pathways that course through the inferior temporal lobe, connect to areas V1 and V2 of the occipital lobe as well.

Brodmann area 39

BA39 includes two unique HCP parcels—areas PGs and PGp.

Area PGs

Area PGs is situated on the superior aspect of the angular gyrus and is flanked anteriorly by area PFm, inferiorly by area PGi, posteriorly by PGp, and superiorly by IP1 and IP0.

Area PGs exhibits functional connectivity with regions in the lateral frontal lobe (areas 8AD, 8AV, 8BL, 8C, s6-8, i6-8, a47r, 47m, 10d, and 9p), the medial frontal lobe (9m, 10v, 10r, 8BM, a24, d32, and s32), the temporal lobe (areas PHA1, PHA2, PreS, EC, the hippocampus, STSva, STSvp, TGd, TE1a, TE1m, TE1p, and TE2a), the lateral parietal lobe (areas PFm, PGi, and IP1), and the medial parietal lobe (areas 7m, 7pm, POS1, POS2, 31a, 31pd, 31pv, d23ab, v23ab, 23d, and RSC.)

Area PGs is structurally connected to parcellations IFJa, IFJp, and IFSp in the inferior frontal sulcus via fibers that project anteriorly from the superior longitudinal fasciculus (SLF), parcellations PHT, FST, and TPOJ2 in the occipitotemporal junction via fibers which project inferiorly from the SLF, and local parcellations PFm, PGi, PGp, IP0, IP1, IP2, and MIP via short association bundle fibers.

Area PGp

Area PGp is a distinct cortical region located on the most posterior part of the inferior parietal lobule. Area PGp is bordered by area PGs anterosuperiorly, area TPOJ3 and LO3 anterioinferiorly, IP0 posterosuperiorly, and area V3cd posteriorly.

Area PGp exhibits functional connectivity with several regions across the brain, including: the premotor areas (areas FEF, PEF, 6a, and 6ma), the middle cingulate regions (areas a24prime, p32prime, 5mv, and 23c), the lateral frontal lobe (areas IFSa and 46), the superior insula opercular regions (areas PFcm and FOP4), the lower opercula and Heschl's gyrus regions (areas PoI1 and PoI2), the temporal lobe (areas PHA1, PHA2, PHA3, TE2p, and PHT), the lateral parietal lobe (areas AIP, MIP, VIP, LIPd, LIPv, PFop, PF, PFt, IP2, IP0, IPS1, 7AL, 7PL, and 7PC), the medial parietal lobe (areas 7AM, 7pm, POS1, POS2, DVT, and PCV), the medial occipital lobe (areas ProS, V1, V2, and V3), the dorsal visual stream (areas V3a, V3b, V6, and V6a), the ventral visual stream (areas FFC, VVC, and VMV2), and the lateral occipital lobe (areas V3cd, LO3, PH, TPOJ2, TPOJ3, and FST).

Area PGp is known to have complex structural connectivity with various brain regions, including areas PH, FFC, and TF within the occipitotemporal junction and fusiform gyrus, areas 7PL and IPS1 within the superior parietal lobe, areas IP0, PGs, TPOJ3, and MT in the surrounding regions, and areas LO3, LO1, and V3CD within the visual processing regions. The connections between area PGp and areas PH, FFC, and TF traverse the inferior longitudinal fasciculus (ILF) and pass beneath the inferior surface of the occipitotemporal junction and fusiform gyrus. Furthermore, the local and visual processing areas are connected to area PGp through short association fiber bundles.

Brodmann area 40 (BA40)

BA40 is divided into 5 unique HCP regions. These HCP divisions include areas PFop, PFt, PFcm, PF, and IP2.

Area PFop

Area PFop is localized to the cortical surface of the anterior inferior parietal lobule at its junction with the anteroinferior postcentral gyrus. Area PFop shares its anterior and posterior borders by area 1 and area PF, respectively. It is also bordered by area PFT and area 2 superiorly, and areas OP4, OP1, and PFcm composing its inferior border. Area PFop exhibits many functional connections with regions.

Area PFop exhibits extensive functional connections to a broad set of regions across the cortex, including: the sensory strip (area 2); the premotor regions (areas SCEF, FEF, PEF, 6ma, 6mp, 6a, 6r, and 6v); the middle cingulate regions (areas a24prime, p24prime, p32prime, 24dv, 5mv, and 23c); the lateral frontal lobe (areas IFSa, 46, and 9-46d); the superior insula opercular regions (areas OP4, OP2-3, OP1, PFcm, FOP1, FOP3, FOP4, and FOP5); the lower opercula and Heschl's gyrus regions (areas LBelt, PBelt, PI, A4, MI, 52, RI, PoI1, and PoI2); the temporal lobe (area TE2p and PHT); the lateral parietal lobe (areas PFt, PF, PGp, intraparietal 0, AIP, MIP, lateral intraparietal, ventral (LIPv), lateral intraparietal, dorsal (LIPd), 7PC, 7PL, and 7AL); the medial parietal lobe (areas 7AM and DVT); the medial occipital lobe (areas V1, V2, and V3); the dorsal visual stream areas (areas V3a and V6); the ventral visual stream areas (areas V8 and FFC); and the lateral occipital lobe (areas LO3, TPOJ2, PH, and FST).

The structural connectivity of area PFop is primarily via short association fibers. These local fibers structurally connect it to nearby areas PF, PFcm, PFt, 4, OP1, and OP4.

Area PFt

Area PFt is topographically situated on the posterior aspect of the postcentral sulcus, at the antero superior margin of the inferior parietal lobule. It is positioned at the intersection of the postcentral sulcus and the inferior end of the intraparietal sulcus. Area PFt is bounded by area AIP superiorly, PFop inferiorly, area 2 anteriorly, and area PF posteriorly.

Area PFt is functionally connected to regions across the cortex, including: the sensory strip (areas 1 and 2); the premotor regions (areas SCEF, FEF, PEF, 6ma, 6mp, 6a, 6d, 6r, and 6v); the middle cingulate regions (areas p32prime, 5mv, and 23c); the lateral frontal lobe (areas IFSa, IFJp, and 46); the superior insula opercular regions (areas OP4, OP1, PFcm, FOP2, and FOP4); the lower opercula and Heschl's gyrus regions (areas MI, PoI1, and PoI2); the temporal lobe (area PHA3 and PHT); the lateral parietal lobe (areas PFop, PF, PGp, intraparietal 2 (IP2), IP0, intraparietal sulcus 1 (IPS1), AIP, ventral intraparietal (VIP), MIP, LIPv, LIPd, 7PC, 7PL, and 7AL); the medial parietal lobe (area 7AM); the ventral visual stream (area FFC); and the lateral occipital lobe (areas PH and FST).

Area PFt is connected to the superior longitudinal fasciculus through white matter projections that extend to nearby areas such as PF and PFop, as well as the inferior parietal lobule and opercular regions. Additionally, white matter pathways that target nonneighboring areas extend from area PFt in two directions: diagonally toward the inferior parietal lobule and anteroinferior to the operculum. These projections terminate at specific areas within the operculum, including OP4, 43, 6r, and 4, and within the inferior parietal lobule at areas 2 and AIP.

Area PFcm

PFcm is cortically positioned on the superior aspect of the supramarginal gyrus, with a significant proportion of its cortical topography concentrated in the regions surrounding the opercular cleft within this area. The anatomical boundaries of PFcm are demarcated by several neighboring regions, with PF and PFop positioned superiorly, OP1 demarcating the anteroinferior border, RI outlining the inferior border, and PSL defining the posterior border.

Area PFcm exhibits broad functional connections to a number of regions, including the sensory strip (areas 1, 2, 3a, and 3b), premotor region (areas SCEF, FEF, PEF, 6mp, 6r, 6a, and 6v) in the premotor regions, areas 24dv, a24prime, p24prime, p32prime, 24dd, 24dv, 5mv, and 23c in the middle cingulate regions, areas 9-46d and 46 in the lateral frontal lobe, areas 43, OP2-3, OP1, PFcm, FOP1, FOP2, FOP3, FOP4, and FOP5 in the superior insula opercular regions, areas LBelt, PBelt, MBelt, A1, TA2, PI, A4, TA2, MI, STV, 52, RI, PoI1, and PoI2 in the lower opercula area and Heschl's gyrus regions, PHT in the temporal lobe, areas PF, PFop, PFt, PGp, AIP, 7PC, and 7AL in the lateral parietal lobe, areas 7am and DVT in the medial parietal lobe, areas V2 and V3 in the medial occipital lobe, area V3a of the dorsal visual stream, areas FFC of the ventral visual stream, and areas LO3, TPOJ1, TPOJ2, TPOJ3, MST, and FST of the lateral occipital lobe.

Area PFcm structural connectivity is characterized by white mattery pathways which extend to a portion of the superior longitudinal fasciculus, bending around the sylvian fissure toward the temporal lobe. Ultimately, these structural connections terminate within the inferior and middle temporal gyrus parcellations TE1a, STSva, and TE2a. It is worth noting that, in contrast to the left hemisphere, the right hemisphere's inferior projections from area PFcm are less consistent. Area PFcm exhibits local connectivity with several nearby cortical regions, including OP1, OP4, PFop, PBelt, PF, and RI, through short association fibers.

Area PF

Area PF (parietal area F) is situated on the lateral surface of the superior aspect of the supramarginal gyrus and is bordered by six neighboring areas. These areas with which it shares its borders include PFcm and PSL inferiorly, PFop and PFt anteriorly, IP2 superiorly, and PFM posteriorly.

Area PF demonstrates functional connectivity to various regions across the cortex, including: the premotor regions (areas SCEF, FEF, 6ma, 6a, and 6r); the middle cingulate regions (areas a24prime, p24prime, p32prime, 5mv, and 23c); the lateral frontal lobe (areas IFSa, IFJp, a9-46v, p9-46v, 46, and 9-46d); the superior insula opercular regions (areas OP4, PFcm, FOP1, FOP3, and FOP5); the lower opercula and Heschl's gyrus regions (areas AVI, MI, 52, PoI1, and PoI2); the temporal lobe (area PHT); the lateral parietal lobe (areas PFt, PFop, PGp, IP0, IP2, AIP, MIP, LIPd, 7PL, and 7AL); the medial parietal lobe (areas 7AM, PCV, and DVT); and the medial occipital lobe (area V1).

Area PF is structurally connected to the superior longitudinal fasciculus and to other nearby areas. Anterior projecting connections to the superior longitudinal fasciculus terminate at somatosensory areas (areas 1, 3a, 4, 43, and OP4), while those projecting inferiorly terminate within the middle and inferior temporal gyrus (areas TE1a, STSva, and TE2a). Area PF also exhibits short association bundles that connect it to other local regions, including AIP, PFcm, PFm, PFop, PFt, PSL, and STV.

Area IP2

Area IP2 is found on the anterior most portion of the inferior bank of the intraparietal sulcus. It shares its borders with several other brain regions. It is anteriorly and inferiorly flanked by area PF and PFm, respectively. Furthermore, it is bounded by IP1 in its posterior aspect, and areas AIP and LIP at its superior border. Area IP2 is found on the anterior most portion of the inferior bank of the intraparietal sulcus. It shares its borders with several other brain regions. It is anteriorly and inferiorly flanked by area PF and PFm, respectively. Furthermore, it is bounded by IP1 in its posterior aspect, and areas AIP and LIP at its superior border.

Area IP2 is functionally connected to several regions across the cortex, including the premotor regions (areas 6ma, 6a, and 6r); the middle cingulate regions (areas 33prime and 8BM); the lateral frontal lobe (areas IFSa, IFJp, a9-46v, p9-46v, 46, 11L, 8C, i6-8, a47r, and p47r); the insula regions (area AVI); the temporal lobe (areas TE1p and PHT); the lateral parietal lobe (areas PFt, PF, PFm, PGp, IP0, IP1, AIP, MIP, LIPd, and 7PL); the medial parietal lobe (areas 7AM, 7pm, 31a, POS2, and RSC); and the occipital lobe (area PH).

Area IP2 is structurally connected to premotor cortex areas 55b and PEF via fibers which project anteriorly from the superior longitudinal fasciculus, and has local connections to PFm, LIPd, AIP, and IP1, via short association fibers. White matter tracts projecting from area IP2 to the inferior frontal gyrus are more extensively connected in the right hemisphere, particularly those located inferiorly.

Brodmann area 41 (BA41)

Area 41 is divided into 3 unique HCP regions. These include areas A1, LBelt, and MBelt.

Area A1

Area A1 is positioned centrally within the superior aspect of Heschl's gyrus. The major proportions of its boundaries abut the neighboring areas of MBelt and LBelt, while its deeper regions interface with the RI region.

Area A1 demonstrates functional connectivity to various regions across the cortex, including areas 2, 3a, and 3b in the sensory strip; areas 43, PFcm, and OP4 in the superior opercular region; areas A4, MBelt, PBelt, LBelt, RI, and 52 in the inferior insula opercular region; and areas V1, V2, V3, and V4 in the medial occipital lobe.

Area A1 exhibits white matter fiber tracts that establish not only local connections but also distant connections with brain regions spanning across the parietal and occipital lobes, as well as the inferior gyrus. Specifically, posterior projecting fiber tracks are thought to involve the middle longitudinal fasciculus and connect area A1 to regions within the parietal lobe, such as areas LIPv and 7PC. Additionally, other posterior projecting fibers connect area A1 to the occipital lobe, namely areas V3B, V3CD, and V4. Further, anteriorly extending fibers connect area A1 to area 44 within the inferior frontal gyrus. Furthermore, short association bundles establish local structural connections between area A1 and areas 52, LBelt, MBelt, PBelt, and MI.

Area LBelt

Situated on the lateral aspect of Heschl's gyrus, Area LBelt shares its medial border with area A1 and a portion of area MBelt, its lateral border with area PBelt, and its deeper structure with area RI.

Area LBelt exhibits functional connectivity with various regions across the cortex, including: the sensory strip regions (areas 1, 2, 3a, and 3b); the superior opercular regions (areas 43, PFcm, OP1, and OP4); the inferior insula opercular regions (areas PoI1, PoI2, A1, A4, A5, MBelt, STV, Ta2, PI, RI, and 52); and the medial occipital lobe (areas V1, V2, V3, and V4).

Area LBelt has white matter tract connections to the superior longitudinal fasciculus, which continue to the ending pole of the sylvian fissure, and finally terminate at the inferior frontal gyrus and insula areas 44 and MI. Additionally, short association bundles connect LBelt to local areas MBelt, PBelt, A4, and A5 via fibers which traverse the temporal terminations of the superior longitudinal fasciculus.

Area MBelt

Area MBelt is localized specifically along the medial aspect of Heschl's gyrus. It is surrounded by several neighboring brain regions, including area A1 laterally, the insular granula and area 52 medially, area RI superiorly, and area TA2 along its inferior edge.

Area MBelt demonstrates functional connectivity to various regions across the cortex, including: the sensory strip (areas 1, 2, 3a, and 3b), the superior opercular region (areas 43, PFcm, and OP4), the inferior insula opercular region (areas PoI1, A1, A4, PBelt, LBelt, Ta2, PI, RI, and 52), and the medial occipital lobe (areas V1, V2, V3, and V4).

Area MBelt exhibits structural connections to occipital and parietal lobe areas, including V3, V3A, V6, V6A, IPS1, and DVT. White matter pathways which connect area MBelt to these areas pass posteriorly through the middle longitudinal fasciculus and extend along the lateral border of the lateral ventricles, ultimately arriving in the parietal and occipital lobes. Additional anterior projecting white matter connections bridge area MBelt to areas A4, TA2, STGa, PBelt, and LBelt.

Brodmann area 42 (BA42)

BA42 is primarily divided into one HCP region—area PBelt.

Area PBelt

Area PBelt is situated on the superior surface of the inferior segment of the supramarginal gyrus and is bordered medially by areas LBelt and MBelt. Laterally, it is bordered by area A4 with its deeper aspect abutting area RI.

Area PBelt demonstrates broad functional connectivity to various regions across the cortex, including: the sensory strip (areas 1, 2, 3a, and 3b), the motor strip (area 4), the paracingulate areas (area 24dd), the premotor areas (areas FEF, 6d, 6v, and 6mp), the superior opercular region (areas 43, PFcm, OP1, OP2-3, and OP4), the inferior insula opercular region (areas A1, A4, A5 MBelt, LBelt, PoI1, PoI2, STV, Ta2, PI, RI, and 52), the parietal lobe (areas PFop and 7PC), the medial occipital lobe (areas V1, V2, V3, and V4), the dorsal visual stream (areas V6, V67a, V7, V3a, and V3b), the ventral visual stream (areas V8, FFC, Pit, and VVC), and the lateral occipital lobe (areas LO2, LO3, V3cd, FST, MT, MST, V4t, TPOJ1, and TPOJ2).

Area PBelt displays notable structural connectivity to two major white fiber bundles, namely the middle longitudinal fasciculus and the superior longitudinal fasciculus. White matter fibers projecting from area PBelt to the superior longitudinal fasciculus traverse the posterior aspect of the sylvian fissure and terminate within the inferior frontal gyrus, specifically at areas 45 and FOP5. On the other hand, white fiber tracts projecting to the middle longitudinal fasciculus extend to the occipital and parietal lobes, with their endpoints specifically located at areas MIP, LIPv, and IP1. These tracts extend in the posterior direction, just lateral to the lateral ventricles. In addition to its long-range connections, area PBelt exhibits local structural connections via short association bundles, which connect it to several other brain regions including areas A1, LBelt, MBelt, PFcm, PSL, A4, A5, and TPOJ1.

Brodmann area 43 (BA43)

BA43 is primarily divided into one HCP region—area 43.

Area 43

Area 43 is situated along the anterior aspects of the subcentral gyrus, specifically at the junction of the pre and postcentral gyri, just inferior to the central sulcus. It encompasses both the lateral and the inferior aspects of the operculum. Area 43 is bounded anteriorly by area 6r, posteriorly by area OP4, superiorly by areas 6v, 4, and 3a, and inferiorly by areas FOP1 and FOP2.

Area 43 is functionally connected to a broad range of cortical regions, including: the sensory strip (areas 1, 2, 3a, and 3b); the motor strip (area 4); the premotor regions (areas SCEF, FEF, PEF, 6ma, 6mp, 6r, and 6v); the middle cingulate regions (areas a24prime, p24prime, p32prime, 24dd, 24dv, 5mv, and 23c); the lateral frontal lobe (areas 46 and 9-46d); the superior insula opercular regions (areas IG, OP4, OP2-3, OP1, PFcm, FOP1, FOP3, FOP4, and FOP5); the lower opercula and Heschl's gyrus regions (areas STV, LBelt, PBelt, MBelt, A1, TA2, PI, A4, MI, STV, 52, RI, PoI1, and PoI2); the temporal lobe (area PHT); the lateral parietal lobe (areas PF and 7AL); in the medial parietal lobe (area DVT); the medial occipital lobe (areas V1, V2, V3, and V4); the dorsal visual stream (areas V3a, V3b, V6, V6a, and V7); the ventral visual stream (areas V8 and FFC); and finally the lateral occipital lobe (areas LO3, TPOJ2, MST, and FST).

Area 43 displays white matter connections to both the superior parietal lobe and nearby areas. Fibers extending posteriorly from area 43 end at areas PFt, PFm, and PFcm within the parietal lobe, while short association bundles link area 43 to neighboring areas MI, 6v, 6r, OP4, 3a, 4, FOP1, and FOP2.

Brodmann area 44 (BA44)

BA44 is primarily divided into one HCP region—area 44.

Area 44

Area 44 is situated on the posterior aspect of the inferior frontal gyrus and comprises the anterior aspect of the pars opercularis. Area 44 shares its borders with a number of HCP parcellations. It is bounded anteriorly by area 45, posteriorly by area 6r, medially by area 8c, inferiorly in between areas 6R and 6V, and superiorly by areas IFSp and IFIa, The opercular surface of area 44 borders area FPO4.

Area 44 is functionally connected with parcellations throughout the cortex, including: the dorsolateral frontal lobe (areas SFL, IFSp, IFJa, 45, 47s, 47L, 9a, 9m, 8AV, 8BL, and 8C); the medial frontal lobe (area 8BM); the premotor areas (area 55b); the insula-opercular region (areas FOP5, AVI, and PSL); the temporal lobe (areas TGd, STSdp, and STSvp); and the inferior parietal lobe (areas PFm and PGi).

In addition to its functional connections, area 44 is structurally linked to several HCP regions via fibers that traverse the superior longitudinal fasciculus and the frontal aslant tract (FAT). The connections between area 44 and the superior longitudinal fasciculus extend in the posterior direction, bend around the sylvian fissure, and terminate at area TE1a within the middle temporal gyrus. Other connections that traverse the superior longitudinal fasciculus exit early and terminate at areas A5 and STSdp. It is worth noting that the connectivity between area 44 and the superior longitudinal fasciculus in the right hemisphere is not consistently observed. The white matter bundle pathways that traverse the FAT extend superiorly to areas SFL, 6ma, and s6-8 of the superior frontal gyrus. In contrast to its long-distance connections, area 44 is also linked to local areas 45 and 8C via short association fiber bundles.

Brodmann area 45 (BA45)

BA45 is primarily divided into one HCP region—area 45.

Area 45

Area 45 is located on the cortical surface of the pars triangularis within the inferior frontal gyrus. It shares its borders with area 47l anteriorly and area 44 posteriorly. In addition, area 44s superior margin adjoins several neighboring regions, namely p47r, IFSa, and IFSp, while its opercular aspect borders area FOP5.

Area 45 demonstrates functional connectivity to a number of regions across the cortex, including: the dorsolateral frontal lobe (areas SFL, IFSp, 44, a47r, 47s, 47L, 9a, 9p, 9m, 8AV, and 8BL); the medial frontal lobe (area 8BM); the premotor area (area 55b); the insula-opercular region (areas FOP5,and PSL); the temporal lobe (areas TGd, TGv, TE1a, STSva, STSdp, and STSvp); the inferior parietal lobe (area PGi); and the medial parietal lobe (area 31pd).

Area 45 displays structural connectivity with both the arcuate/SLF and IFOF fiber tracts, although the consistency of arcuate/SLF connections can vary between individuals. Area 45's structural connections to the superior longitudinal fasciculus further extend in the posterior direction, bending around sylvian fissure and terminate at parcellation TE1p in middle temporal gyrus. Area 45 is also connected to areas A4 and PBelt through fibers which exit the superior longitudinal fasciculus before it ends. In contrast, the IFOF connections originating at area 45 pass through the external capsule, traverse the temporal lobe, and end at areas V1, V2, V3, and V4 within the occipital lobe. Additionally, area 45 is also connected to neighboring areas 44s and FOP4 via short association fibers.

Brodmann area 46 (BA46)

BA46 is divided into 4 unique HCP regions. These include areas 46, p9-46v, a9-46v, IFSp, and IFSa.

Area 46

Area 46 is located within the ventral prefrontal cortex and is parallel to area 9-46d in an oblique direction, commencing from the depths of the superior frontal sulcus in the posterior region, with its anterior aspect extending toward the middle frontal gyrus. Area 46 is flanked by a number of areas identified by the HCP. It is bounded medially by area 9-46d, laterally by area p9-46v, anterolaterally along the lateral aspect of the middle frontal gyrus with IFSa, and anteriorly with a9-46v, with its posterior portion coming to a point and being bordered by areas 8AV and 8AD.

Area 46 shows functional connectivity with various regions across the cortex, including: the dorsolateral frontal lobe (areas a9-46v, p9-46v, and IFSa); the medial frontal lobe (areas SCEF, a32prime, p24prime, and a24prime); the premotor regions (areas 6ma, 6a, and 6r); the orbitofrontal region (area 11L); the insula-opercular region (areas FOP1, FOP3, FOP4, FOP5, 52 PFop, PFcm, 43 PoI2, PoI1, and MI); the temporal lobe (area PHT); the parietal lobe (areas PF, PFt, PGp, AIP, MIP, 7AL, 7PL, LIPd, IP2, and IP0); the medial parietal lobe (areas 23c, PCV, POS2, PCV, 7am, 7pm, and DVT); and to the occipital lobe (areas V1, V2, V3, V4, V3a, and V6).

Area 46 exhibits structural connectivity to a number of local areas and to the contralateral hemisphere. White matter fiber tracts that connect Area 46 to the contralateral hemisphere traverse the corpus callosum and terminate at areas 9-46d and p9-47v. Local short association bundles connect it to areas 9-46d, a9-46v, p9-46v, IFSp, and IFSa.

Area p9-46v

Area p9-46v is a smaller parcellation with a delta shaped configuration situated on the cortical surface of the middle frontal gyrus. Its boundaries include area 8c posteriorly, area 46 medially, and IFSp and IFJa laterally.

Area p9-46v demonstrates functional connectivity to various regions across the cortex, including: the dorsolateral frontal lobe (areas 46, 9-46d, a9-46v, a10p, p10p, 6ma, i6-8, a47r, p47r, 8C, IFJp, IFJa, IFSp, and IFSa), the medial frontal lobe (areas 8BM and 33prime), the premotor area (areas 6a and 6r), the orbitofrontal region (area 11L), the temporal lobe (areas TE1p, TE2p, PH, and PHT), the insula (area AVI), the parietal lobe (areas LIPd, AIP, MIP, PFm, 7PL, IP0, IP1, and IP2), the medial parietal lobe (area 7pm), and the occipital lobe (area V1).

Area p9-46d exhibits structural connectivity to the superior longitudinal fasciculus and local parcellations. Area 9-46d is connected to area TE2a through white matter tracts that traverse the superior longitudinal fasciculus, pass through the Sylvian fissure, and terminate at the inferior temporal gyrus. Short association bundles provide local structural connections between area 9-46d and several other brain regions, including areas 46, a9-46v, IFJa, IFSa, IFSp, 8C, and 9-46d.

Area a9-46v

Area a9-46v is located on the cortical surface of the anterior aspect of the middle frontal gyrus, with its posterior boundary marked by area 46. Medially, it is adjacent to area 9-46d, while laterally, it is bordered by area p47r and IFSa to a lesser extent. The anterior aspect of a9-46v forms a point and is demarcated by the boundary between p10p and a47r.

Area a9-46v is functionally connected to various regions across the cortex, including: the dorsolateral frontal lobe (areas p9-46v, 46, 9-46d, a10p, p10p, 6ma, i6-8, a47r, p47r, 8C, and IFSa); the medial frontal lobe (areas 8BM and a32prime); the orbitofrontal region (area 11L); the temporal lobe (area TE1p); the insula (area AVI); the parietal lobe (areas LIPd, PF, IP1, and IP2); and the medial parietal lobe (areas 7pm, 31a, and RSC).

Area a9-46v exhibits interindividual variability in terms of its structural connectivity to neighboring regions. It has been reported that some individuals display structural connections between area a9-46v and nearby parcellations, such as 8C, 9-46d, 46, IFSa, and p47r.

Area IFSp

Area IFSp is situated at the anterior aspect of the IFS and includes the lower portion of the MFG's inferior bank in its upper region. It is positioned approximately above the IFG's pars triangularis segment. Area IFSp is bounded anteriorly by area IFSa and posteriorly by IFJa. Its inferior point borders areas 45 and 44, while its superior flank comprises areas p9-46v and 8C.

Area IFSp exhibits functional connectivity with several regions across the cortex, including the dorsolateral frontal lobe (areas a47r, p47r, IFSa, IFJa, IFJp, p9-46v, 47l, 44, 45, i6-8, and 8C), the medial frontal lobe (area 8BM), the premotor areas (area 55b), the orbitofrontal region (area 47m), the temporal lobe (areas PH, TE1p, STSdp, and STSvp), and the inferior parietal lobe (areas IP0, IP1, TPOJ1, and LIPd).

Area IFSp also exhibits structural connects with the superior longitudinal fasciculus (SLF) and local areas. Its connections to the SLF extend posteriorly, bending around the Sylvian fissure and ultimately end in areas TE1a, TE1m, and TE2a of the inferior temporal gyrus. Additional local connections to areas 46, IFJa, IFSa, IFSp, TE2a, TE1m, TE1a, 9-46d, p9-46v, 8C, and 8Av via short association fibers are also appreciated.

Area IFSa

Area IFSa is situated within the anterior part of the inferior frontal sulcus and occupies the lower regions of the middle frontal gyrus. More specifically, it is located superior to the pars orbitalis component of the inferior frontal gyrus. Area IFSa shares its anterior and posterior boundary with area p47r and area IFSp, respectively. Furthermore, area IFSa's inferior border abuts area 45, while its superior border is flanked by areas a9-46v, 46, and p9-46v.

Area IFSa exhibits functional connectivity with several regions across the cortex, including the dorsolateral frontal lobe (areas a47r, p47r, IFSp, IFJa, IFJp, a9-46v, p9-46v, and 46), the medial frontal lobe (areas 8BM and 33prime), the premotor areas (areas 6ma, 6a, 6r, PEF, FEF), the insula opercular area (areas FOP4, FOP5, PFop, MI, and PoI2), the orbitofrontal region (area 11L), the temporal lobe (areas PH, PHT, TE1p, TE2p, PeEc, and PHA3), the inferior parietal lobe (areas 7PL, IP0, IP1, IP2, PF, PGp, PFt, AIP, MIP, and LIPd), and the medial parietal lobe (areas 7am and 23c).

Area IFSa exhibits structural connectivity to area TE2a of the inferior temporal gyrus via fibers which extend posteriorly through the superior longitudinal fasciculus, taking a circular route around the Sylvian fissure. Additional connections from the superior longitudinal fasciculus connect area IFSa to area 4. Local structural connections exist to areas p47r, 46, p9-46v, 45, IFSa, 8C, 44, IFSp, and p47r, a9-46v via short association fibers.

Brodmann area 47 (BA47)

BA47 is divided into 5 unique HCP regions. These include areas a47r, p47r, 47L, 47m, and 47s.

Area a47r

Area a47r is situated in the anterior inferior aspect of the pars orbitalis within the inferior frontal gyrus. As a crescent-shaped region, area a47r curves minimally around its anterior-most aspect and contains a small portion in the anterior middle frontal gyrus. Area a47r is flanked superiorly between areas p47r and a9-46v, medially by area a10p, and posteriorly with 47l. Additionally, surrounding area a47r is area 11r adjacent to its orbitofrontal aspect.

Area a47r is functionally connected to various regions across the cortex, including: the dorsolateral frontal lobe (areas p47r, 8C, 8av, 8BL, 8BM, i6-8, a9-46v, and p9-46v); the inferior frontal gyrus (areas IFSa, IFSp, 47l, and 45); the inferior parietal lobule (areas IP1, IP2, PGi, PGs, and PFm); the lateral temporal lobe (areas TE1m, TE1p, TE2a, and STSv); and the posterior cingulate region (area d23ab).

A major fiber bundle connecting area a47r to the occipital lobe is the IFOF, a prominent white matter tract known to be the largest in the brain (Conner et al., 2018). The IFOF connects area a47r to areas within the occipital lobe (V1, V2, V3, V3a, V6, V6a, 7Am, and 7PL) via posterior projecting tracts that traverse posteriorly through the external capsule. Additionally, locally projecting fibers structurally connect area a47r to areas 47m and 11l.

Area p47r

Area p47r is situated on the cortical surface of anterior aspect of the inferior frontal sulcus, specifically within the extreme anterosuperior aspect of the inferior frontal gyrus. Additionally, area p47r shares its anterior and posterior boundaries with a47r and IFSa, respectively. Its smaller lateral side is flanked by area 45, while its bigger medial side is adjacent to area a9-46v.

Area p47r demonstrates functional connectivity to regions across the cortex, including: the dorsolateral frontal lobe (areas a47r, a9-46v, p9-46v, 8BM, 8C, and i6-8); the inferior frontal lobe (areas IFSa, IFSp, IFJp, and 6r); the insula (area anterior ventral insula, AVI); the temporal lobe (areas TE1p and PHT); the inferior parietal lobe (area PFm); and the intraparietal area (areas IP2, IP1, and LIPd).

Area p47r is primarily connected to other brain regions through locally projecting white matter tracts, which terminate at areas a47r, p10p, 45, IFSa, s6-8, 9-47d, and 9a. While some individuals exhibit long-distance structural connections to area p47r via the IFOF, this finding has been inconsistent across different patients.

Area 47L

Area 47L is situated in the inferior frontal gyrus, more precisely within the posterior aspect of the pars orbitalis, at its inferior and lateral boundary. Area 47L is flanked by area 45 superioposteriorly, FOP5 posteriorly, 47s and 47m inferiorly, and a47r and p47r superiorly.

Area 47L exhibits functional connectivity with various regions across the cortex, including: the dorsolateral frontal lobe (areas SFL, IFSp, 45, 44, a47r, 47s, 9a, 9p, 9m, 8AV, and 8BL); the medial frontal lobe (areas 8BM and 10v); the premotor areas (area 55b); the temporal lobe (areas TGd, TE1a, STSda, STSdp, and STSva); the inferior parietal lobe (area PGi); and the medial parietal lobe (areas 31pd and 31pv).

Area 47L is structurally connected to two major fiber bundles, the IFOF and the uncinate fasciculus. Connections which parse the IFOF link area 47L to areas within the occipital lobe (V1, V2, V3, and V4) via tracts which pass through the external capsule, while fibers that traverse the uncinate fasciculus link area 47L to areas within the temporal pole (TGd and STGa) via tracts which pass through the limen insulae. Interestingly, structural connections from area 47L to the uncinate fasciculus is not consistent in the right hemisphere. Lastly, area 47L has additional local association bundles which structurally connect it to areas 45 and FOP5.

Area 47m

Area 47m is situated on the surface of the posterior aspect of the lateral orbitofrontal cortex and is bounded anteriorly by area 11l, laterally by areas 47l and a47r, posteriorly by area 47s, and medially by area 13l.

Area 47m′s functional connectivity spans the cortex and includes connections to the orbitofrontal region (area 13L), the frontal lobe (areas 8AD and IFSp), the temporal lobe (areas TE1p and PHA2), and the parietal lobe (areas PGs and POS1).

Area 47m is structurally connected to the IFOF and other local areas. Connections projecting from area 47m to the inferior frontooccipital fasciculus continue posteriorly, pass through the external capsule, and terminate at area V1 within the occipital lobe. Additional connections from area 47m project to local areas 13l, 47l, 47s, AAIC, and Pir, via short association bundles.

Area 47s

Area 47s is situated in the inferior frontal gyrus, specifically in the posterior aspect of the pars orbitalis. Area 47s is flanked posteriorly by areas AVI and Pir, medially by area 13r, anteriorly by area 47l, and laterally by area 47m.

Area 47s exhibits functional connectivity with various regions across the cortex, including: the dorsolateral frontal lobe (areas SFL, 45, 44, 47L, 9a, 9p, 9m, 8AV, and 8BL); the medial frontal lobe (area d32); the insula (area AAIC); the temporal lobe (areas TGd, TE1a, STSdp, STSvp, and STSva); the inferior parietal lobe (area PGi); and the medial parietal lobe (areas 7M, 31pd, and 31pv).

Area 47s, like area 47L, shares structural connections with the IFOF and the uncinate fasciculus. Connections which parse the IFOF link area 47s to areas within the occipital lobe (V1, V2, and V3) via tracts which parse the external capsule, while fibers that traverse the uncinate fasciculus link area 47s to the temporal lobe (TGd) via tracts, which pass inferiorly through limen insulae. Additionally, there are white matter fibers that project anteriorly from the uncinate and end in the frontal pole areas 10v and 10pp.

Brodmann area 52 (BA52)

BA52 is divided into two unique HCP regions—areas PI and 52.

Area PI

Area PI (para-insular) is situated on the surface of the anterior aspect of the inferior circular sulcus of the insula. This region is bordered by PoI1 medially and areas TA2, STGa, and MBelt laterally. Furthermore, the anterior border of area PI is adjacent to areas Pir and TGd.

Area PI exhibits functional connectivity with several regions within the insula, such as areas PoI1, PoI2, MBelt, PBelt, 43, PFcm, RI, 52, FOP4, OP4, TA2, and STGa. Moreover, it has functional connections with PFop, which is situated in the parietal lobe.

Area PI exhibits structural connectivity to several brain regions. However, the proximity of the white matter tracts that project from this this area to the extreme and external capsule, has made it challenging to define is connectivity. Area PI exhibits structural connectivity to area V1, via white matter fibers which traverse the temporal lobe. Another subset of structural projections passes inferiorly from area PI to area TGd of the temporal lobe. Moreover, short association bundles project to local parcellations, including areas Pol1, Pol2, Ig, and OP2-3.

Area 52

Area 52 is located between Heschl's gyrus and the long gyri of the insula, specifically in the posterior aspect of the lower limb of the circular sulcus of the insula. Area 52 shares its boundaries with area MBelt laterally, POL1 and IG medially, OP2-3 posterosuperiorly, and PI anteroinferiorly.

Area 52 exhibits functional connectivity to various regions across the cortex, including: the sensory strip (areas 1, 2, 3a, and 3b); the middle cingulate areas (areas 24dv, p32prime, a24prime, 23c, and 5mv); the premotor regions (areas SCEF and FEF); the lateral frontal lobe (area 46); the superior insula opercular regions (areas 43, IG, FOP1, FOP3, FOP4, OP4, OP1, and PFcm); the lower opercula and Heschl's gyrus regions (areas MI, PoI1, PoI2, LBelt, PBelt, MBelt, A1, TA2, PI, A4, and STV); the temporal lobe (area PHT); the parietal lobe (areas DVT, PF, and PFop); the medial occipital lobe (areas V2, V3, and V4); the dorsal visual stream areas (areas V6 and V3a); and the lateral occipital lobe (area TPOJ2).

Area 52 exhibits structural connections exclusively to local areas. One subset of white matter fibers that originate from area 52 extend anteriorly, ultimately terminating at Pol1 and Pol2. Another subset of fibers connects to surrounding areas such as A1, Ig, MBelt, OP1, OP2-3, PoI1, PoI2, and TA2. Additionally, it is worth noting that connectivity between area 52 and the occipital and parietal lobes varies among individuals.

Conclusion

While admittedly a long chapter, this work covers a significant amount of material which is necessary as we move toward a connectomic-based thinking and framework when examining and studying human brain anatomy. Starting with Brodmann's atlas, a significant map was provided for cortical anatomy. In this chapter, we provide additional insight to Brodmann's map by bridging that work with that of the HCP authors as well as providing data on the underlying structural and functional connectivity in and across each region.

Acknowledgments

The authors would like to thank a number of additional contributors for their insight and work provided for this chapter. Specifically, we would like to thank Courtney Abbriano, Daniel Brenner, Daniel Valdivia, and the authors of the Connectomic Atlas of the Human Cerebrum (Operative Neurosurgery).

References

Aggleton, J. P., & Christiansen, K. (2015). The subiculum: The heart of the extended hippocampal system. *Progress in Brain Research, 219*, 65–82. https://doi.org/10.1016/bs.pbr.2015.03.003

Aggleton, J. P., Saunders, R. C., Wright, N. F., & Vann, S. D. (2014). The origin of projections from the posterior cingulate and retrosplenial cortices to the anterior, medial dorsal and laterodorsal thalamic nuclei of macaque monkeys. *European Journal of Neuroscience, 39*(1), 107–123. https://doi.org/10.1111/ejn.12389

Alexander, A. S., Place, R., Starrett, M. J., Chrastil, E. R., & Nitz, D. A. (2023). Rethinking retrosplenial cortex: Perspectives and predictions. *Neuron, 111*(2), 150–175. https://doi.org/10.1016/j.neuron.2022.11.006

Amunts, K., Lenzen, M., Friederici, A. D., Schleicher, A., Morosan, P., Palomero-Gallagher, N., & Zilles, K. (2010). Broca's region: Novel organizational principles and multiple receptor mapping. *PLoS Biology, 8*(9). https://doi.org/10.1371/journal.pbio.1000489

Amunts, K., & Zilles, K. (2012). Architecture and organizational principles of Broca's region. *Trends in Cognitive Sciences, 16*(8), 418–426. https://doi.org/10.1016/j.tics.2012.06.005

Amunts, K., & Zilles, K. (2015). Architectonic mapping of the human brain beyond Brodmann. *Neuron, 88*(6), 1086–1107. https://doi.org/10.1016/j.neuron.2015.12.001

Bailey, P. (1951). The isocortex of man. *Urbana, 3*.

Baker, C. M., Burks, J. D., Briggs, R. G., Conner, A. K., Glenn, C. A., Sali, G., McCoy, T. M., Battiste, J. D., O'Donoghue, D. L., & Sughrue, M. E. (2018a). A connectomic atlas of the human cerebrum-chapter 1: Introduction, methods, and significance. *Oper Neurosurg (Hagerstown), 15*(Suppl. 1_1), S1–S9. https://doi.org/10.1093/ons/opy253

Baker, C. M., Burks, J. D., Briggs, R. G., Conner, A. K., Glenn, C. A., Morgan, J. P., Stafford, J., Sali, G., McCoy, T. M., Battiste, J. D., O'Donoghue, D. L., & Sughrue, M. E. (2018b). A connectomic atlas of the human cerebrum-chapter 2: The lateral frontal lobe. *Oper Neurosurgery (Hagerstown), 15*(Suppl. 1_1), S10–s74. https://doi.org/10.1093/ons/opy254

Baker, C. M., Burks, J. D., Briggs, R. G., Stafford, J., Conner, A. K., Glenn, C. A., Sali, G., McCoy, T. M., Battiste, J. D., O'Donoghue, D. L., & Sughrue, M. E. (2018c). A connectomic atlas of the human cerebrum-chapter 4: The medial frontal lobe, anterior cingulate gyrus, and orbitofrontal cortex. *Oper Neurosurgery (Hagerstown), 15*(Suppl. 1_1), S122–s174. https://doi.org/10.1093/ons/opy257

Beckmann, M., Johansen-Berg, H., & Rushworth, M. F. (2009). Connectivity-based parcellation of human cingulate cortex and its relation to functional specialization. *Journal of Neuroscience, 29*(4), 1175–1190. https://doi.org/10.1523/JNEUROSCI.3328-08.2009

Bludau, S., Eickhoff, S. B., Mohlberg, H., Caspers, S., Laird, A. R., Fox, P. T., Schleicher, A., Zilles, K., & Amunts, K. (2014). Cytoarchitecture, probability maps and functions of the human frontal pole. *NeuroImage, 93 Pt 2*(Pt 2), 260–275. https://doi.org/10.1016/j.neuroimage.2013.05.052

Burgess, P. W., Gilbert, S. J., & Dumontheil, I. (2007). Function and localization within rostral prefrontal cortex (area 10). *Philosophical Transactions of the Royal Society of London B Biological Sciences, 362*(1481), 887–899. https://doi.org/10.1098/rstb.2007.2095

Bzdok, D., Heeger, A., Langner, R., Laird, A. R., Fox, P. T., Palomero-Gallagher, N., Vogt, B. A., Zilles, K., & Eickhoff, S. B. (2015). Subspecialization in the human posterior medial cortex. *NeuroImage, 106*, 55–71. https://doi.org/10.1016/j.neuroimage.2014.11.009

Carmichael, S. T., & Price, J. L. (1995). Sensory and premotor connections of the orbital and medial prefrontal cortex of macaque monkeys. *The Journal of Comparative Neurology, 363*(4), 642–664. https://doi.org/10.1002/cne.903630409

Cavanna, A. E., & Trimble, M. R. (2006). The precuneus: A review of its functional anatomy and behavioural correlates. *Brain, 129*(Pt 3), 564–583. https://doi.org/10.1093/brain/awl004

Chahine, G., Diekhof, E. K., Tinnermann, A., & Gruber, O. (2015). On the role of the anterior prefrontal cortex in cognitive "branching": An fMRI study. *Neuropsychologia, 77*, 421–429. https://doi.org/10.1016/j.neuropsychologia.2015.08.018

Choi, M. H., Kim, H. S., Baek, J. H., Lee, J. C., Park, S. J., Jeong, U. H., Gim, S. Y., Kim, S. P., Lim, D. W., & Chung, S. C. (2015). Differences in activation area within Brodmann area 2 caused by pressure stimuli on fingers and joints: In case of male subjects. *Medicine (Baltimore), 94*(38), e1657. https://doi.org/10.1097/MD.0000000000001657

Chouinard, P. A., & Paus, T. (2006). The primary motor and premotor areas of the human cerebral cortex. *The Neuroscientist, 12*(2), 143–152. https://doi.org/10.1177/1073858405284255

Coco, M., Perciavalle, V., Cavallari, P., & Perciavalle, V. (2016). Effects of an exhaustive exercise on motor skill learning and on the excitability of primary motor cortex and supplementary motor area. *Medicine (Baltimore), 95*(11), e2978. https://doi.org/10.1097/MD.0000000000002978

Conner, A. K., Briggs, R. G., Sali, G., Rahimi, M., Baker, C. M., Burks, J. D., Glenn, C. A., Battiste, J. D., & Sughrue, M. E. (2018). A connectomic atlas of the human cerebrum-Chapter 13: Tractographic description of the inferior fronto-occipital fasciculus. *Operative neurosurgery (Hagerstown), 15*(1), S436–S443. https://doi.org/10.1093/ons/opy267. PMID: 30260438. PMCID: PMC6890527.

Dadario, N. B., & Sughrue, M. E. (2023). The functional role of the precuneus. *Brain.* https://doi.org/10.1093/brain/awad181

Dadario, N. B., Tanglay, O., Stafford, J. F., Davis, E. J., Young, I. M., Fonseka, R. D., Briggs, R. G., Yeung, J. T., Teo, C., & Sughrue, M. E. (2023). Topology of the lateral visual system: The fundus of the superior temporal sulcus and parietal area H connect nonvisual cerebrum to the lateral occipital lobe. *Brain Behaviour, 13*(4), e2945. https://doi.org/10.1002/brb3.2945

Dadario, N. B., Tanglay, O., & Sughrue, M. E. (2023). Deconvoluting human Brodmann area 8 based on its unique structural and functional connectivity. *Frontiers in neuroanatomy, 17*, 1127143. https://doi.org/10.3389/fnana.2023.1127143.

Devinsky, O., Morrell, M. J., & Vogt, B. A. (1995). Contributions of anterior cingulate cortex to behaviour. *Brain, 118*(Pt 1), 279–306. https://doi.org/10.1093/brain/118.1.279

Drevets, W. C., Savitz, J., & Trimble, M. (2008). The subgenual anterior cingulate cortex in mood disorders. *CNS Spectrums, 13*(8), 663–681. https://doi.org/10.1017/s1092852900013754

Fecteau, J. H., & Munoz, D. P. (2006). Salience, relevance, and firing: A priority map for target selection. *Trends in Cognitive Sciences, 10*(8), 382–390. https://doi.org/10.1016/j.tics.2006.06.011

Field, D. T., Biagi, N., & Inman, L. A. (2020). The role of the ventral intraparietal area (VIP/pVIP) in the perception of object-motion and self-motion. *NeuroImage, 213*, 116679. https://doi.org/10.1016/j.neuroimage.2020.116679

Fogassi, L., Gallese, V., Buccino, G., Craighero, L., Fadiga, L., & Rizzolatti, G. (2001). Cortical mechanism for the visual guidance of hand grasping movements in the monkey: A reversible inactivation study. *Brain, 124*(Pt 3), 571–586. https://doi.org/10.1093/brain/124.3.571

Furlan, M., & Smith, A. T. (2016). Global motion processing in human visual cortical areas V2 and V3. *Journal of Neuroscience, 36*(27), 7314–7324. https://doi.org/10.1523/JNEUROSCI.0025-16.2016

Galletti, C., & Fattori, P. (2018). The dorsal visual stream revisited: Stable circuits or dynamic pathways? *Cortex, 98*, 203–217. https://doi.org/10.1016/j.cortex.2017.01.009

Gasquoine, P. G. (2013). Localization of function in anterior cingulate cortex: From psychosurgery to functional neuroimaging. *Neuroscience & Biobehavioral Reviews, 37*(3), 340–348. https://doi.org/10.1016/j.neubiorev.2013.01.002

Glasser, M. F., Coalson, T. S., Robinson, E. C., Hacker, C. D., Harwell, J., Yacoub, E., Ugurbil, K., Andersson, J., Beckmann, C. F., Jenkinson, M., Smith, S. M., & Van Essen, D. C. (2016). A multi-modal parcellation of human cerebral cortex. *Nature, 536*(7615), 171–178. https://doi.org/10.1038/nature18933

Gottfried, J. A. (2007). What can an orbitofrontal cortex-endowed animal do with smells? *Annals of the New York Academy of Sciences, 1121*, 102–120. https://doi.org/10.1196/annals.1401.018

Grabenhorst, F., & Rolls, E. T. (2011). Value, pleasure and choice in the ventral prefrontal cortex. *Trends in Cognitive Sciences, 15*(2), 56–67. https://doi.org/10.1016/j.tics.2010.12.004

Grefkes, C., Ritzl, A., Zilles, K., & Fink, G. R. (2004). Human medial intraparietal cortex subserves visuomotor coordinate transformation. *NeuroImage, 23*(4), 1494–1506. https://doi.org/10.1016/j.neuroimage.2004.08.031

Grefkes, C., & Fink, G. R. (2005). The functional organization of the intraparietal sulcus in humans and monkeys. *Journal of Anatomy, 207*(1), 3–17. https://doi.org/10.1111/j.1469-7580.2005.00426.x

Hazem, S. R., Awan, M., Lavrador, J. P., Patel, S., Wren, H. M., Lucena, O., Semedo, C., Irzan, H., Melbourne, A., Ourselin, S., Shapey, J., Kailaya-Vasan, A., Gullan, R., Ashkan, K., Bhangoo, R., & Vergani, F. (2021). Middle frontal gyrus and area 55b: Perioperative mapping and language outcomes. *Frontiers in Neurology, 12*, 646075. https://doi.org/10.3389/fneur.2021.646075

Hein, G., & Knight, R. T. (2008). Superior temporal sulcus–it's my area: Or is it? *Journal of Cognitive Neuroscience, 20*(12), 2125–2136. https://doi.org/10.1162/jocn.2008.20148

Huffman, K. J., & Krubitzer, L. (2001). Area 3a: Topographic organization and cortical connections in marmoset monkeys. *Cerebral Cortex, 11*(9), 849–867. https://doi.org/10.1093/cercor/11.9.849

Hyvarinen, J., & Poranen, A. (1978). Receptive field integration and submodality convergence in the hand area of the post-central gyrus of the alert monkey. *The Journal of Physiology, 283*, 539–556. https://doi.org/10.1113/jphysiol.1978.sp012518

Jitsuishi, T., & Yamaguchi, A. (2023). Characteristic cortico-cortical connection profile of human precuneus revealed by probabilistic tractography. *Scientific Reports, 13*(1), 1936. https://doi.org/10.1038/s41598-023-29251-2

Kruger, B., Bischoff, M., Blecker, C., Langhanns, C., Kindermann, S., Sauerbier, I., Reiser, M., Stark, R., Munzert, J., & Pilgramm, S. (2014). Parietal and premotor cortices: Activation reflects imitation accuracy during observation, delayed imitation and concurrent imitation. *NeuroImage, 100*, 39–50. https://doi.org/10.1016/j.neuroimage.2014.05.074

Kubler, A., Dixon, V., & Garavan, H. (2006). Automaticity and reestablishment of executive control-an fMRI study. *Journal of Cognitive Neuroscience, 18*(8), 1331–1342. https://doi.org/10.1162/jocn.2006.18.8.1331

Kuehn, E., Haggard, P., Villringer, A., Pleger, B., & Sereno, M. I. (2018). Visually-driven maps in area 3b. *Journal of Neuroscience, 38*(5), 1295–1310. https://doi.org/10.1523/JNEUROSCI.0491-17.2017

Mars, R. B., Jbabdi, S., Sallet, J., O'Reilly, J. X., Croxson, P. L., Olivier, E., Noonan, M. P., Bergmann, C., Mitchell, A. S., Baxter, M. G., Behrens, T. E., Johansen-Berg, H., Tomassini, V., Miller, K. L., & Rushworth, M. F. (2011). Diffusion-weighted imaging tractography-based parcellation of the human parietal cortex and comparison with human and macaque resting-state functional connectivity. *Journal of Neuroscience, 31*(11), 4087–4100. https://doi.org/10.1523/JNEUROSCI.5102-10.2011

Martuzzi, R., van der Zwaag, W., Dieguez, S., Serino, A., Gruetter, R., & Blanke, O. (2015). Distinct contributions of Brodmann areas 1 and 2 to body ownership. *Social Cognitive and Affective Neuroscience, 10*(11), 1449–1459. https://doi.org/10.1093/scan/nsv031

Naito, E., Scheperjans, F., Eickhoff, S. B., Amunts, K., Roland, P. E., Zilles, K., & Ehrsson, H. H. (2008). Human superior parietal lobule is involved in somatic perception of bimanual interaction with an external object. *Journal of Neurophysiology, 99*(2), 695–703. https://doi.org/10.1152/jn.00529.2007

Naya, Y. (2016). Declarative association in the perirhinal cortex. *Neuroscience Research, 113*, 12–18. https://doi.org/10.1016/j.neures.2016.07.001

Neubert, F. X., Mars, R. B., Thomas, A. G., Sallet, J., & Rushworth, M. F. (2014). Comparison of human ventral frontal cortex areas for cognitive control and language with areas in monkey frontal cortex. *Neuron, 81*(3), 700–713. https://doi.org/10.1016/j.neuron.2013.11.012

Nitschke, K., Kostering, L., Finkel, L., Weiller, C., & Kaller, C. P. (2017). A Meta-analysis on the neural basis of planning: Activation likelihood estimation of functional brain imaging results in the Tower of London task. *Human Brain Mapping, 38*(1), 396–413. https://doi.org/10.1002/hbm.23368

Ongur, D., Ferry, A. T., & Price, J. L. (2003). Architectonic subdivision of the human orbital and medial prefrontal cortex. *The Journal of Comparative Neurology, 460*(3), 425–449. https://doi.org/10.1002/cne.10609

Paus, T. (1996). Location and function of the human frontal eye-field: A selective review. *Neuropsychologia, 34*(6), 475–483. https://doi.org/10.1016/0028-3932(95)00134-4

Petit, L., Clark, C. P., Ingeholm, J., & Haxby, J. V. (1997). Dissociation of saccade-related and pursuit-related activation in human frontal eye fields as revealed by fMRI. *Journal of Neurophysiology, 77*(6), 3386–3390. https://doi.org/10.1152/jn.1997.77.6.3386

Petrides, M. (2005). Lateral prefrontal cortex: Architectonic and functional organization. *Philosophical Transactions of the Royal Society of London B Biological Sciences, 360*(1456), 781–795. https://doi.org/10.1098/rstb.2005.1631

Petrides, M., & Pandya, D. N. (2012). Chapter 26 - the frontal cortex. In J. K. Mai, & G. Paxinos (Eds.), *The human nervous system* (3rd ed., pp. 988–1011). San Diego: Academic Press. https://doi.org/https://doi.org/10.1016/B978-0-12-374236-0.10026-4.

Pierrot-Deseilligny, C. (1994). Saccade and smooth-pursuit impairment after cerebral hemispheric lesions. *European Neurology, 34*(3), 121–134. https://doi.org/10.1159/000117025

Pierrot-Deseilligny, C., Gaymard, B., Muri, R., & Rivaud, S. (1997). Cerebral ocular motor signs. *Journal of Neurology, 244*(2), 65–70. https://doi.org/10.1007/s004150050051

Pierrot-Deseilligny, C., Muri, R. M., Nyffeler, T., & Milea, D. (2005). The role of the human dorsolateral prefrontal cortex in ocular motor behavior. *Annals of the New York Academy of Sciences, 1039*, 239–251. https://doi.org/10.1196/annals.1325.023

Ploner, M., Schmitz, F., Freund, H. J., & Schnitzler, A. (2000). Differential organization of touch and pain in human primary somatosensory cortex. *Journal of Neurophysiology, 83*(3), 1770–1776. https://doi.org/10.1152/jn.2000.83.3.1770

Pouget, P. (2015). The cortex is in overall control of "voluntary" eye movement. *Eye, 29*(2), 241–245. https://doi.org/10.1038/eye.2014.284

Qiu, F. T., & von der Heydt, R. (2005). Figure and ground in the visual cortex: v2 combines stereoscopic cues with gestalt rules. *Neuron, 47*(1), 155–166. https://doi.org/10.1016/j.neuron.2005.05.028

Rajimehr, R., Young, J. C., & Tootell, R. B. (2009). An anterior temporal face patch in human cortex, predicted by macaque maps. *Proceedings of the National Academy of Sciences of the United States of America, 106*(6), 1995–2000. https://doi.org/10.1073/pnas.0807304106

Ranganath, C. (2006). Working memory for visual objects: Complementary roles of inferior temporal, medial temporal, and prefrontal cortex. *Neuroscience, 139*(1), 277–289. https://doi.org/10.1016/j.neuroscience.2005.06.092

Rolls, E. T. (2007). The representation of information about faces in the temporal and frontal lobes. *Neuropsychologia, 45*(1), 124–143. https://doi.org/10.1016/j.neuropsychologia.2006.04.019

Rolls, E. T. (2011). Taste, olfactory and food texture reward processing in the brain and obesity. *International Journal of Obesity, 35*(4), 550–561. https://doi.org/10.1038/ijo.2010.155

Roux, F., Wibral, M., Mohr, H. M., Singer, W., & Uhlhaas, P. J. (2012). Gamma-band activity in human prefrontal cortex codes for the number of relevant items maintained in working memory. *Journal of Neuroscience, 32*(36), 12411–12420. https://doi.org/10.1523/JNEUROSCI.0421-12.2012

Roux, A., Lemaitre, A. L., Deverdun, J., Ng, S., Duffau, H., & Herbet, G. (2021). Combining electrostimulation with fiber tracking to stratify the inferior fronto-occipital fasciculus. *Frontiers in Neuroscience, 15*, 683348. https://doi.org/10.3389/fnins.2021.683348

Sarkissov, S., Filimonoff, I., Kononowa, E., Preobraschenskaja, I., & Kukuew, L. (1955). *Atlas of the cytoarchitectonics of the human cerebral cortex* (p. 20). Moscow: Medgiz.

Schall, J. D. (2013). *Production, control, and visual guidance of saccadic eye movements. ISRN Neurol* (p. 752384), 2013 https://doi.org/10.1155/2013/752384, 2013.

Scheperjans, F., Palomero-Gallagher, N., Grefkes, C., Schleicher, A., & Zilles, K. (2005). Transmitter receptors reveal segregation of cortical areas in the human superior parietal cortex: Relations to visual and somatosensory regions. *NeuroImage, 28*(2), 362–379. https://doi.org/10.1016/j.neuroimage.2005.06.028

Scheperjans, F., Eickhoff, S. B., Homke, L., Mohlberg, H., Hermann, K., Amunts, K., & Zilles, K. (2008). Probabilistic maps, morphometry, and variability of cytoarchitectonic areas in the human superior parietal cortex. *Cerebral Cortex, 18*(9), 2141–2157. https://doi.org/10.1093/cercor/bhm241

Schmahmann, J. D., & Pandya, D. N. (2007). The complex history of the fronto-occipital fasciculus. *Journal of the History of the Neurosciences, 16*(4), 362–377. https://doi.org/10.1080/09647040600620468

Shahab, Q. S., Young, I. M., Dadario, N. B., Tanglay, O., Nicholas, P. J., Lin, Y. H., Fonseka, R. D., Yeung, J. T., Bai, M. Y., Teo, C., Doyen, S., & Sughrue, M. E. (2022). A connectivity model of the anatomic substrates underlying Gerstmann syndrome. *Brain Commun, 4*(3), fcac140. https://doi.org/10.1093/braincomms/fcac140

Sheets, J. R., Briggs, R. G., Young, I. M., Bai, M. Y., Lin, Y. H., Poologaindran, A., Conner, A. K., O'Neal, C. M., Baker, C. M., Glenn, C. A., & Sughrue, M. E. (2021). Parcellation-based modeling of the supplementary motor area. *Journal of the Neurological Sciences, 421*, 117322. https://doi.org/10.1016/j.jns.2021.117322

Tsao, D. Y., Moeller, S., & Freiwald, W. A. (2008). Comparing face patch systems in macaques and humans. *Proceedings of the National Academy of Sciences of the United States of America, 105*(49), 19514–19519. https://doi.org/10.1073/pnas.0809662105

Vann, S. D., Aggleton, J. P., & Maguire, E. A. (2009). What does the retrosplenial cortex do? *Nature Reviews Neuroscience, 10*(11), 792–802. https://doi.org/10.1038/nrn2733

Vierck, C. J., Whitsel, B. L., Favorov, O. V., Brown, A. W., & Tommerdahl, M. (2013). Role of primary somatosensory cortex in the coding of pain. *Pain, 154*(3), 334–344. https://doi.org/10.1016/j.pain.2012.10.021

Vogt, C., & Vogt, O. (1919). *Allgemeine ergebnisse unserer hirnforschung* (JA Barth).

Vogt, B. A., & Vogt, L. (2003). Cytology of human dorsal midcingulate and supplementary motor cortices. *Journal of Chemical Neuroanatomy, 26*(4), 301–309. https://doi.org/10.1016/j.jchemneu.2003.09.004

von Economo, C. F., & Koskinas, G. N. (1925). *Die cytoarchitektonik der hirnrinde des erwachsenen menschen*. J. Springer.

Wandell, B. A., Dumoulin, S. O., & Brewer, A. A. (2007). Visual field maps in human cortex. *Neuron, 56*(2), 366–383. https://doi.org/10.1016/j.neuron.2007.10.012

Wang, J., Yang, Y., Fan, L., Xu, J., Li, C., Liu, Y., Fox, P. T., Eickhoff, S. B., Yu, C., & Jiang, T. (2015). Convergent functional architecture of the superior parietal lobule unraveled with multimodal neuroimaging approaches. *Human Brain Mapping, 36*(1), 238–257. https://doi.org/10.1002/hbm.22626

Whitsel, B. L., Vierck, C. J., Waters, R. S., Tommerdahl, M., & Favorov, O. V. (2019). Contributions of nociresponsive area 3a to normal and abnormal somatosensory perception. *The Journal of Pain, 20*(4), 405–419. https://doi.org/10.1016/j.jpain.2018.08.009

Zilles, K., & Amunts, K. (2010). Centenary of Brodmann's map — conception and fate. *Nature Reviews Neuroscience, 11*(2), 139–145. https://doi.org/10.1038/nrn2776

CHAPTER

5 The transdiagnostic model of mental illness and cognitive dysfunction

Introducing a transdiagnostic hypothesis

The mechanistic origin of emotional and cognitive problems is only beginning to become well-articulated. Here is what we know for sure: most cognitive, emotional, or neurological symptoms have been linked to various focal lesions like strokes, brain tumors, trauma, etc., by various groups. However, the exact anatomy of what areas are affected is not well delineated. For example, we know that judgment issues result from injury to the frontal lobe, especially on the right side, however we have yet to see a clear definition of what specifically is injured in these patients which is all encompassing, suggesting that the lesion pattern is complex (Hormovas et al., 2023). Depression and other psychiatric illnesses are known to result from various focal lesions, but again, the anatomy is well known to be complex.

The understanding of neurological and psychocognitive functions of the brain started in a deductive manner. Historically, the understanding of brain functions by anatomical regions happened by the ways of survivable injuries to the brain. A significant example in the understanding of functional neuroanatomy was the case of Phineas Gage in the 1800s. Gage was a railroad construction worker who had survived after a large iron rod was driven completely through his head, destroying much of his brain's left frontal lobe. The change in his personality in the subsequent years of his life dictated much of what we understood as the functional importance of the left frontal lobe. He was "gross, profane, coarse, and vulgar, to such a degree that his society was intolerable to decent people"(1851). He was generally considered as behaviorally disinhibited with intact memory formation. This observation dominated much of our understanding of what we considered as left "dominant" frontal lobe function. However, by this time, we know that our brain is more complicated than that.

The fact that clinical syndromes meeting diagnostic criteria for depression can be caused by a specific unifocal lesion clearly demonstrates that the disease is not widely diffuse and does not involve the whole brain, but given that a variety of lesions have been demonstrated to cause this suggests that the normal system damaged in these patients is multifocal. This lines up with the neuroimaging literature which clearly shows that mental illnesses display functional connectivity problems in specific areas, and not the entire brain, making these diseases fundamentally multifocal.

Further, complicating this issue is that the patterns of different people meeting the diagnostic criteria for mental illnesses often are different in some ways and similar in others. This aligns well with the clinical phenotype of mental illnesses which both overlap across diagnostic categories and diverge within diagnostic categories.

Connectomic Medicine. https://doi.org/10.1016/B978-0-443-19089-6.00002-1

This chapter builds a framework for understanding the transdiagnostic hypothesis: the idea that symptoms link to dysfunction in specific circuits, that regardless of exact biologic mechanism of various diseases causing a symptom, the symptom caused is circuit-specific, not disease-specific, and that a mechanistic explanation of cognitive and emotional problems can be described in an anatomically specific way. While this is a work in progress, the potential explanatory power of this approach is profound and gives us a path toward more intelligent brain therapies.

Brain function as a series of states

The previous chapter outlined brain anatomy and the canonical brain networks and subdivisions. These are useful in that they provide some basic structure to the connectome, and in many ways provide a useful simplifying framework for understanding the brain that provides specific functional insights. Having said this, one cannot explain all of brain function with networks alone. Higher cognition and emotional function involves combinations of networks working together to achieve combined states specific to an emotion or cognitive task.

One way to think of this is as a symphony orchestra. In this metaphor, the DMN is the string section, the CEN the flutes, the salience network the percussion, the sensorimotor the wood winds, etc. While it is possible for a single section to carry a portion of the piece as a solo part, we all acknowledge that a very different song results from mixing of these instruments, creating something that is not possible alone. Also, it is common for different sections to play complementary yet distinct parts in sync with other sections for short parts of the piece, but to not play identical parts. This is what makes a symphony rewarding.

The brain is quite similar. When we look at brain states, such as emotions, they involve complex combinations of areas off and on for a short stretch. Fig. 5.1 shows an example of this, specifically, that two emotions result from different combinations of salience, DMN and CEN firing at a specific time. As time passes, the brain transits through various states and what results are our experiences.

How does functional connectivity become disturbed in brain disease?

The evidence that functional connectivity is disturbed in mental illnesses of various types is extensive and hard to refute. While specific pathofunctional relationships are still under intense study, what is clear is that parts of the brain of mentally ill people show abnormalities between region correlations when measured with BOLD, EEG, or other similar modalities (Dadario & Sughrue, 2022; Gupta et al., 2021; Menon, 2011; Uddin et al., 2013).

The reason why is not entirely known, but groundbreaking work from several groups has shown us some possible mechanisms how this might occur. For example, it is known that the functional connectivity patterns of a given connectome result from a complex interaction between signals which run through the entire system causes some areas to fire together frequently and others to fire opposite each other (Sporns, 2013). Changing the overall basal brain rhythm, the structural connectivity pattern, or the electrochemical gain between areas, all changes which pattern of functional connectivity emerges from this complex system. Note that the electrochemical gain is likely modified by ascending neurotransmitter systems like the serotonin, dopamine, or norepinephrine systems well known in psychopharmacology.

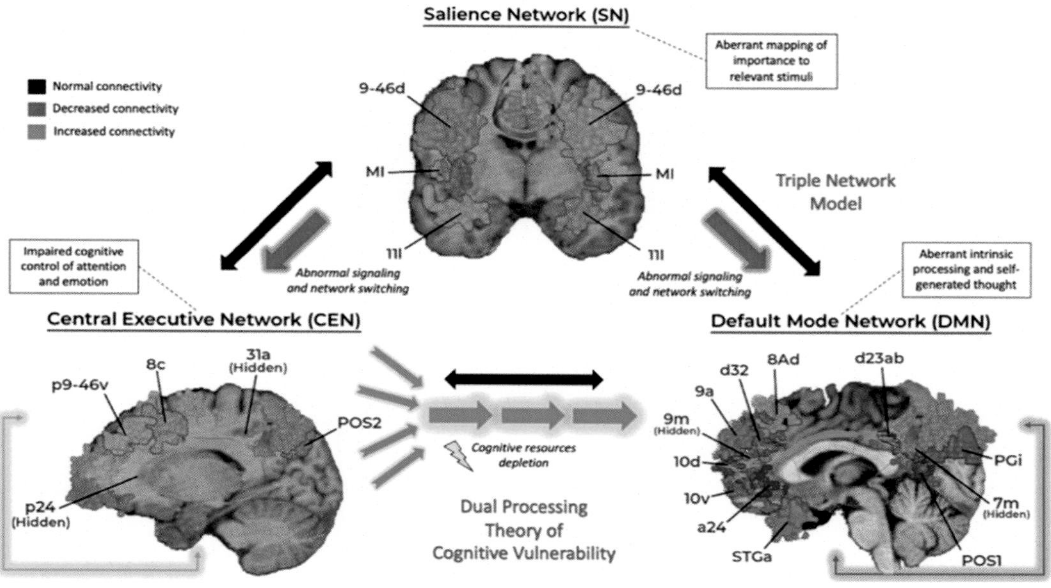

FIGURE 5.1

Different combinations of firing between the cognitive control networks determines specific emotional expressions. Dysfunction or disruption in these interactions forms the basis for many psychiatric and neurological disorders, most commonly elucidated in the context of depression.

Figure adapted from Young et al. (2023).

Pioneering work by Danielle Bassett's group has given us some insight into reasons why some brains develop a tendency for mental symptoms while others do not (Gu et al., 2015; Medaglia et al., 2017). Specifically, network control theory tells us that given a specific set of wiring, an electrical system has states which are energetically easier to get the system into than others, because certain state transitions are easier than others in a given system. Fig. 5.2 shows why this is the case. Taken *en toto*, we can estimate an energy landscape showing that some states are simpler than others. Her group has shown that networks like the CEN play a key role in controlling the brain and achieving energetically difficult brain states. The maturation of the CEN during adolescence and early adulthood (Sherman et al., 2014) suggests that this plays a role in the social maturation process and the development of abstract thinking and judgment that occurs during this period.

In this theory, mental illness occurs when undesirable brain states become energetically favorable and/or desirable states become unfavorable. It is easy to hypothesize that differences in wiring patterns (due to genetic or experiential causes) or changes in the level or receptor function of various neurotransmitters could change this landscape and cause states to happen or not happen that causes the symptoms of disease. The role of the CEN in exploring the energy landscape suggests an answer to the question of why genetically mediated psychiatric risks do not cause symptoms until adolescence or later, as opposed to childhood. Additionally, injury to the brain could change the wiring pattern to make the landscape different and unfavorable. Therapies which work do so by changing the landscape

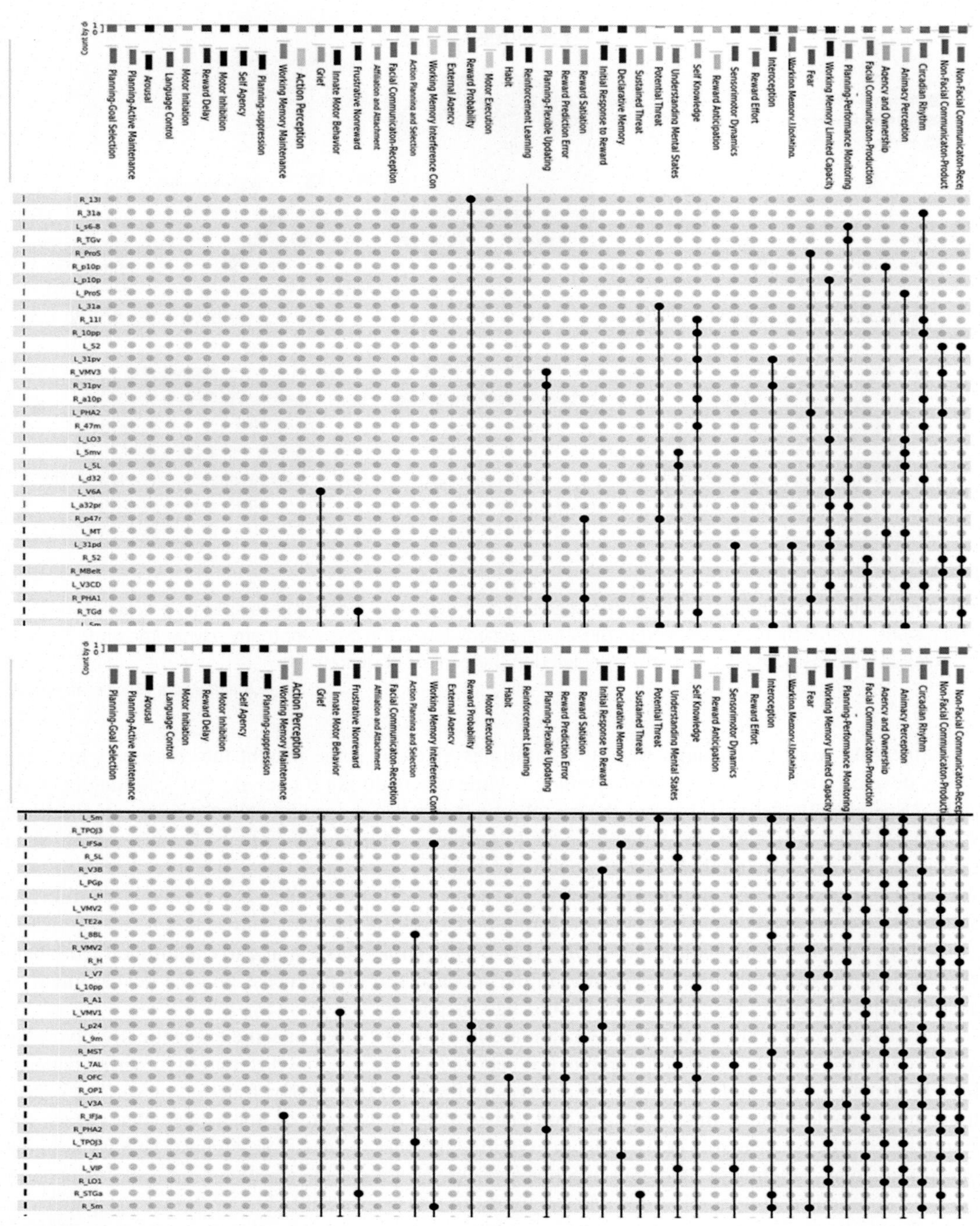

FIGURE 5.2

An upset plot demonstrating the intersections of specific areas belonging to particular domains.

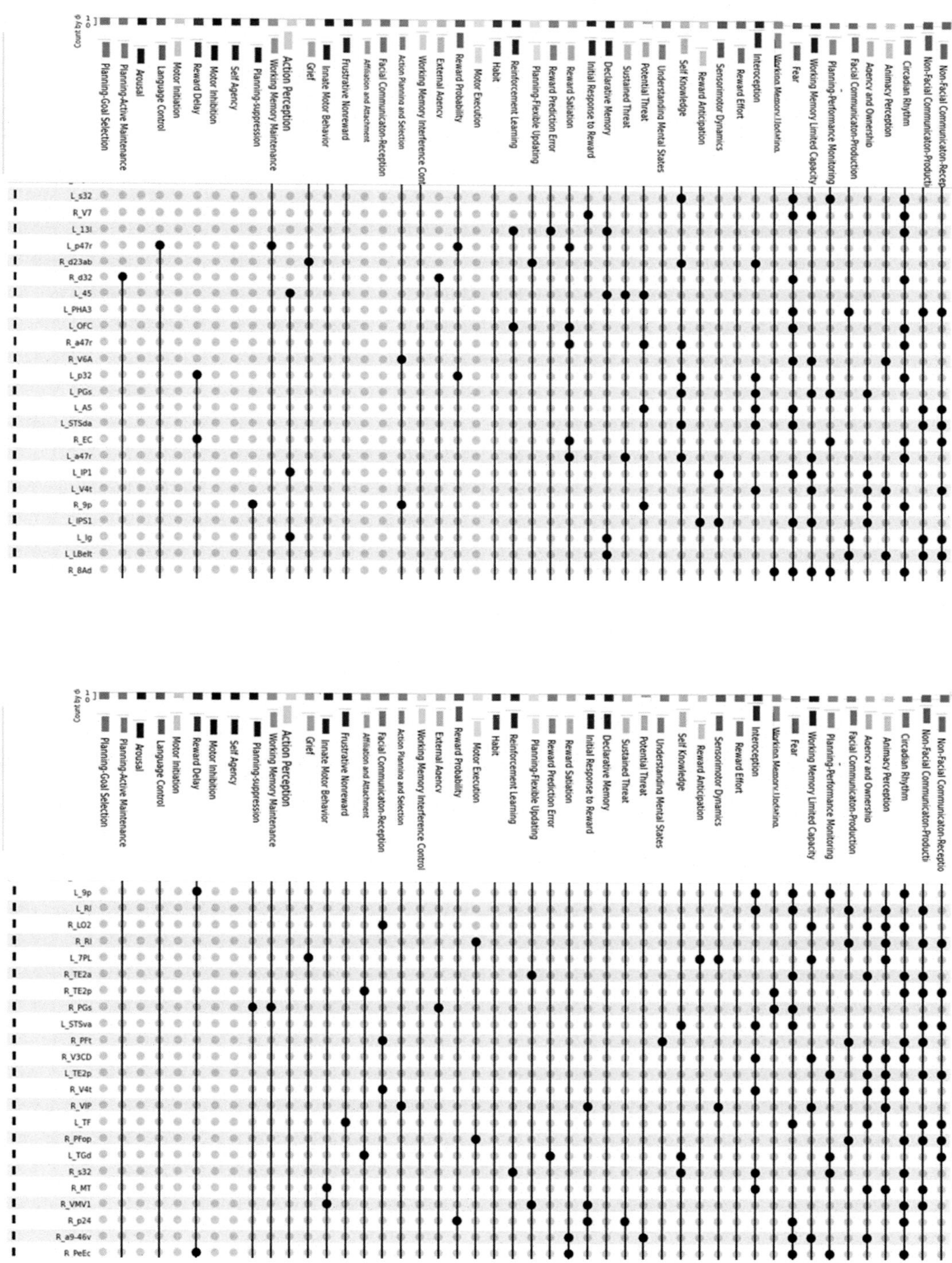

FIGURE 5.2 Cont'd

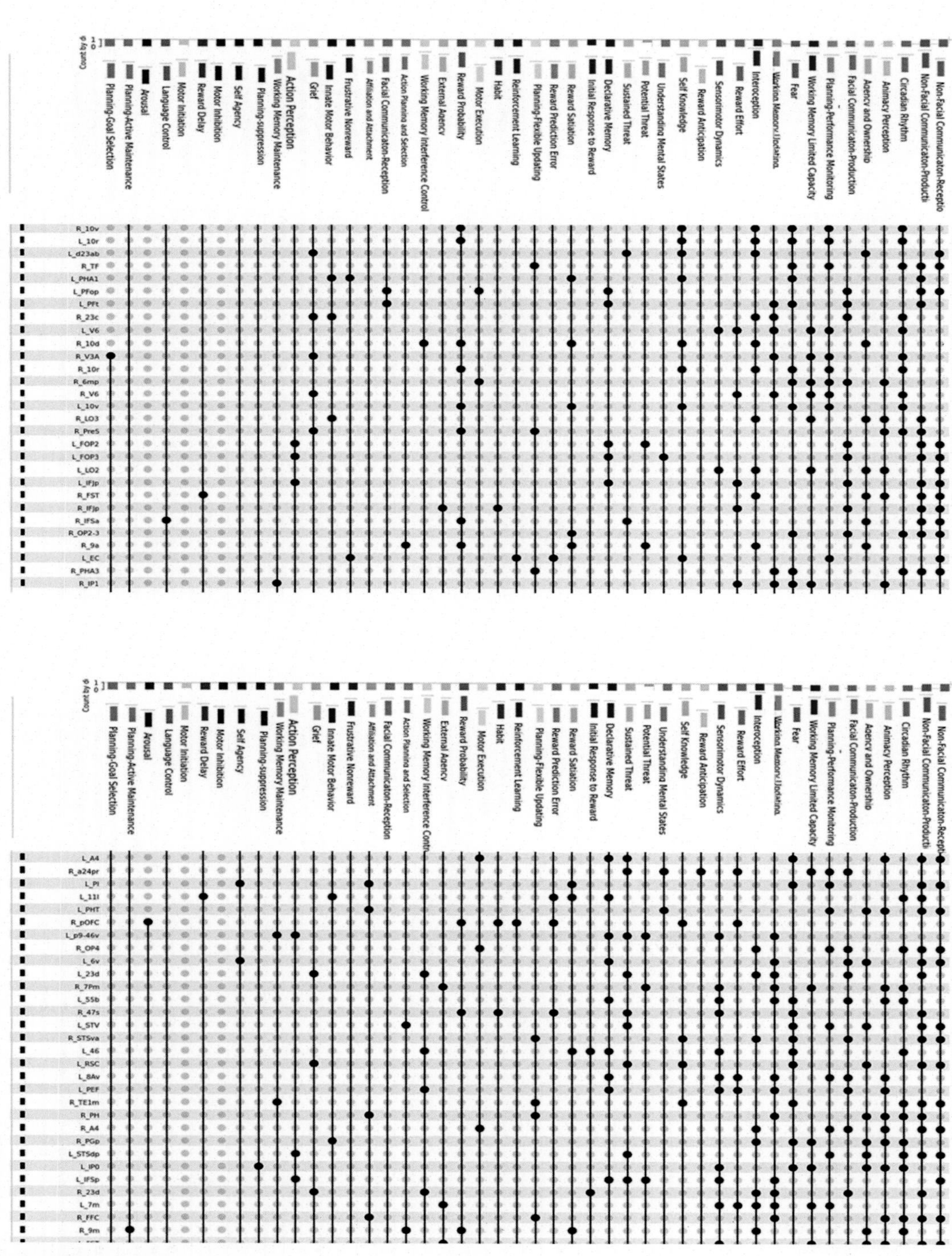

FIGURE 5.2 Cont'd

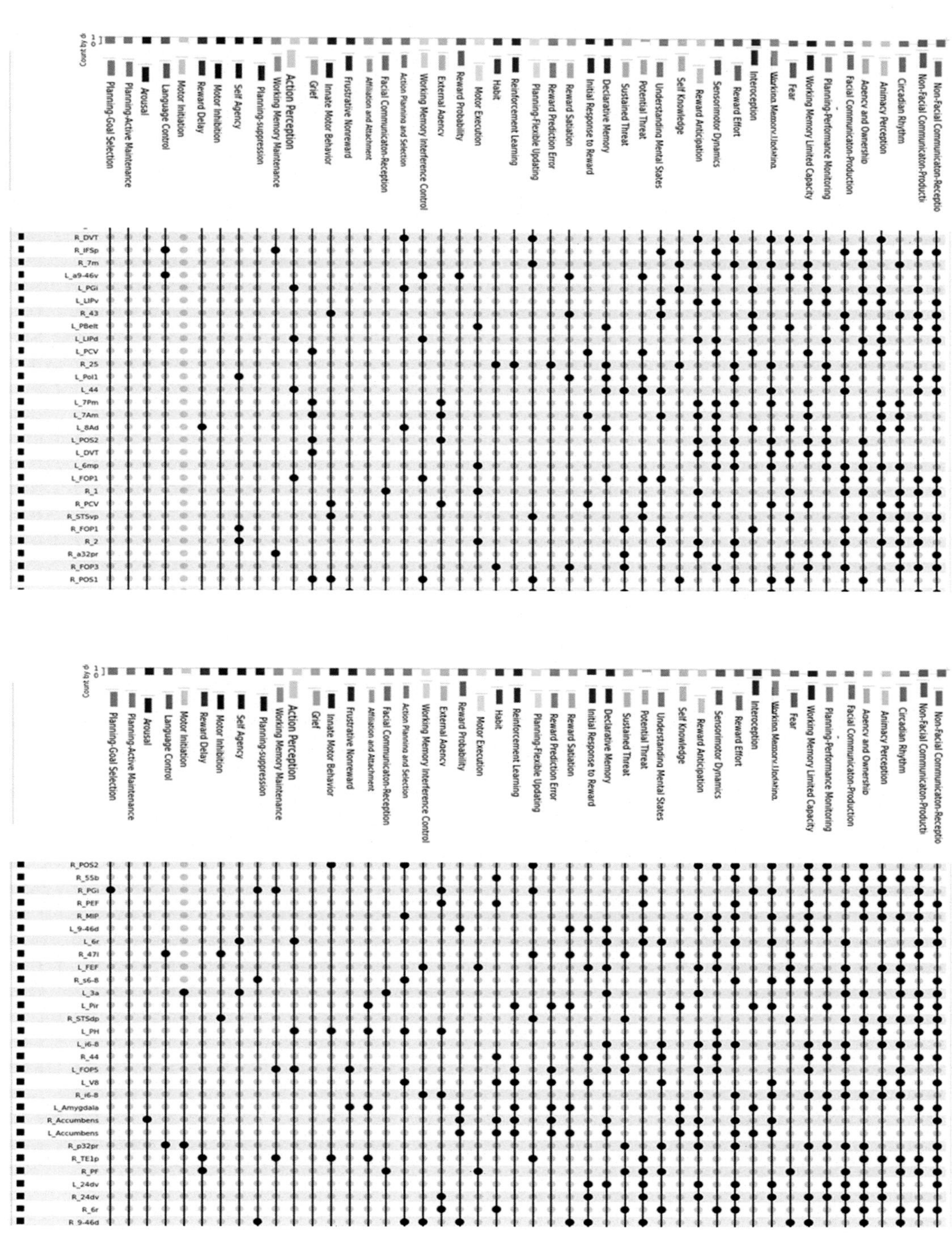

FIGURE 5.2 Cont'd

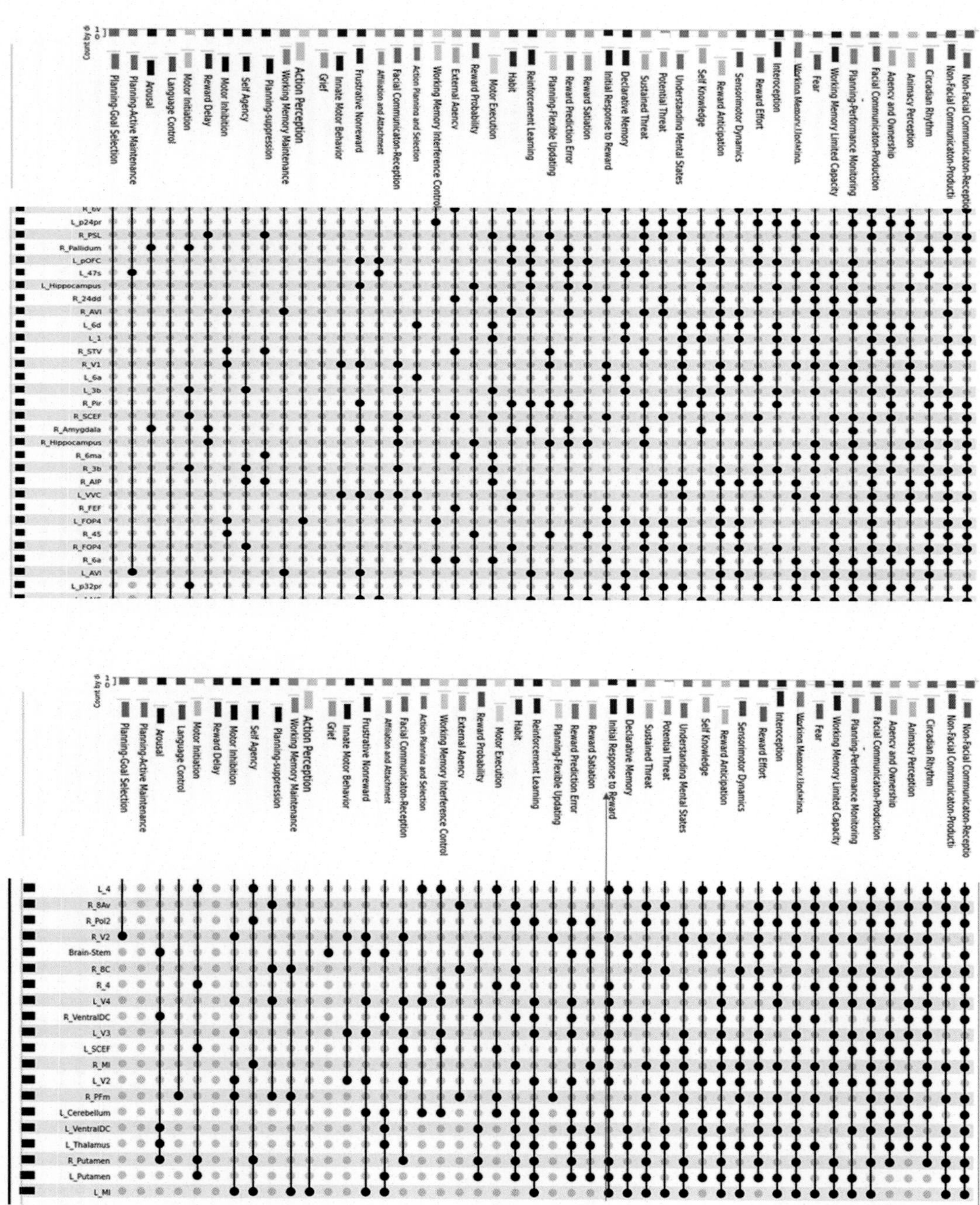

FIGURE 5.2 Cont'd

in some way or another, for example, psychopharmacologic therapies change the between area gain which changes the landscape.

However, the most important idea to come out of this science is the idea that while diseases like anxiety, depression, and schizophrenia have diverse origins, if they change the energetic landscape, they change how circuits fire, and as a result, cause decreased function, loss of function, or overactive function, and thus the symptom. Given the complexity of wiring, and the complex relationship between genes, life experiences, and neuronal wiring, it is unrealistic to expect that energetic landscapes will fit inside of easy diagnostic boxes. Some circuit problems will be common given similarities in human connectivity, but there should be great deals of heterogeneity which has long been noted clinically.

It differs from traditional Diagnostic and Statistical Manual of Mental Disorders (DSM) framework that requires us to "force-fit" a patient's symptoms into categories and classifications that are not based on biophysiological principles. We once again use MDD as an example of a disease that can have variations in presentations and have various biotypes. It is a highly heterogenous disorder that is associated with multiple other psychiatric disorders and have major crossover in symptomology with other separate diagnoses, such as bipolar disorder. MDD diagnosis can be broken down further by multiple modifiers based on comorbidities of phobias, panic, substance use disorders, posttraumatic stress disorder, personality disorder, schizotypal, borderline, antisocial, and generalized anxiety disorder (GAD) (Hasin et al., 2018). Resting state functional MRI studies on patients with anxiety disorders demonstrated similar connectivity abnormalities in DMN and the salience network as those found in depression (Gupta et al., 2021). The REST-meta-MDD Project found that there are two subtypes of MDD with differing functional connectivity profiles (Liang et al., 2020). Other studies found four biotypes of MDD (Drysdale et al., 2017; Fu et al., 2019).

The RDoc framework

The heterogeneity problem and the need for transdiagnostic frameworks which could meaningfully tackle this issue prompted the NIMH to organize a framework called the Research Domain Criteria (RDoC) (https://www.nimh.nih.gov/research/research-funded-by-nimh/rdoc). RDoc aims at guiding research toward studying illnesses in terms of normal brain mechanisms which are dysfunctioning, as opposed to basing research around symptoms, which are often poorly defined, hard to reproduce, and subjective (Insel et al., 2010).

Despite the word "research" in the acronym, we suggest that RDoc provides a useful way to think about symptoms in terms of brain mechanisms gone wrong. This provides a powerful way to tie a dysfunction to a set of brain regions which are involved in that function, and grounds mental illness in anatomy and physiology in a way not previously possible. Thus, while it is currently not possible to entirely predict the result of a treatment on an energetic landscape of the brain, it is possible to know where in the brain we would expect to matter for this function, and thus likely know where the psychopathology is likely occurring, even if we do not totally know the proximal cause of why the functional connectivity is off.

RDoc views mental and emotional function as a set of six main domains: negative valence (systems which avoid unpleasant or dangerous experiences), positive valence (the reward system), the cognitive system (the decision making, sensory, and memory systems), the social process system (systems for understanding and communicating with others, and understanding relationships between actions and consequences), the arousal system (sleep-wake, alertness, and circadian rhythms), and the sensorimotor system (executing planned and overlearned motor behaviors). Each system has a number of subdomains which we cover below, and which are the primary unit of functional analysis.

In such a framework, it is possible to often explain symptoms in terms of mechanisms in RDoc, and because we have combed the literature and linked the functions of the RDoc domains to specific places in the connectome (and even describe novel networks such as in Dadario and Sughrue (2023)), it is worth spending the following pages articulating this system in detail, and demonstrating a symptom to circuit mapping. This brings reality to this complex system and our hope is it makes it useful clinically.

Negative valence domains

Acute threat (fear)

The acute threat system causes a physiologic response to threats. Dysfunction of this system is fundamental to phobias, panic attacks, and aspects of PTSD among other diseases.

Potential threat (anxiety)

The potential threat system scans the environment for potential threats. Overactivity of this system is fundamental to anxiety disorders, as it involves the threat response occurring in otherwise benign situations.

Sustained threat

Sustained threat refers to processes which are activated by long-standing psychological trauma like abuse, and overactivity of this system is a risk factor for suicide.

Loss (grief)

Loss is a system activated when something someone is accustomed to is removed from their life. It is commonly an exacerbating factor in numerous mental illnesses, and it is a known suicide risk factor.

Frustrative nonreward

Frustrative nonreward refers to irritation and anger when one does not achieve a specific expected outcome. Problems with this system are common in ADHD, conduct disorder, antisocial personality disorder, among other problems.

Positive valence domains

Reward responsiveness systems

Reward anticipation

Reward anticipation refers to the process of looking forward to receiving a reward. It is underactive in many diseases, notably depression, where it leads to negative thinking. It likely contributes to mania in that it causes a feeling of imminent reward in otherwise meaningless situations.

Reward response

Reward response is abnormal in anhedonic patients as they do not feel good when they receive a reward.

Reward satiation

This system provides a signal that someone has had enough reward. Underactivity of the reward satiation system is fundamental to eating disorders like bulimia and overeating.

Reward learning systems

Reward prediction error

Reward prediction error refers to the system that adjusts future expectations when an expected reward is not received (or vice versa). It is abnormal in many diseases, but most strongly in schizophrenia where it causes the well-known failure of these patients to change rule sets in the Wisconsin cart sort task.

Habit

Habit in the reward system (note there is a motor habit subdomain later) refers to strongly engrained behavior patterns generated by past experiences. Pathologic habits are common to many diseases, most notably compulsive disorders like OCD.

Reward valuation systems

Reward probability

Reward probability refers to the cognitive process of evaluating the likelihood that something will result from one's actions. Negativity in diseases like depression result from failure of this system. Mania also can result, in part, from inappropriate overestimates of reward in certain context.

Reward effort

Reward effort refers to the cognitive systems which estimate how difficult a reward will be to obtain and makes a judgment call whether this is worthwhile. Depression shows obvious problems with motivation related to this system.

Reward delay

Reward delay refers to the cognitive ability to estimate how long a reward will take to be realized, and to make a judgment of whether this is valuable and worth pursuing. Failure of the reward delay system is common in impulsive behavior in bipolar disorder, bulimia, various personality disorders, schizophrenia, and possibly ADHD. Overactive reward delay can also be problematic in diseases like anorexia nervosa, where patients can choose severe deprivation in the short term for the perception of future benefits in the long term.

Cognitive system domains

Attention

Attention in this context, refers to the ability to maintain focus on a specific object, conversation, or task without unintentionally changing focus to a distractor. It is disturbed across a wide spectrum of diseases, most famously ADHD.

Perception

Visual

Visual abnormalities in mental illnesses most dramatically present as visual hallucinations in schizophrenia spectrum disorders. Oversensitivity to visual stimuli or light is also common in many diseases.

Auditory

Auditory hallucinations are linked to problems with the auditory system. Also, sensory oversensitivity is also common with many diseases.

Enteroception

Enteroception refers to the sensation of the inner organs. A wide spectrum of mental disorders cause somatization disorders of various types, and this is presumably due to problems with these circuits of some type.

Declarative memory

Declarative memory refers to the ability to recall people, places, and events. It is most prominently disturbed in degenerative diseases, but dysfunction can also be a part of mental illnesses.

Language

Language refers to a broad spectrum of cognitive abilities related to using the ability of speech in productive ways, specifically syntax, appropriate sematic content, etc. A broad variety of subdisorders in this group are well known. Note that speech is distinct from motor in this definition, and refers to a specific palatal-lingual-labial-facial motor abilities needed to produce meaningful speech, and it can be disturbed in combination with, or separately from speech.

Cognitive control

Goal selection

The goal selection system chooses between specific goals and is key to making decisions. Problems with this system lie at the root of compulsions, as the system is directed at selecting inappropriate goals. Additionally, indecisiveness in various diseases also likely results from problems with the goal selection system.

Goal updating

Goal updating allows someone to change goals in response to changing situations. The well-known failure of set switching of schizophrenics on the Wisconsin Card sort test is a good example of the goal rigidity caused when this system fails.

Goal maintenance

The goal maintenance system helps maintain a plan or goal for a period of time. As such, it is critical to accomplishing goals, and failure of maintenance is common in ADHD. Additionally, problems with this system play a role in the cognitive disorganization seen in schizophrenia.

Goal inhibition

This system is critical for behavioral and impulse control, and failure of this system causes impulsiveness or poor decision making.

Performance monitoring

The performance monitoring system assesses how closely one's performance matches intended goals and assesses the quality of performance. Problems with this system are central to cognitive disorganization and poor performance seen in schizophrenia and ADHD. As well, poor insight also results in part from a failure of performance monitoring.

Working memory

Active maintenance

The active maintenance memory system helps to keep ideas and/or facts in working memory. Problems with this system cause forgetfulness, lack of concentration, and poor working memory performance. In addition, these problems are common in learning disabilities.

Flexible updating

Flexible updating refers to the ability of the working memory system to eliminate a previously held memory and move on to holding new information in working memory. One well-known example is cognitive rigidity and cognitive dysfunction in schizophrenia.

Limited capacity

Working memory has a fixed capacity, as evidenced by the well-known seven-digit span of normal adults. This function seems to be actively managed and problems with this process can reduce working memory capacity. Additionally, this system can be disturbed in learning disabilities.

Interference control

The interference control system prevents competing thoughts and ideas from pushing an actively maintained memory out of the working memory system. This causes distractibility and decreased ability to multitask. These issues are common in disease processes like ADHD.

Social process system domains

Affiliation and attachment

This system helps to form relationships, bonds, and feelings of attachment to other people. This system is abnormal in a variety of diseases, notably autism and schizophrenia.

Social communication

Facial communication: production and reception

Production of facial communication and reception of facial communication can be disturbed in numerous diseases notably autism spectrum diseases and schizophrenia type diseases.

Nonfacial communication: production

This term is better viewed as a collection of things like body language, prosody, phrasing sentences in an appropriate way, and correctly conveying understood meaning behind communication. This is obviously a problem for people on the higher functioning end of the autism spectrum.

Nonfacial communication: reception

Failure to appropriately interpret social context and the flow of conversation leads to unintentionally saying things that anger others, interrupting others, taking an argument too far, and other problematic behaviors.

Perception and understanding of self

Agency

The agency system makes someone aware of being in control of themselves and their actions. Failures of this system can lead to delusions of being controlled by an outside force, or other delusions.

Self-knowledge

Self-knowledge systems help in metacognitive tasks such as being aware of your own thoughts and behavior. Overactivity of this system in depression is thought to lead to rumination on negative thoughts. Failure of this system leads to problems with insight and judgment.

Perception and understanding of others

Animacy perception

Animacy perception refers to the ability to detect that objects in the environment are independent actors and living beings. More profound disturbances of this system in autism can lead to problems with interacting with other people.

Action perception

This system helps individuals link cause and effect in the surrounding environment and to determine the source of an action. Problems with this domain in schizophrenia and other mental illnesses can lead to problems with insight.

Understanding mental states

This system specifically refers to the ability of people to imply intentions of others, and to predict the emotions and opinions of others in reaction to events including their own behavior. This is obviously an issue for patients with autistic spectrum disorder, as well as diseases like schizophrenia.

Alertness system domains

Arousal

Difficulties with arousal are common, particularly in depression and neurological diseases. This system does not seem to result as much from a focal set of regions as from a global brain rhythm and the emergent properties therein, though admittedly there are likely regions more important than others in this process though we have not mentioned specifically.

Sleep-wake

Problems with the sleep-wake cycle lie at the root of numerous mental illnesses both as a symptom and an exacerbating factor. Again, this does not seem to be strictly locatable to specific brain regions as other factors in the transdiagnostic approach.

Circadian rhythms

Circadian rhythm disturbances can best be differentiated from difficulties with sleep wake cycles by the ability to keep a regular timing of sleep. Certain structures in the brain seem to be specifically involved in this process, making it more focal than other aspects of wakefulness and arousal.

Sensorimotor system domains

Motor actions

Action planning and selection

Action planning is the cognitive control of motor function and determines motor sequences needed to achieve an action. Problems with this system are related to forms of apraxia.

Sensorimotor dynamics

Problems with sensorimotor dynamics tend to end in coordination problems, which sometimes are congenital and not related to an obvious neurologic disorder.

Initiation

Problems of the motor initiation system can cause apathy in diseases like depression, Parkinson disease, and schizophrenia.

Execution

Dysfunction of the motor execution system is thought to be at the heart of psychomotor retardation seen in diseases like major depression.

Inhibition

Problems with motor inhibition can lead to such problems in schizophrenia as perseveration, catatonic rituals, automatic obedience where people feel compelled to follow instructions without question.

Agency and ownership

Agency and ownership refers to the ability to recognize that one is in control of their own body parts. In addition to well-known neuropsychologic deficits like alien hand or somatoparaphrenia, patients with borderline personality often suffer from similar problems.

Sensorimotor habits

Sensorimotor habits refer to overlearned behaviors. Problems with this system are linked to behaviors like tics.

Innate motor patterns

Innate motor patterns work at a somewhat lower level than habits, and when this system is abnormal, behaviors like stereotypical behavior of autistic spectrum disorders, and explosive inappropriate effects can result.

Conclusion

In this chapter, we introduced the transdiagnostic hypothesis and how specific symptoms can be linked to disruption or dysfunction within specific circuits, regardless of the mechanistic reason for this abnormality. Further, we introduce how the RDoC can provide powerful way to tie a dysfunction to a set of brain regions which are involved in that function, and grounds mental illness in anatomy and physiology in a way not previously possible. In the next chapter, we introduce the idea of considering neurocognitive functions as emergent phenomena.

References

Dadario, N. B., & Sughrue, M. E. (2022). Should neurosurgeons try to preserve non-traditional brain networks? A systematic review of the neuroscientific evidence. *Journal of Personalized Medicine, 12*(4), 587.

Dadario, N. B., & Sughrue, M. E. (2023). The functional role of the precuneus. *Brain: A Journal of Neurology, 146*(9), 3598–3607. https://doi.org/10.1093/brain/awad181

Drysdale, A. T., Grosenick, L., Downar, J., Dunlop, K., Mansouri, F., Meng, Y., Fetcho, R. N., Zebley, B., Oathes, D. J., Etkin, A., Schatzberg, A. F., Sudheimer, K., Keller, J., Mayberg, H. S., Gunning, F. M., Alexopoulos, G. S., Fox, M. D., Pascual-Leone, A., Voss, H. U., … Liston, C. (2017). Resting-state connectivity biomarkers define neurophysiological subtypes of depression. *Nature Medicine, 23*(1), 28–38. https://doi.org/10.1038/nm.4246

Fu, C. H. Y., Fan, Y., & Davatzikos, C. (2019). Addressing heterogeneity (and homogeneity) in treatment mechanisms in depression and the potential to develop diagnostic and predictive biomarkers. *Neuroimage Clinical, 24*, 101997. https://doi.org/10.1016/j.nicl.2019.101997

Gu, S., Pasqualetti, F., Cieslak, M., Telesford, Q. K., Yu, A. B., Kahn, A. E., Medaglia, J. D., Vettel, J. M., Miller, M. B., Grafton, S. T., & Bassett, D. S. (2015). Controllability of structural brain networks. *Nature Communications, 6*(1), 8414. https://doi.org/10.1038/ncomms9414

Gupta, A., Wolff, A., & Northoff, D. G. (2021). Extending the "resting state hypothesis of depression" — dynamics and topography of abnormal rest-task modulation. *Psychiatry Research: Neuroimaging, 317*, 111367. https://doi.org/10.1016/j.pscychresns.2021.111367

Hasin, D. S., Sarvet, A. L., Meyers, J. L., Saha, T. D., Ruan, W. J., Stohl, M., & Grant, B. F. (2018). Epidemiology of adult DSM-5 major depressive disorder and its specifiers in the United States. *JAMA Psychiatry, 75*(4), 336–346. https://doi.org/10.1001/jamapsychiatry.2017.4602

Hormovas, J., Dadario, N. B., Tang, S. J., Nicholas, P., Dhanaraj, V., Young, I., Doyen, S., & Sughrue, M. E. (2023). Parcellation-based connectivity model of the judgement core. *Journal of Personalized Medicine, 13*(9), 1384. https://doi.org/10.3390/jpm13091384

Insel, T., Cuthbert, B., Garvey, M., Heinssen, R., Pine, D. S., Quinn, K., Sanislow, C., & Wang, P. (2010). Research domain criteria (RDoC): Toward a new classification framework for research on mental disorders. *The American Journal of Psychiatry, 167*(7), 748–751. https://doi.org/10.1176/appi.ajp.2010.09091379

Liang, S., Deng, W., Li, X., Greenshaw, A. J., Wang, Q., Li, M., Ma, X., Bai, T. J., Bo, Q. J., Cao, J., Chen, G. M., Chen, W., Cheng, C., Cheng, Y. Q., Cui, X. L., Duan, J., Fang, Y. R., Gong, Q. Y., Guo, W. B., … Li, T. (2020). Biotypes of major depressive disorder: Neuroimaging evidence from resting-state default mode network patterns. *NeuroImage: Clinical, 28*, 102514. https://doi.org/10.1016/j.nicl.2020.102514

Medaglia, J. D., Pasqualetti, F., Hamilton, R. H., Thompson-Schill, S. L., & Bassett, D. S. (2017). Brain and cognitive reserve: Translation via network control theory. *Neuroscience and Biobehavioral Reviews, 75*, 53–64. https://doi.org/10.1016/j.neubiorev.2017.01.016

Menon, V. (2011). Large-scale brain networks and psychopathology: A unifying triple network model. *Trends in Cognitive Sciences, 15*(10), 483–506. https://doi.org/10.1016/j.tics.2011.08.003

Sherman, L. E., Rudie, J. D., Pfeifer, J. H., Masten, C. L., McNealy, K., & Dapretto, M. (2014). Development of the default mode and central executive networks across early adolescence: A longitudinal study. *Developmental Cognitive Neuroscience, 10*, 148–159. https://doi.org/10.1016/j.dcn.2014.08.002

Sporns, O. (2013). Structure and function of complex brain networks. *Dialogues in Clinical Neuroscience, 15*(3), 247–262. https://doi.org/10.31887/DCNS.2013.15.3/osporns

Uddin, L. Q., Supekar, K., Lynch, C. J., Khouzam, A., Phillips, J., Feinstein, C., Ryali, S., & Menon, V. (2013). Salience network-based classification and prediction of symptom severity in children with autism. *JAMA Psychiatry, 70*(8), 869–879. https://doi.org/10.1001/jamapsychiatry.2013.104

Young, I. M., Dadario, N. B., Tanglay, O., Chen, E., Cook, B., Taylor, H. M., Crawford, L., Yeung, J. T., Nicholas, P. J., Doyen, S., & Sughrue, M. E. (2023). Connectivity model of the anatomic substrates and network abnormalities in major depressive disorder: A coordinate meta-analysis of resting-state functional connectivity. *Journal of Affective Disorders Reports*, 100478. https://doi.org/10.1016/j.jadr.2023.100478

Further reading

A most remarkable case. (1851). *American Phrenological Journal and Repository of Science, Literature, and General Intelligence, 13*(4), 89. col. 3.

CHAPTER

6 Reimagining neurocognitive functions as emergent phenomena: What resting state is really showing us

Introduction

From the earlier chapters' description of brain anatomy (Chapter 4), the careful reader will find the previous Chapter 5 a bit confusing. If networks are the fundamental building blocks of brain function, why do most functions not really map into a single network very well. For example, none of the RDoC functions are a single network. Complex behaviors like math and judgment are not either. Also, many parts of the brain demonstrate functional connectivity patterns which show they do not fit well into a single network, and may facilitate cross-talk between them. So are networks important, or are they not?

It turns out that while not immediately clinically accessible, the explanations behind these issues are both intellectually rewarding, as well as full of profound implications for future directions in determining new therapeutic approaches.

What do we mean by the term network dynamics?

The network concept arises ultimately from a statistical perspective on firing patterns of various cortical and subcortical regions during various paradigms which measure brain activation, such as EEG or fMRI BOLD (Haufe et al., 2018; Portnova et al., 2018). People imaged in the resting state have been repeatedly found to have specific patterns of coactivation between different brain regions which are very reproducible. For example, the primary motor and sensory cortices show high degree of correlation of firing patterns with each other, as do the visual cortices in the occipital lobe, or parts of the default mode network, etc. In other words, it is implied that brain areas with highly correlated activity patterns are very likely to be intercommunicating and participating in similar functions, a fact that has been confirmed by many different ways (Ip et al., 2019; Moon et al., 2007; Zhang et al., 2020).

Having said this, no network has regions that are 100% correlated with each other. It is hard to see how a network which only communicated internally would have any impact on brain function. Thus, parts of the sensorimotor or default mode networks, sometimes are firing with components of other networks. This is how cognitive processes likely arise, as an emergent phenomenon of the sum of these interactions.

Connectomic Medicine. https://doi.org/10.1016/B978-0-443-19089-6.00008-2

Emergent phenomena refer to when varying elementary elements, such as individual neurons or cortical regions, collectively produce a complex function that cannot be 100% explained by their individual counterparts alone. Emergent phenomena are located all around us in the human world, ranging from natural disasters to the New York stock market (Turkheimer et al., 2019). A useful analogy again is an orchestra. Sections like the violins, the woodwinds, the percussion, or the brass have similar instruments that commonly play the same or similar tunes, but together when a violin and flute part align for a short section, the orchestra produces a sound that is greater than the sum of the parts. Violins are all similar to each other, as are flutes. Elimination of the violin section would make some functions impossible, others different, but sound would still go on. This is similar to the brain, if you think of each section as a network. Networks are the building blocks of cognition, and play a key role in cognitive and emotional function, but they are not all of the story.

Thus, when we think of how parts of the brain talk both inside their core network, and occasionally form brief collaborations with other regions, we can imagine the brain transiting through numerous intermediate states of what areas are talking or not talking to each other at a given time point. The collective trajectory of these states over time creates the total sum of all episodes in our conscious experience (Fig. 6.1).

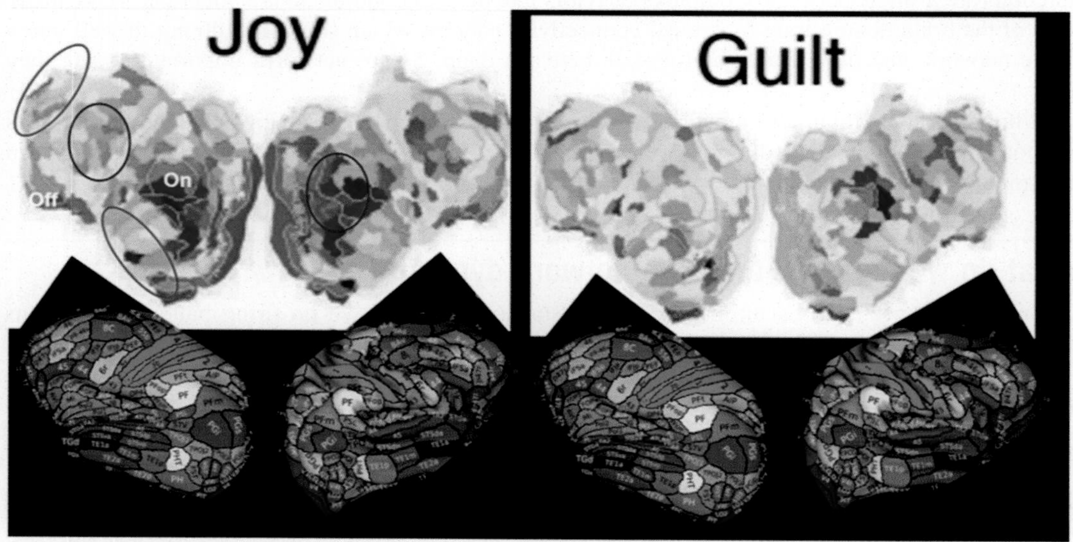

FIGURE 6.1 Brain States.

These two images demonstrate flat maps of fMRI studies showing subjects experiencing two emotions. Parcellations from the HCP are provided between in similar position for comparison. Regions while experiencing this emotion are indicated in red and regions off are indicated in blue. Note that while the DMN, CEN, and Salience networks are involved in both the emotional states, the exact set of what is on and off in these states differs, a fact that has been demonstrated for numerous other emotions.

Adapted from Horikawa et al., (2020). The neural representation of emotion is high-dimensional, categorical, and distributed across transmodal brain regions.

Where do network dynamics come from?

Network dynamics result from two basic processes: internal state dynamics and the effect of external stimuli on perturbing the internal dynamics. While most readers likely have some concept of the effects of external stimulation on the brain, internal dynamics likely are less obvious, and probably play a greater role in diseased thought, as even the effect of external stimuli is processed through the internal brain dynamic state. Thus, to build a brain model which is understandable, we will outline what is the best current thinking about how these dynamics occur, in part by building this model in steps.

The structural connectome

The first component driving dynamics is the structural connectome, the sum of all wires in the brain. This wiring pattern is complex, with multiple parallel pathways, redundant intercommunication, and many areas interconnected to a large number of other areas. While all of these things make visual inspection inadequate for understanding the structural connectome, a few things have been found with mathematical approaches. First, while areas of the brain connect to many other areas, every part of the brain is not connected to every other. Second, some areas, usually called *hubs*, are more widely interconnected than others (Sporns et al., 2007; Yeung et al., 2021). Third, the wiring pattern is not random, but instead follows a "small world" type pattern, where most of the brain is mainly connected to its neighboring areas, but some areas are connected more distantly (Watts & Strogatz, 1998). These balances not only seem to minimize energy costs, but also seem necessary for producing the complexity of brain dynamics seen in the actual brain (Bullmore & Sporns, 2012). For example, overly connected or randomly connected networks seem to produce dynamic patterns incompatible with conscious thought, like synchronized oscillation.

A key consequence of these wiring traits is that some paths in the brain are shorter than others, and thus areas of the brain are more likely to strongly couple with some areas as opposed to others. This has obvious implications for what brain states are likely to occur and which ones are not. It also provides some link between genetic changes which affect axon stability, synapse formation, or other similar processes, which appear involved in some inherited brain diseases, and specific symptoms which result from these genes.

Subcortical and other nonelicited rhythmic dynamics

Structures like the thalamus and basal ganglia are well known to generate pulsatile rhythms of varying frequencies (Cannon et al., 2014). When these rhythms are projected through the cortical connectome, a highly complex set of interactions occurs which causes some areas to cooscillate and others to inhibit or coinhibit each other. The sum of all these interactions, interacting across the entire set of the structural connectome, creates the patterns of functional connectivity seen in resting state brain activity, and has likely implications for how brain functions occur or do not occur, and which brain states are possible, impossible, or unlikely in a given brain.

Computational modeling of connectome dynamics from models based on structural connectivity have shown that changes in the overall frequency of rhythmic dynamics changes the functional connectivity of the brain significantly, namely changing which parts of the brain cooscillate (Preti et al., 2017).

Neurotransmitters and other things that change the gain

It is well known that not all neural firing causes an action potential in the target: many neurons serve to modulate the excitability of the target area. In other words, they change the gain of the target area, making inputs to that area more or less likely to cause activity in that area. Taken over the totality of the connectome, changing the gain in a few places in the brain would be expected to have an impact on what area is firing with what, and thus change how the overall brain functions cognitively, neurologically, and emotionally on the global scale.

Neurotransmitters in the well-known ascending serotonin, dopamine, acetylcholine, and norepinephrine systems have long been linked to cognitive and emotional function, but recent evidence suggests that they have these effects by changing the gain of their target areas of cortex (Ferguson & Cardin, 2020). While they are widely distributed, they are not universally synapsing around the brain, and thus they can have complex effects on the functional connectome (Leiser et al., 2015). We could hypothesize that some genetic changes might also exert an effect at the gain modulation level in their link between genotype and phenotype.

Network control theory

The above discussion should make it clear that structural wiring patterns, global rhythms, and gain play a key role in the internal dynamics of what the brain is doing. More importantly, they cumulatively dictate what brain states are easy to obtain, and which are difficult. This has been termed the energetic landscape, which refers to the complexity of getting the brain to a given state given its intrinsic structure and internal dynamics.

Brilliant work by Danielle Bassett has elucidated a likely mechanism by which energy landscapes arise from structural connectomes, and how this dictates spontaneous behavioral patterns and tendencies in normal and pathologic states (Gu et al., 2015; Medaglia et al., 2017). Fig. 6.2 highlights the concept of state changes and how structure leads to favorability or unfavorability of a transition between two states. These changes can be induced by external stimuli, or from internal processes. The latter method of moving the brain into energetically difficult states seems to be a key function of the central executive network, which interestingly matures primarily in adolescence and early adulthood when mental illnesses begin to appear more commonly.

A different structural connectome, set of gains or other factors, changes the landscape making some brain states more or less easy to reach. If undesirable states become easy to reach, and desirable states become difficult, spontaneous and/or context-specific cognitive or emotional responses deviate from expected, normal function (Fig. 6.3). This deep theory provides a possible explanation about why genetic diseases like schizophrenia fail to present until adolescence: an underdeveloped CEN does not make full realization of the landscape possible, and thus pathologic brain states, like delusions, while wired to happen, cannot be reached until later in life.

What does this tell us about the clinical utility of resting state fMRI?

Abnormal patterns of functional connectivity are a part of mental illness and other diseases. While every observed pattern has not been completely characterized, it is really beyond doubt that abnormal correlations occur in sick people that do not manifest frequently in normal people. Thus, a resting-state

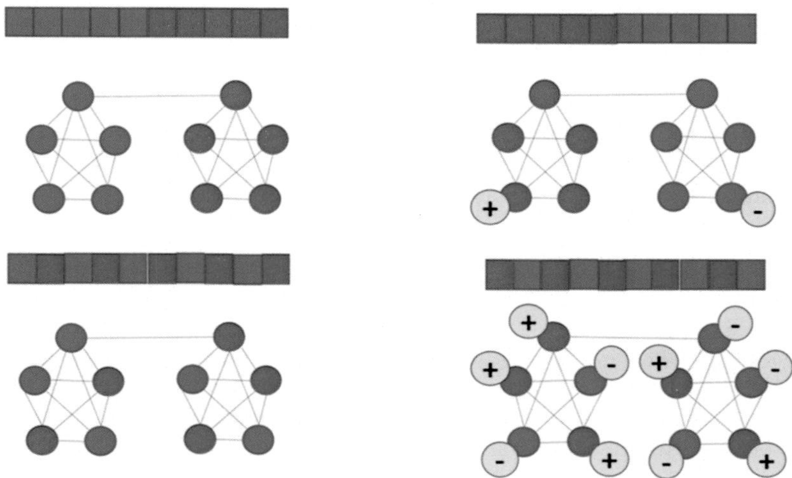

FIGURE 6.2 Energy Demands for Brain State Transitions.

These schematic diagrams show conceptually how differences in network wiring can affect the energetic demands of different brain state transitions. Note that in the lower diagram, more inputs in more places are necessary to put the network into the changed state than in the upper diagram.

Adapted from talk by Danielle Bassett, Cognitive Neuroscience Society meeting, 2020.

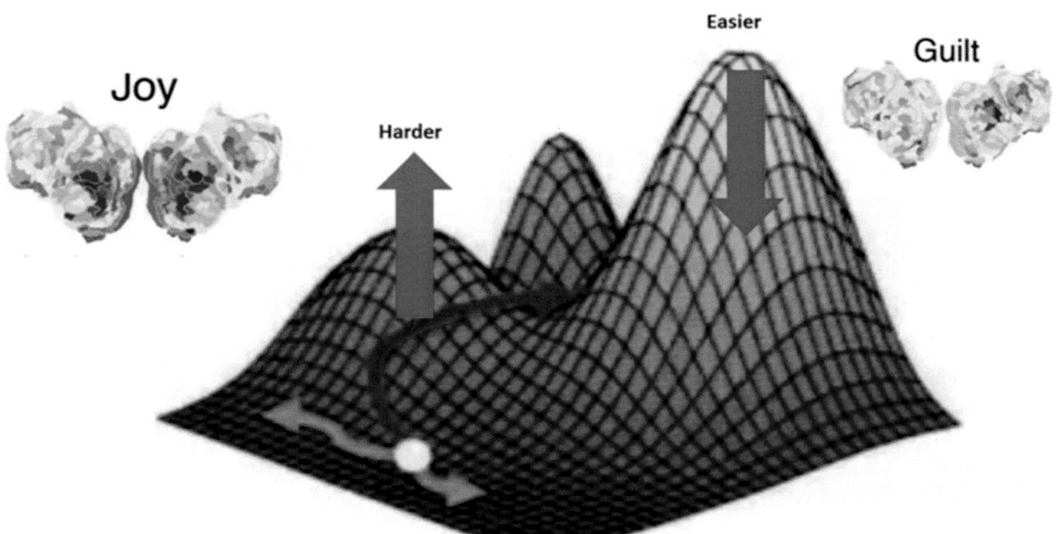

FIGURE 6.3 Energy Landscapes and How They Relate to Pathologic Brain States.

(A) We can estimate the energy required to put the brain into every possible brain state from a structural connectome. Higher peaks in this diagram indicate brain states in that person which are difficult to obtain compared to lower peaks. As such we would expect higher peaks to happen less often or not be possible at all, while lower points in the landscape should occur more frequently or spontaneously occur, (B) a specific structural connectivity pattern might lead to a different landscape where certain bad brain states might become easier, or good states harder to get the brain into. Given this plausible path toward how mental and cognitive changes occur, it seems possible that manipulating this landscape for therapeutic benefit might be a reasonable approach.

scan provides a window into what patterns of connectivity are emerging from the complex dynamics of brain function, which is quite plausibly related to the brain transitioning between undesirable states or failing to pass through normal states. It seems promising that future approaches which are aimed at altering these landscapes will provide us better insight into where to target therapies so that bad brain states happen less and good states happen more often.

Conclusion

In this chapter, we introduced a way to think about neurocognitive functioning as emergent phenomenon. Specifically, these complex functions arise from the cumulative interactions of various dynamically interacting brain networks. More specifically, network dynamics, consisting of internal state dynamics and the effect of external stimuli on perturbing the internal dynamics, are important in dictating which brain states are more or less energetically favorable to achieve. Together, structural wiring patterns, global rhythms in deep structures, and electrochemical gain from neurotransmitters play a key role in the internal dynamics of what the brain is doing. In this context, if undesirable states become easy to reach, and desirable states become difficult to reach, spontaneous and/or context-specific cognitive or emotional responses deviate from expected, normal function. This energetic landscape is strongly related to underlying structural connections, and these structural connections are also important in determining overall efficiency of a network and its brain functional connectivity at rest, thus allowing resting-state fMRI to provide key insights into these processes in healthy and pathologic states. In the next chapter, we focus on both invasive and noninvasive techniques that allow us to modulate brain activity.

References

Bullmore, E., & Sporns, O. (2012). The economy of brain network organization. *Nature Reviews Neuroscience, 13*(5), 336–349. https://doi.org/10.1038/nrn3214

Cannon, J., McCarthy, M. M., Lee, S., Lee, J., Börgers, C., Whittington, M. A., et al. (2014). Neurosystems: Brain rhythms and cognitive processing. *European Journal of Neuroscience, 39*(5), 705–719. https://doi.org/10.1111/ejn.12453

Ferguson, K. A., & Cardin, J. A. (2020). Mechanisms underlying gain modulation in the cortex. *Nature Reviews Neuroscience, 21*(2), 80–92. https://doi.org/10.1038/s41583-019-0253-y

Gu, S., Pasqualetti, F., Cieslak, M., Telesford, Q. K., Yu, A. B., Kahn, A. E., et al. (2015). Controllability of structural brain networks. *Nature Communications, 6*(1), 8414. https://doi.org/10.1038/ncomms9414

Haufe, S., DeGuzman, P., Henin, S., Arcaro, M., Honey, C. J., Hasson, U., et al. (2018). Elucidating relations between fMRI, ECoG, and EEG through a common natural stimulus. *NeuroImage, 179*, 79–91. https://doi.org/10.1016/j.neuroimage.2018.06.016

Ip, I. B., Emir, U. E., Parker, A. J., Campbell, J., & Bridge, H. (2019). Comparison of neurochemical and BOLD signal contrast response functions in the human visual cortex. *Journal of Neuroscience, 39*(40), 7968–7975. https://doi.org/10.1523/jneurosci.3021-18.2019

Leiser, S. C., Li, Y., Pehrson, A. L., Dale, E., Smagin, G., & Sanchez, C. (2015). Serotonergic regulation of prefrontal cortical circuitries involved in cognitive processing: A review of individual 5-HT receptor mechanisms and concerted effects of 5-HT receptors exemplified by the multimodal antidepressant vortioxetine. *ACS Chemical Neuroscience, 6*(7), 970–986. https://doi.org/10.1021/cn500340j

Medaglia, J. D., Pasqualetti, F., Hamilton, R. H., Thompson-Schill, S. L., & Bassett, D. S. (2017). Brain and cognitive reserve: Translation via network control theory. *Neuroscience & Biobehavioral Reviews, 75*, 53–64. https://doi.org/10.1016/j.neubiorev.2017.01.016

Moon, C.-H., Fukuda, M., Park, S.-H., & Kim, S.-G. (2007). Neural interpretation of blood oxygenation level-dependent fMRI maps at submillimeter columnar resolution. *Journal of Neuroscience, 27*(26), 6892–6902. https://doi.org/10.1523/jneurosci.0445-07.2007

Portnova, G. V., Tetereva, A., Balaev, V., Atanov, M., Skiteva, L., Ushakov, V., et al. (2018). Correlation of BOLD signal with linear and nonlinear patterns of EEG in resting state EEG-informed fMRI. *Frontiers in Human Neuroscience, 11*. https://doi.org/10.3389/fnhum.2017.00654

Preti, M. G., Bolton, T. A. W., & Van De Ville, D. (2017). The dynamic functional connectome: State-of-the-art and perspectives. *NeuroImage, 160*, 41–54. https://doi.org/10.1016/j.neuroimage.2016.12.061

Sporns, O., Honey, C. J., & Kotter, R. (2007). Identification and classification of hubs in brain networks. *PLoS One, 2*(10), e1049. https://doi.org/10.1371/journal.pone.0001049

Turkheimer, F. E., Hellyer, P., Kehagia, A. A., Expert, P., Lord, L.-D., Vohryzek, J., et al. (2019). Conflicting emergences. Weak vs. strong emergence for the modelling of brain function. *Neuroscience & Biobehavioral Reviews, 99*, 3–10. https://doi.org/10.1016/j.neubiorev.2019.01.023

Watts, D. J., & Strogatz, S. H. (1998). Collective dynamics of 'small-world'networks. *Nature, 393*(6684), 440–442.

Yeung, J. T., Taylor, H. M., Young, I. M., Nicholas, P. J., Doyen, S., & Sughrue, M. E. (2021). Unexpected hubness: A proof-of-concept study of the human connectome using pagerank centrality and implications for intracerebral neurosurgery. *Journal of Neuro-Oncology, 151*(2), 249–256. https://doi.org/10.1007/s11060-020-03659-6

Zhang, X., Pan, W.-J., & Keilholz, S. D. (2020). The relationship between BOLD and neural activity arises from temporally sparse events. *NeuroImage, 207*, 116390. https://doi.org/10.1016/j.neuroimage.2019.116390

CHAPTER

Brain stimulation techniques

7

Introduction

Stimulating the brain with some kind of electrical pulse is not new to neurosurgeons, neurologists, and psychiatrists. Direct cortical stimulation (DCS) was first demonstrated in dogs in the 19th century and has been widely used in brain surgery for many decades (Hagner, 2012). Deep brain stimulation is commonly used by neurosurgeons to treat movement disorders, such as Parkinson's disease or essential tremor, as part of a team with dedicated subspecialized neurologists. In recent years, there has been various clinical trials aiming to target deep brain structures to treat psychiatric disorders. Transcranial magnetic stimulation (TMS) was pioneered by psychiatrists, but has gained traction in many other fields.

However, we would argue that the pace at which we are integrating the expertise of the three different fields is slower than it ought to be because there is limited cross-pollination of the disciplines. We do not think that this is a product of the lack of will to collaborate, but, rather, there lacks a common platform by which we can communicate concerning neuropsychiatric diseases due to the different angles from which we approach these conditions. In addition, there is lack of understanding of the various discipline string to training of each subspecialty, thus limiting the communication and sharing of ideas. We strongly feel that this is going to change now with our ability to process large datasets of functional connectivity at an industrial scale—as long as we understand the principles behind alterations in the functional connectome as the common language among neurological and psychiatric diseases (van den Heuvel & Sporns, 2019).

In this chapter, we will focus on both invasive and noninvasive techniques that allow us to modulate brain activity. We will not be focusing on destructive brain lesioning techniques, such as radiosurgery or ablations of various types. We will briefly review the roles of deep brain stimulation, transcranial direct electrical stimulation, and do a deep dive into transcranial magnetic stimulation, which we personally think will be the most common method for brain stimulation using connectomic medicine.

Deep brain stimulation (DBS)

Deep brain stimulation is an invasive neurosurgical procedure that involves the implantation of electrodes into specific targets within the deep brain (Lozano et al., 2019). The electrodes are then connected to an adjustable and implanted battery source away from the cranium. In neurosurgery, DBS is a favorable approach as it allows for demodulation of the deep brain structure as opposed to

Connectomic Medicine. https://doi.org/10.1016/B978-0-443-19089-6.00013-6

lesioning of the deep brain structures, which was common in the beginning of functional neurosurgery. It is most commonly used in the United States for the treatment of Parkinson's disease (subthalamic nucleus or globus pallidus internus), dystonia, and essential tremor (thalamic ventralis intermedius nucleus) (Iorio-Morin et al., 2020). Other therapeutic potentials include disorders that not only affect motor functions, but limbic and other cognitive functions as well, many of which are under current clinical trial investigations. Recently, cerebral stimulation via DBS has been approved for the treatment of refractory epilepsy as well (Ellis & Stevens, 2008). The only psychiatric indication to receive FDA approval to date is severe obsessive compulsive disorder (Visser-Vandewalle et al., 2022). Other indications that are currently under investigation include major depressive disorder, tinnitus, Tourette syndrome, schizophrenia, Alzheimer's disease, chronic pain, addiction, and anorexia nervosa. However, it is still not definitive by what mechanism DBS allows for neurostimulation and the clinical correlates of this mechanism. Nonetheless, it is not the point of this book to dwell in the various hypotheses focusing on the mechanisms of DBS and how that alters brain activity. What we do know is that high frequency (~100 Hz) trains of pulses are likely inhibitory as opposed to low frequency (~10 Hz) stimulation (Farokhniaee & McIntyre, 2019).

DBS is a very established technique for neural modulation and a very familiar approach for functional neurosurgeons. Its ability to modulate essentially any deep brain structure is unique among various brain simulation strategies. However, a number of concerns remain at the current time which are worth mentioning. It is an invasive neurosurgical procedure that carries the risk of surgical morbidity, although it is minimally invasive by nature. It is highly resource-intensive as it requires a multidisciplinary team to first arrive at a consensus to choose this treatment, usually taking months if not years, and the surgical portion of this treatment strategy incurs large capital costs, which bleeds into subsequent cost for device programming. Device programming is time-consuming and labor-intensive. It is usually a trial-and-error experience. While it may be easy to program a device to titrate to motor symptoms, as in the case of Parkinson's disease, it is likely difficult in the cases of psychiatric diseases where patients experience episode events and will prove to be difficult for symptom monitoring and subsequent titration. The battery implants have a limited lifetime and require subsequent replacements as well. Patients can theoretically become refractory to DBS as well if they form scar tissue around the electrodes. Lastly, with all of the hindrances mentioned above, DBS is not a readily available resource to community practices.

Transcranial direct current stimulation (tDCS)

This is a noninvasive and painless brain stimulation technique that uses direct electrical currents to stimulate specific parts of the brain. We should begin by saying that tDCS is not an approved treatment by the FDA. It is an experimental form of brain stimulation that has had research applications in depression, anxiety, Parkinson's disease, and chronic pain, among other experimental indications (Tu et al., 2022). This type of stimulation utilizes this weak electrical current, usually at low strengths (1–2 mA) applied for several minutes, passed between two electrodes attached to the scalp of the subject. There exists two methods of stimulation, positive anodal current and negative cathodal currents. It is noted that nodal current stimulation encourages communication between the cortical regions under the target electrodes by theoretically depolarizing neurons and increasing the probability of action potentials, and cathodal current does the opposite by hyperpolarizing the neurons and decreasing the likelihood of neuronal firing (Thair et al., 2017). However, it is likely that this is an oversimplification of tDCS. Nonetheless, it is without question that tDCS has the ability to modulate brain activity whether it is to modify behavior, accelerate learning, or increase memory acquisition in an experimental setting. Targeted electrode placement is traditionally performed using a 10:20 EEG system and anatomical measurement from the inion to the nasion, and from the left to the right pre-auricular areas. Target selection is dependent on the hypothesis and the task in question. If the investigative action is regarding motor outputs, then dual target electrodes may be placed on both motor cortices, identified using either transcranial magnetic stimulation or neuronavigation. Stimulation variables may include the location of the target electrodes, reference electrodes, size of the electrodes, intensity of the simulation, and duration of the stimulation. In the research setting, the behavioral effects of the subject can be accessed using an online method, during which the patient is monitored for the behavioral tasks during stimulation, or the offline method, where the participant participates in a task before and after the stimulation to allow for comparison. This method of stimulation has shown tremendous safety in the research setting. However, at this point in time, tDCS is considered completely experimental and there are no established parameters in the treatment of patients with neurocognitive disorders. The dose-response relationship is also unknown; in fact, higher current may be detrimental in experimental setting. One of the potential benefits of this treatment modality may be in its ability to be carried out using preset headsets that can be worn in an outpatient basis.

Transcranial magnetic stimulation (TMS)

Transcranial magnetic stimulation is another form of noninvasive brain stimulation technique that uses a magnetic field to generate a subsequent electrical field that acts on the neuron. It is our opinion that this simulation technique holds the most promise in the near future for noninvasive brain stimulation due to its ease of use and established track record. When the stimulation is subthreshold to the depolarization potential, it is possible to enable inhibition of the cortex. Conversely, if the stimulation is suprathreshold, then neuronal depolarization can result in neurotransmitter release and affect subsequent synaptic communications (Pelletier & Cicchetti, 2014). Studies using TMS suggest that it might have a role in promoting neuronal plasticity (Kozyrev et al., 2018). Currently, the magnetic field can be delivered by using a Figure 8 coil or an H coil. There are benefits and tradeoffs in using each coil. The Figure 8 coil is more precise and focal but has limited penetration deep into the cortex. The H coil on the other hand can reach deeper targets but has more electrical spread along the cortex such that the stimulation is less focal. An example of a Figure 8 coil is provided in Fig. 7.1.

Localization. Traditionally, TMS localization in depression studies are based on targeting of the dorsal lateral prefrontal cortex. This is estimated using probabilistic localization, also known as the 5 cm (in actuality 5.5 cm) rule. This rule utilizes prefrontal sites that are 5.5 cm anterior to the motor cortex, as determined by TMS activation. This method is fast and simple but does not take into account individual variances in anatomy and skull sizes. The other method of localization of the DLPFC is performed using the 10:20 EEG method. Using this system, the DLPFC is believed to be near the F3 electrode. The Beam F3 method is developed to estimate the location of the F3 electrode and is thought to be more accurate than the 5 cm rule. It requires the input of three measurement values, including the distance from the nasion to inion, from the left preauricular point to the right preauricular point, and

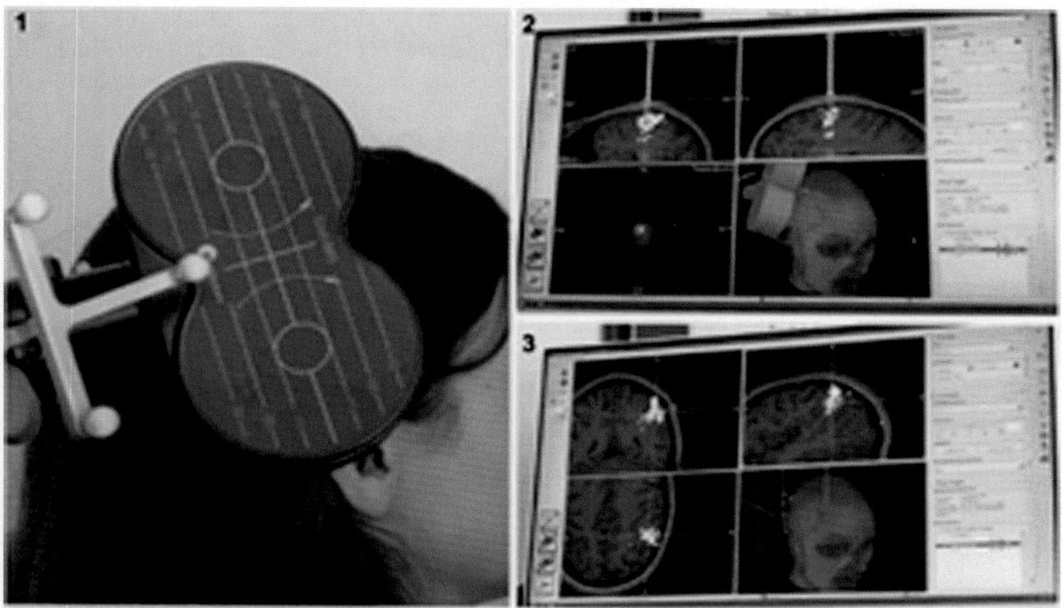

FIGURE 7.1

Demonstration of image-guided TMS treatment using a Figure 8 coil. Here, the coil is placed with computer image guidance to ensure accurate placement over the target (panel 2). An example of a target area defined by a precise anatomic cortical parcellation is shown in the panel 3, in which the red area is the target, green is the entry zone, and white is the parcellation.

This figure has been adapted from Einstein et al. (2022b).

the head circumference measured at the level of the eyebrow and the inion. Outside of the treatment of depression, at a functional location can be determined using TMS activation until a certain behavior or perception occurs. Strictly anatomical or structural locations may also be selected based on a structural MRI focusing on an anatomical region of interest. Functional targeting can also be done in a theoretical fashion while obtaining concurrence from functional task-based MRI. In our continued discussion of implementing connectomics medicine and noninvasive stimulation using TMS, we will be solely focused on the adjunctive use of frameless stereotaxy or neuronavigation. The targets will be selected based on our study of a patient's functional connectome which we will discuss in detail in the next chapter (Dadario et al., 2022; Young et al., 2023). The limitation of TMS targeting is approximately 3.5 cm, but we believe that this depth limitation can be overcome if we adopt the philosophy that TMS stimulates an entire network, not just a single cortical site (Rosen et al., 2021).

Dosage. When we talk about dosage for transcranial magnetic stimulation, there are many possible parameters, unlike a medication that only varies by amount and frequency. TMS can vary by intensity, frequency (Hz), number of total pulses, number of trains, number of sessions within a day, and a specific duration (Chail et al., 2018). These parameters are demonstrated in Fig. 7.2. There are also special parameters to consider such as the shape, size, position, and orientation of the coil. Unlike

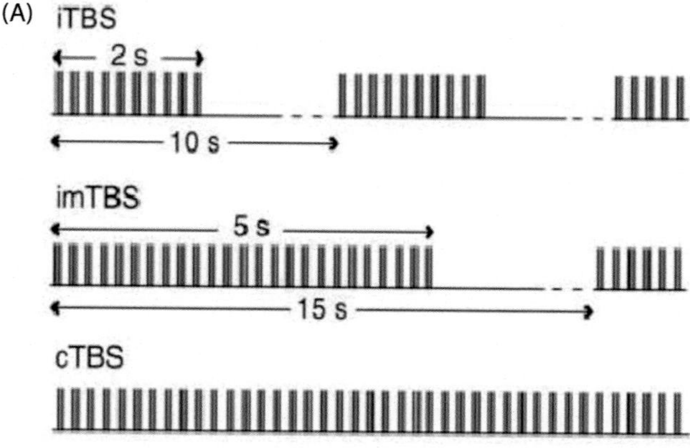

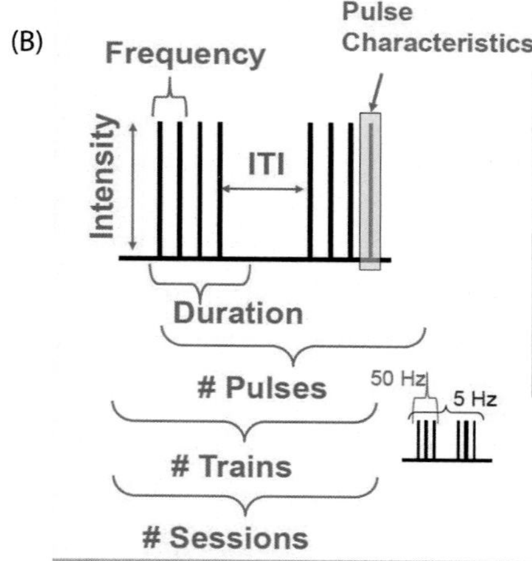

FIGURE 7.2

Different forms of theta burst stimulation paradigms. (A) In the intermittent TBS (iTBS), 20 2 s periods (10 bursts) of TBS are applied at a rate of 0.1 Hz. In the intermediate theta burst stimulation paradigm (imTBS), a 5 s train of TBS is repeated every 15 s for a total of 110 s (600 pulses). In continuous TBS (cTBS), bursts of three pulses at 50 Hz are applied at a frequency of 5 Hz for either 20 s (100 bursts) or 40 s (200 bursts). (B) Explanation of different parameters discussed in regards to the stimulation treatment (Huang et al., 2005; Oberman et al., 2011).

tCDS, there is a clear dose-response relationship with TMS. The greater the intensity usually equates greater depth and spread. Lower intensity stimulation may also have an effect on brain connectivity as well (Siebner et al., 2009). The way to determine prescribed intensity (mA) is commonly determined by the percentage of motor threshold, which is the minimum current required to activate the motor cortex such that movement can be observed in the corresponding body part (Herbsman et al., 2009). As a rule of thumb, high frequency (>3 Hz) can upregulate cortical excitability and low frequency (1Hz) producing cortical inhibition (Speer et al., 2000). When repetitive trains of TMS (rTMS) pulses are applied to the brain with a short interval between stimuli, the effects are thought to be long-lasting cortical excitability beyond the time of stimulation (Klomjai et al., 2015). Repetitive TMS protocol utilizes pulses that are applied in bursts of three at 50 Hz with an inter burst interval of 5 Hz. The idea behind that is this firing protocol may mix the pyramidal cells in the hippocampus and the hippocampus is an anatomical structure that is analogous to short-term learning and is associated with plasticity. Theta burst stimulation (TBS) appears to induce sustained changes in cortical activity, changing the initial paradigm that TMS only induces acute effects. This forms the basis of connectomic medicine and how we can induce long lasting changes in the brain through the study of functional connectivity, once we can visualize and quantify to connectome (Einstein et al., 2022a).

Safety. TMS treatment has been shown to be extremely safe. The most obvious risk of brain stimulation involves the induction of seizures. According to the clinical TMS society consensus statement, the overall risk of seizure is estimated to be less than one in 30,000 treatment sessions or less than one in 1000 patient exposures (Perera et al., 2016). There have not been any reported adverse severe seizure events using TMS (Einstein et al., 2022a). Other risks include excessive heat production if a patient has metallic cranial implants or other electronic implants and potential discomfort caused by stimulation of cranial musculature.

This form of brain stimulation has a long track record in efficacy and safety, and has a deep repertoire of clinical trial data to back up its efficacy in neural modulation. The first FDA approved indication for transcranial magnetic stimulation was in the treatment of major depressive disorder. FDA approval has since then been extended to obsessive compulsive disorder and short-term smoking cessation. However, all of these indications include the targeting of the dorsolateral prefrontal cortex. Hopefully we have conveyed by now that everyone's connectome is different and every disorder may involve different aberrant foci. It is thus unreasonable to expect to maximize efficacy by using the same target for everyone.

How does TMS work?

While this is not entirely understood, what we do know has implications for clinical practice. Notably, it seems to reduce the GABA-ergic tone of the stimulated cortex which puts the brain in a state prone to the induction of neuroplasticity (Dubin et al., 2016). In other words, it seems to set the brain up to be malleable, such as for superlearning.

As such, we have structured our practice to be able to take advantage of this superlearning period. This involves performing therapies aimed at the symptom of interest during the period between TMS treatments.

Conclusions

In this chapter we covered a variety of clinical tools which are used to modulate underlying brain activity. While DBS, tDCS, and TMS have only been FDA approved for a small number of diseases, their ability to study and treat a wide variety of patients is clear. In particular for connectome-based medicine, TMS has significant potential given its noninvasive nature, safety profile, and significant body of literature supporting its efficacy. In the next chapter, we will introduce how high dimensional patient data can be broken down with machine learning techniques so as to better understand patient connectomic features, such as for neuromodulatory targets discussed in this chapter.

References

Chail, A., Saini, R. K., Bhat, P. S., Srivastava, K., & Chauhan, V. (2018). Transcranial magnetic stimulation: A review of its evolution and current applications. *Indian Psychiatry Journal, 27*(2), 172–180. https://doi.org/10.4103/ipj.ipj_88_18

Dadario, N. B., Young, I. M., Zhang, X., Teo, C., Doyen, S., & Sughrue, M. E. (2022). Prehabilitation and rehabilitation using data-driven, parcel-guided transcranial magnetic stimulation treatment for brain tumor surgery: Proof of concept case report. *Brain Network and Modulation, 1*(1), 48.

Dubin, M. J., Mao, X., Banerjee, S., Goodman, Z., Lapidus, K. A., Kang, G., Liston, C., & Shungu, D. C. (2016). Elevated prefrontal cortex GABA in patients with major depressive disorder after TMS treatment measured with proton magnetic resonance spectroscopy. *Journal of Psychiatry and Neuroscience, 41*(3), E37–E45. https://doi.org/10.1503/jpn.150223

Einstein, E. H., Dadario, N. B., Khilji, H., Silverstein, J. W., Sughrue, M. E., & D'Amico, R. S. (2022a). Transcranial magnetic stimulation for post-operative neurorehabilitation in neuro-oncology: A review of the literature and future directions. *Journal of Neuro-Oncology, 157*(3), 445 (Apr, 10.1007/s11060-022-03987-9, 2019).

Einstein, E. H., Dadario, N. B., Khilji, H., Silverstein, J. W., Sughrue, M. E., & D'Amico, R. S. (2022b). Transcranial magnetic stimulation for post-operative neurorehabilitation in neuro-oncology: A review of the literature and future directions. *Journal of Neuro-Oncology*, 1–9.

Ellis, T., & Stevens, A. (2008). Deep brain stimulation for medically refractory epilepsy. *Neurosurgical Focus, 25*, E11. https://doi.org/10.3171/FOC/2008/25/9/E11

Farokhniaee, A., & McIntyre, C. C. (2019). Theoretical principles of deep brain stimulation induced synaptic suppression. *Brain Stimulation, 12*(6), 1402–1409. https://doi.org/10.1016/j.brs.2019.07.005

Hagner, M. (2012). The electrical excitability of the brain: Toward the emergence of an experiment. *Journal of the History of the Neurosciences, 21*(3), 237–249. https://doi.org/10.1080/0964704x.2011.595634

Herbsman, T., Forster, L., Molnar, C., Dougherty, R., Christie, D., Koola, J., Ramsey, D., Morgan, P. S., Bohning, D. E., George, M. S., & Nahas, Z. (2009). Motor threshold in transcranial magnetic stimulation: The impact of white matter fiber orientation and skull-to-cortex distance. *Human Brain Mapping, 30*(7), 2044–2055. https://doi.org/10.1002/hbm.20649

Huang, Y.-Z., Edwards, M. J., Rounis, E., Bhatia, K. P., & Rothwell, J. C. (2005). Theta burst stimulation of the human motor cortex. *Neuron, 45*(2), 201–206. https://doi.org/10.1016/j.neuron.2004.12.033

Iorio-Morin, C., Fomenko, A., & Kalia, S. K. (2020). Deep-brain stimulation for essential tremor and other tremor syndromes: A narrative review of current targets and clinical outcomes. *Brain Sciences, 10*(12). https://doi.org/10.3390/brainsci10120925

Klomjai, W., Katz, R., & Lackmy-Vallée, A. (2015). Basic principles of transcranial magnetic stimulation (TMS) and repetitive TMS (rTMS). *Annals of Physical and Rehabilitation Medicine, 58*(4), 208–213. https://doi.org/10.1016/j.rehab.2015.05.005

Kozyrev, V., Staadt, R., Eysel, U. T., & Jancke, D. (2018). TMS-induced neuronal plasticity enables targeted remodeling of visual cortical maps. *Proceedings of the National Academy of Sciences, 115*(25), 6476–6481. https://doi.org/10.1073/pnas.1802798115

Lozano, A. M., Lipsman, N., Bergman, H., Brown, P., Chabardes, S., Chang, J. W., Matthews, K., McIntyre, CC, Schlaepfer, T. E., Schulder, M., Temel, Y., Volkmann, J., & Krauss, J. K. (2019). Deep brain stimulation:

Current challenges and future directions. *Nature Reviews Neurology, 15*(3), 148–160. https://doi.org/10.1038/s41582-018-0128-2

Oberman, L., Edwards, D., Eldaief, M., & Pascual-Leone, A. (2011). Safety of theta burst transcranial magnetic stimulation: A systematic review of the literature. *Journal of Clinical Neurophysiology, 28*(1), 67–74. https://doi.org/10.1097/WNP.0b013e318205135f

Pelletier, S. J., & Cicchetti, F. (2014). Cellular and molecular mechanisms of action of transcranial direct current stimulation: Evidence from in vitro and in vivo models. *International Journal of Neuropsychopharmacology, 18*(2). https://doi.org/10.1093/ijnp/pyu047

Perera, T., George, M. S., Grammer, G., Janicak, P. G., Pascual-Leone, A., & Wirecki, T. S. (2016). The clinical TMS society consensus review and treatment recommendations for TMS therapy for major depressive disorder. *Brain Stimulation, 9*(3), 336–346. https://doi.org/10.1016/j.brs.2016.03.010

Rosen, A. C., Bhat, J. V., Cardenas, V. A., Ehrlich, T. J., Horwege, A. M., Mathalon, D. H., Roach, B. J., Glover, G. H., Badran, B. W., Forman, S. D., George, M. S., Thase, M. E., Yurgelun-Todd, D., Sughrue, M. E., Doyen, S. P., Nicholas, P. J., Tian, L., & Yesavage, J. A. (2021). Targeting location relates to treatment response in active but not sham rTMS stimulation. *Brain Stimulation, 14*(3), 703–709. https://doi.org/10.1016/j.brs.2021.04.010

Siebner, H. R., Hartwigsen, G., Kassuba, T., & Rothwell, J. C. (2009). How does transcranial magnetic stimulation modify neuronal activity in the brain? Implications for studies of cognition. *Cortex, 45*(9), 1035–1042. https://doi.org/10.1016/j.cortex.2009.02.007

Speer, A. M., Kimbrell, T. A., Wassermann, E. M., Repella, J. D., Willis, M. W., Herscovitch, P., & Post, R. M. (2000). Opposite effects of high and low frequency rTMS on regional brain activity in depressed patients. *Biological Psychiatry, 48*(12), 1133–1141. https://doi.org/10.1016/s0006-3223(00)01065-9

Thair, H., Holloway, A. L., Newport, R., & Smith, A. D. (2017). Transcranial direct current stimulation (tDCS): A beginner's guide for design and implementation. *Frontiers in Neuroscience, 11*. https://doi.org/10.3389/fnins.2017.00641

Tu, Y., Zhang, L., & Kong, J. (2022). Placebo and nocebo effects: From observation to harnessing and clinical application. *Translational Psychiatry, 12*(1), 524. https://doi.org/10.1038/s41398-022-02293-2

van den Heuvel, M. P., & Sporns, O. (2019). A cross-disorder connectome landscape of brain dysconnectivity. *Nature Reviews Neuroscience, 20*(7), 435–446. https://doi.org/10.1038/s41583-019-0177-6

Visser-Vandewalle, V., Andrade, P., Mosley, P. E., Greenberg, B. D., Schuurman, R., McLaughlin, N. C., Voon, V., Krack, P., Foote, K. D., Mayberg, H. S., Figee, M., Kopell, B. H., Polosan, M., Joyce, E. M., Chabardes, S., Chabardes, K., Chabardes, J. C., Tyagi, H., Holtzheimer, P. E. ..., & Okun, M. S. (2022). Deep brain stimulation for obsessive–compulsive disorder: A crisis of access. *Nature Medicine, 28*(8), 1529–1532. https://doi.org/10.1038/s41591-022-01879-z

Young, I. M., Dadario, N. B., Tanglay, O., Chen, E., Cook, B., Taylor, H. M., Crawford, L., Yeung, J., Nicholas, P. J., Doyen, S., & Sughrue, M. E. (2023). Connectivity model of the anatomic substrates and network abnormalities in major depressive disorder: A coordinate meta-analysis of resting-state functional connectivity. *Journal of Affective Disorders Reports, 11*, 100478. https://doi.org/10.1016/j.jadr.2023.100478

CHAPTER

Machine learning and its utility in connectomic medicine

8

Introduction

This is not a chapter about the technical aspects of machine learning. There is no new math for clinicians already versed in basic medical epidemiology. There is nothing about coding and computers. Instead, this chapter aims to help you grasp the rationale and concepts of why this is useful, how to have basic familiarity with these techniques, and to not be afraid of the language used.

The umbrella of AI

The reader will note that the term Artificial intelligence "AI" is often loosely utilized for most applications within the clinical neuroscience community. A more strict definition of "AI" refers to the broad idea of computers being able to independently process data and execute tasks in an intelligent way which is similar to the human brain. Instead, most brain analysis applications currently available in the field more appropriately fall under the category of machine learning (ML), a subset of AI (Fig. 8.1). ML refers to using mathematical algorithms to allow a computer to learn about data and improve in the ability to make decisions. ML-based analyses are becoming an integral way to address complex topics in the neuroscience community as the amount of available data continues to increase

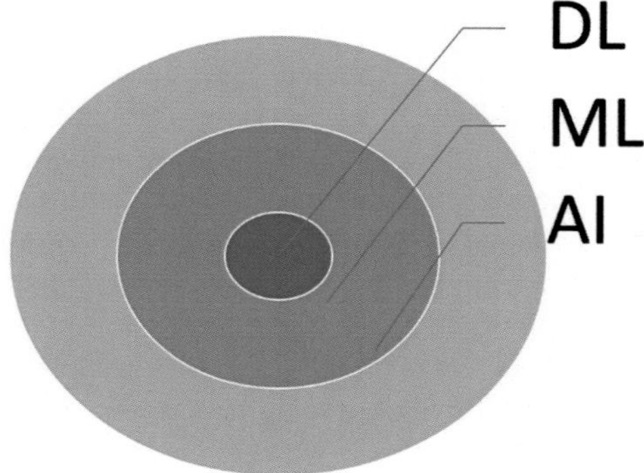

FIGURE 8.1

The AI umbrella. This includes, deep learning (DL), machine learning (ML), and artificial intelligence (AI).

Connectomic Medicine. https://doi.org/10.1016/B978-0-443-19089-6.00004-5

along with complex patient pathophysiology and treatments. Thus, it is imperative for those in the field to have a general understanding of the field of machine learning in order to understand its applications in brain analyses, however, without avoiding all of its unnecessary nuances.

Machine learning in medicine

ML is a buzz word in medicine, but in many cases it comes off as a tool being forced to solve problems we do not really have. So before we dive into the technical details about how to do ML, we need to be clear of what this basically is used for in connectomic medicine and what kinds of questions we should use this to answer. In our experience, this is due to the fact that nonclinicians are left largely alone to define what machine learning tools need to be made, as clinicians view the topic as too specialized to meaningfully engage with the area and to ensure the solutions fit our actual problems. I call these inventions "lasers which cut butter," which are basically expensive solutions for nonproblems. So being literate is important, but we will refrain from trying to cover all topics in this field as they are unnecessary for most people reading this book.

What ML is good for is finding signal inside noisy and/or high dimensional data. In other words, if you have a big dataset, with a lot of potentially relevant variables with potential complex effects on the outcome, machine learning approaches are good at learning this relationship and finding a reasonable statistical approximation of the outcome (aka dependent variable, aka the thing we want to predict) using the feature variables (aka independent variables, aka covariates) you show it. There is nothing more or less to this than that.

Another way to think of this is using the paradigm of facial recognition (Alturki et al., 2022). To teach the model, we provide features, in this case pixel values from the images of peoples' *faces*, and labels, in this case peoples' *names*. The machine learning algorithm then learns which combinations of pixels are useful in making an estimate of which name goes with which face, and which pixels to ignore as they do not provide useful predictive power to differentiate peoples' faces.

If one is familiar with the concept of regression, note that everything in machine learning can basically be thought of as increasingly complex, increasingly nonlinear forms of regression. In regression, the operation basically aims to plot a line in the form of

$$\mathbf{Y = aX_1 + bX_2 + cX_3 + dX_4 + intercept}$$

Where Y is the variable one is attempting to predict (i.e., the variable we care about knowing in advance), X_{1-4} represent the feature variables, or covariates (essentially things that we can measure that might predict the value of Y because they might in theory affect or cause Y) and a, b, c, and d are the feature weights (called co-efficient in regression speak, but feature weights in machine learning). The feature weights are variables changed by the modeling process to assign relative importance to the variables in the model to attempt to minimize the cost-function, a function which measures the prediction error of the model in its prediction of Y. In linear regression, this cost function is measured as the mean squared error of the distance of actual values of Y from the prediction line (Fig. 8.2). In other words, the weights are changed until the error of predicting Y is minimized. The feature weights tell us how to predict Y from the X variables, and these weights are the "machine learning model." If you provide the weights of a good model to someone over the phone, they should be able to make a reasonable prediction of Y if they have the X variables and know how to use the weights.

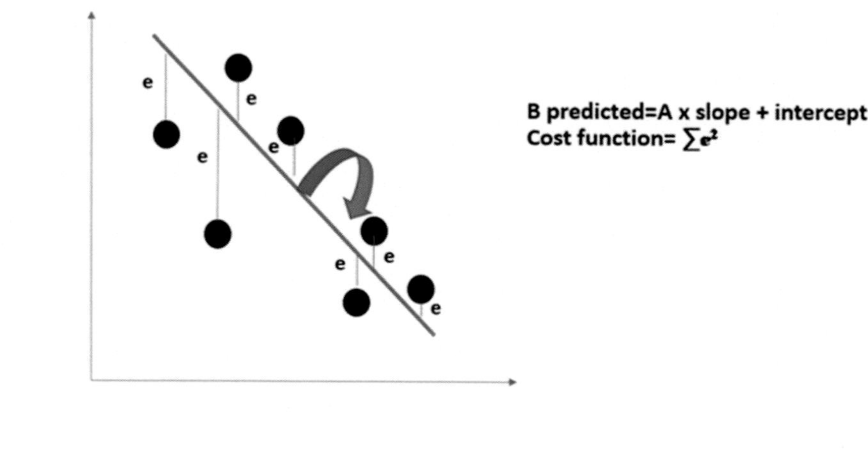

FIGURE 8.2

Simple schematic of a linear regression.

Unless you are going to go deep into this area, for most clinicians, it is best to not waste too much time trying to grasp the dizzying array of machine learning approaches as ultimately they work in similar ways to this. There are differences in how weights are allowed to cointeract with each other and the feature variables, there are differences in what kinds of predictions these models can make (binary versus continuous data for example), there are differences how the cost function is calculated, the nature of hyperparameters (specific ways to alter the relationship about how feature weights work in that model and how it is trained, this of this line turning knobs on an amplifier, and getting slightly different sound for the same maneuver) and how the feature weights are changed to improve the performance of the model to minimize the cost function (Doyen & Dadario, 2022). Ultimately, most of these models are a commodity these days: they are packaged into simple routines in languages like Python, they can be executed with a few lines of code by anyone with basic computer skills, and in modern machine learning hyperclusters, it is possible to run all sorts of combinations in parallel, with many hyperparameters tried (Fig. 8.3).

What is necessary to know for ML competency are two topics: the black box concept and the issue of area under the curve. The black box concept is a major issue in this book (Petch et al., 2022). In short, it refers to the fact that many modeling approaches, notably deep learning techniques, model the feature weights in complex ways, which really make understanding how the model works and what it uses to make predictions nearly impossible for humans. Open box methods are comparatively less common but allow feature weights to be read in meaningful ways (Wood & Choubineh, 2020).

Area under the curve involves testing the ability of the model, specifically its sensitivity and specificity in predicting new data (Bradley, 1997). This is usually done by holding a portion of the training dataset out and not showing it to the algorithm during the training phase (when you run the computer to assign weights). These holdout data are then tested using the model to determine its

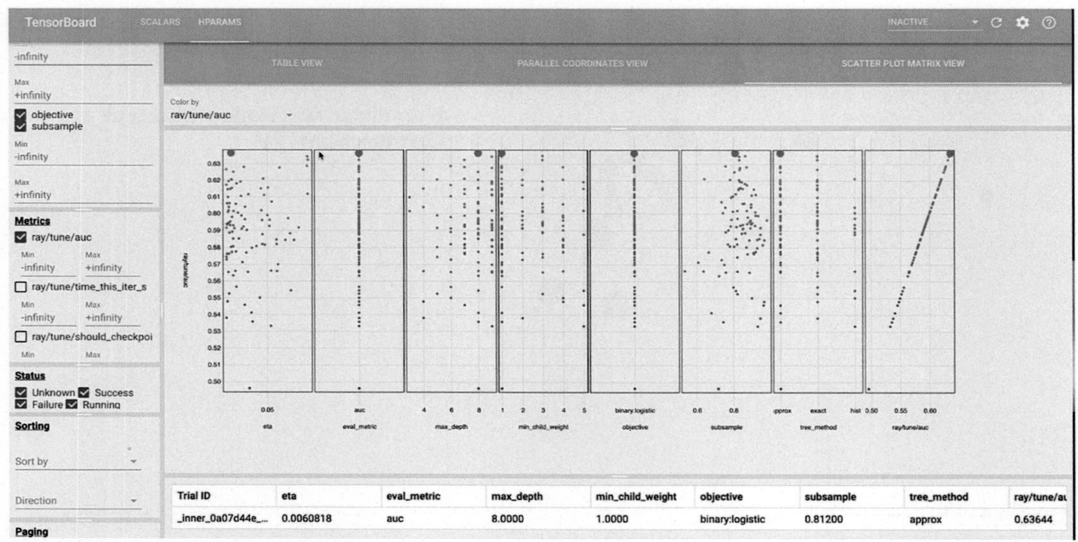

FIGURE 8.3

Building a machine learning model. This figure presents an example on TensorFlow of attempting to build an ML model and choosing or testing various hyperparameters at once, all of which affect various metrics of the model, such as its accuracy (https://www.tensorflow.org/tensorboard/hyperparameter_tuning_with_hparams).

performance. The term "overfitting" (Fig. 8.4) refers to when a model trains precisely to idiosyncratic features of the training data in a way that no other data set would be expected to conform to. Testing a model with holdout data reduces this risk if the AUC remains high for the test set.

While no hard guidelines of a "good AUC" are definitive, a good rule of thumb is that AUC around 0.5 is basically a coin-flip, AUC > 0.7 is pretty promising, AUC > 0.8 is generally excellent performance, and AUC > 0.95 usually involves overfitting or some kind of other methodologic error, for example, training a model to predict a variable from a basically identical variable in the same form, such as predicting height in inches from height in centimeters.

A key benefit of machine learning over regression or other standard biostatistical methods is that it is more able to handle nonlinear interactions between variables as well as being more able to effectively scale to massive numbers of variables. In comparison, it is very, very difficult for a linear regression to fit a meaningful hyperplane equation to model 140,000 variables at once, but newer machine methods can scale to this dimensionality given enough training data.

What do we need machine learning for in the brain?

The fundamental question in medical epidemiology can be framed as the following: What factors (denoted as A, C, D, etc.) relate to an outcome of interest (denoted as B)?

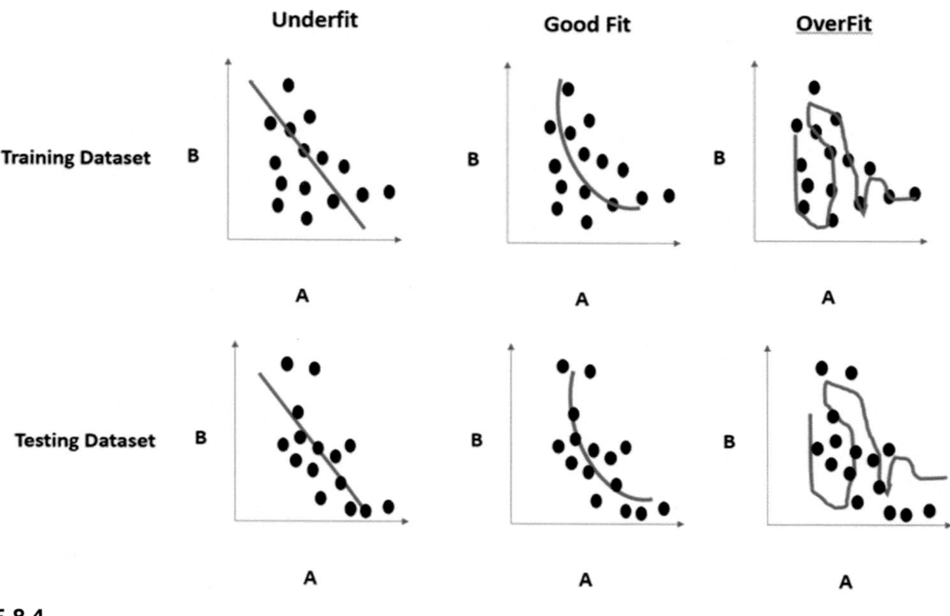

FIGURE 8.4

Overfitting in mathematical modeling. Overfitting refers to when a statistical model corresponds too closely or exactly to its training data set.

In other words we want to study if A causes B to happen.

Usually, we expect multiple factors to affect B and need to control for the potential confounding effects of other variables like C and D.

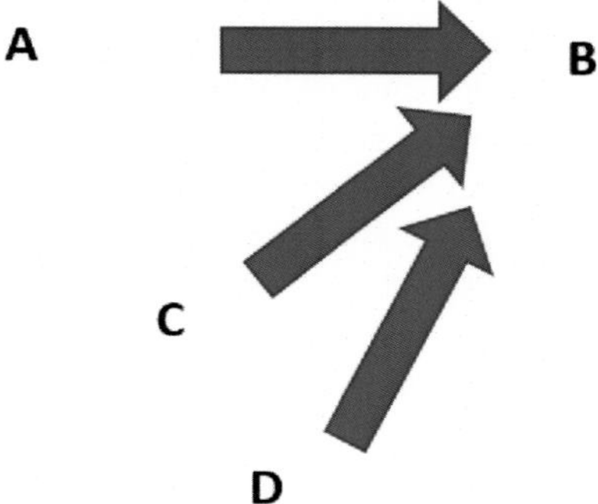

The purpose of this brief description is to point out that what we usually want to know from an epidemiologic study is a set of causes or the outcome of interest. We want to know causes because this helps us get an idea of how to fix the disease. While some situations may benefit from a blind prediction of the likelihood of an event happening, we often are more interested in knowing why a specific patient has a specific problem. But this is the form of the solution machine learning methods often provide: *a prediction or a probability.*

This should cause everyone to immediately note the key issue in the connectome is to figure how to set up the question so the answer we get is useful. In other words, if we ask the question the wrong way, we may end up with a prediction of something we do not care about or one we gain little from being able to predict.

For example, it is possible to train a model to predict the presence of a clinical diagnosis of schizophrenia from the connectome with AUC around 0.7 generally in datasets which are achievable. However, not only is this diagnosis not difficult to make accurately by an experienced psychologist, but given the heterogeneity of this disease, a key question is should we be training models to predict a disease class which may not really exist as such. Alternately, predicting a response to a drug or other treatment might be useful in certain contexts. The key point is we need to model the question correctly.

What does a connectome question look like?

Usually the A in the relationship described in the diagram in the past section in a connectome study is a feature in the connectome, say the function connectivity between Area 44 and 4 (aka Broca's area and the primary motor cortex), and C, D, and the huge number of other potential variables in the model are other connectome features, perhaps functional connectivity between other pairs of areas, tract number between areas, or even clinical covariates which might confound the data, like age, or a gene marker of interest. Machine learning can fix confounding relationships pretty easily by adding more features to the list of model candidates.

B is an outcome of interest. Most commonly, this is an outcome scale, the result of a neuropsych test, or some other endpoint we care about. It is particularly interesting to model multiple subparts of an outcome scale to gain insight of the individual components also. However, how this B is expressed matters, as some modeling techniques require the use of a cutoff to binarize the data, so this can obviously impact how the model works and needs to be considered.

Finally, it is key to determine what you care about: the weights or the model predictions.

Questions that use model prediction are generally ones where the decision being made is not adaptable. For example, you either give an SSRI or you do not. You make a prediction about recovery or you do not. If the question you are trying to answer does not involve you making a nuanced decision about how to treat, you can use model predictions to give a useful answer.

Alternately, sometimes we care about what the model learns, not what it predicts. For example, if you are interested in finding a new target for a symptom of depression, a machine learning output which tells you whether the patient has the symptom already is basically useless as you can ask the patient. You need to know how the model learns the differences between people who have that symptom and those who do not (Fig. 8.5).

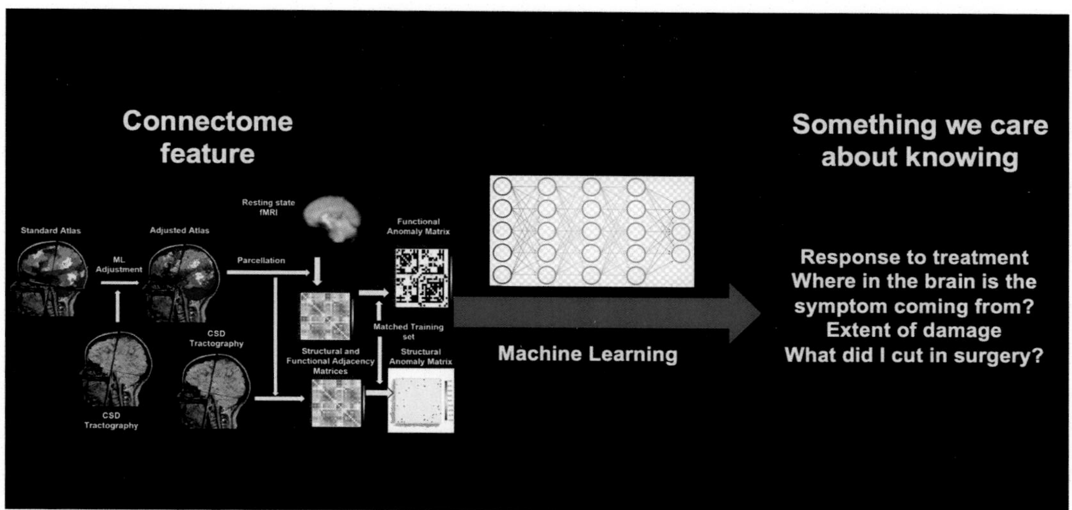

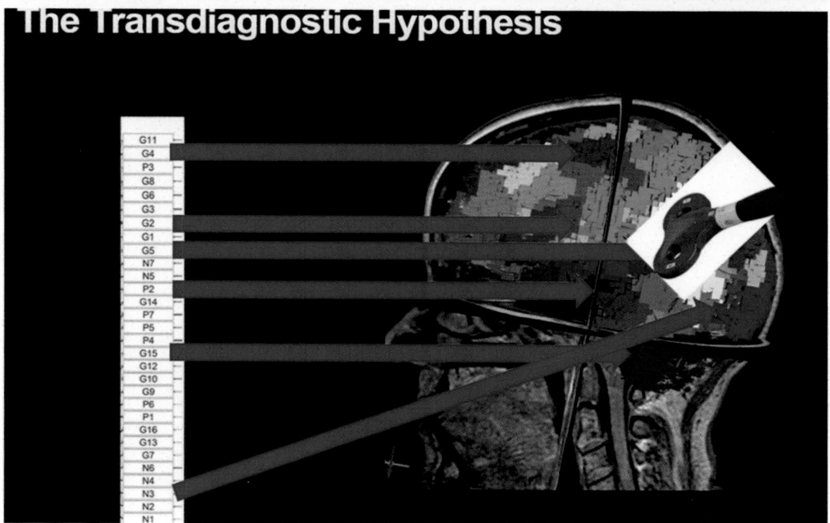

FIGURE 8.5

ML approach to teach us about disease. Instead of simply predicting if a patient has a symptom, which we can gather through a thorough history or exam, ML based analysis can teach us about a lot of other important features, such as what makes two people different who have and do not have this symptom (A). By breaking down specific diseases into individual symptoms and finding their underlying neural correlates (B), then we can better understand the importance of specific regions in causing specific deficits between individuals (C). Such ML based analyses allow for dimensionality reduction to in turn better characterize and make sense of a patient's complex individual pathology for subsequent treatments, especially on an indvidiualized basis (D).

Panels (C) and (D) adapted from Taylor et al. (2023).

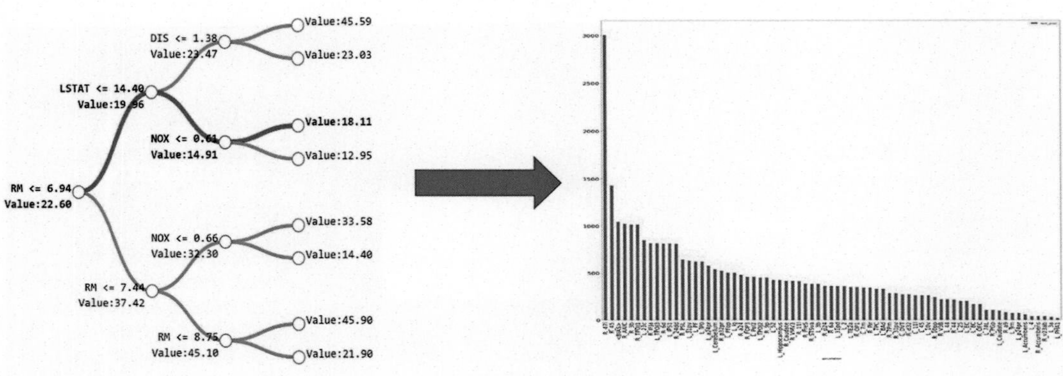

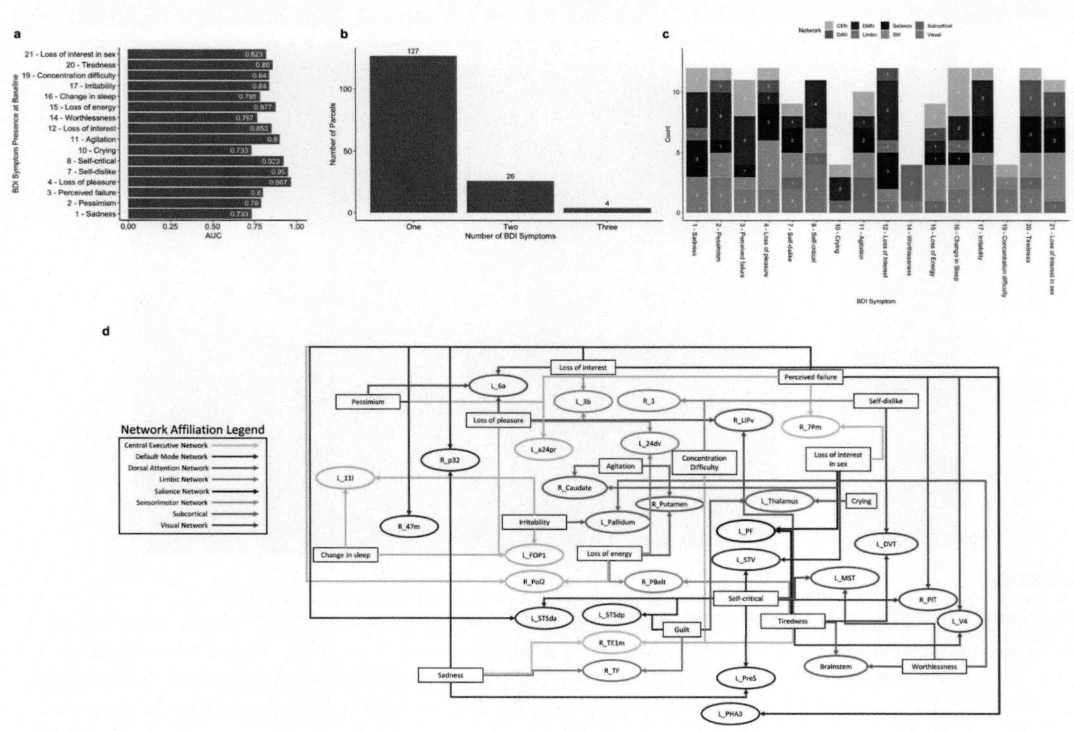

FIGURE 8.5 Cont'd

What is machine learning's real promise in connectomic medicine?

The human brain is a vast and complex structure. Searching blindly in the connectome for possible answers is fraught with either false positives, or missing the truth due to the very high statistical thresholds needed to avoid the extreme issue of multiple comparisons in large datasets.

The best approach to using a big dataset is to learn from the standard Bayesian thinking we learn early in medical school: a test administered to a patient with a high pretest probability will have a better posttest probability. Thus, if we look in areas we know are relevant to the patient's symptoms, we by definition have a higher pretest probability on the tests we are running because we are not looking randomly.

ML's key promise is teaching us where to look for answers to our questions.

Conclusions

In this chapter we outline some of the nuances of ML for the practicing clinician when analyzing brain data. ML-based brain software has provided us tremendous advancements in our understanding of brain structural-functional relationships and thus novel insight into how these relationships may be distorted by pathology and targeted for treatment. In the next chapter, we discuss how to leverage these tools and connectomic information to identify patient -specific targets for neuromodulation.

References

Alturki, R., Alharbi, M., AlAnzi, F., & Albahli, S. (2022). Deep learning techniques for detecting and recognizing face masks: A survey. *Frontiers in Public Health, 10.* https://doi.org/10.3389/fpubh.2022.955332

Bradley, A. P. (1997). The use of the area under the ROC curve in the evaluation of machine learning algorithms. *Pattern Recognition, 30*(7), 1145–1159. https://doi.org/10.1016/S0031-3203(96)00142-2

Doyen, S., & Dadario, N. B. (2022). 12 plagues of AI in healthcare: A practical guide to current issues with using machine learning in a medical context. *Frontiers in Digital Health, 4.*

Petch, J., Di, S., & Nelson, W. (2022). Opening the black box: The promise and limitations of explainable machine learning in cardiology. *Canadian Journal of Cardiology, 38*(2), 204–213. https://doi.org/10.1016/j.cjca.2021.09.004

Taylor, H., Nicholas, P., Hoy, K., Bailey, N., Tanglay, O., Young, I. M., Dobbin, L., Doyen, S., Sughrue, M. E., & Fitzgerald, P. B. (2023). Functional connectivity analysis of the depression connectome provides potential markers and targets for transcranial magnetic stimulation. *Journal of Affective Disorders, 329*, 539–547. https://doi.org/10.1016/j.jad.2023.02.082

Wood, D. A., & Choubineh, A. (2020). Transparent open-box learning network and artificial neural network predictions of bubble-point pressure compared. *Petroleum, 6*(4), 375–384. https://doi.org/10.1016/j.petlm.2018.12.001

CHAPTER

Fundamentals of connectome based decision making and targeting

9

Introduction

While the basic approach to treating patients with TMS using standard targeting approaches is well documented, it is limited to a finite and small number of targets and does not possibly work in all patients who might benefit from TMS. If we want to expand our armamentarium to be able to treat a wider range of patients, we need to be able to think through cases systematically, logically, and efficiently. The principles of reading a connectomic image set do not differ radically from reading labs or other images, though the ideas are new. However, using best practices for interpreting medical data and making decisions from it provides a good starting point.

This is a critical chapter in this book as it provides a framework for how we think about our patients. The rest of the book is focused on applying these principles to specific cases.

This chapter aims to build a framework for approaching patients with brain disease being evaluated for brain stimulation. These kinds of ideas can be applied as an approach to the patient more generally, even if a target selection is not part of the continuum of care. Sound thinking principles are useful regardless of actual therapy.

Our four-step approach to any patient

Step 1: Define the problem.

Step 2: Look at the relevant parts of Structural connectome.

Step 3: Look at the relevant parts of the Functional connectome.

Step 4: Determine a treatment which addresses these issues and integrates into the overall care plan.

Our rationale for these steps is outlined in Fig. 9.1. In short, it is common that patients present with more than one symptom. Both mental illnesses and focal lesions, like strokes and brain tumors, commonly affect multiple circuits and cause a diverse range of neurologic, cognitive, and emotional problems (Dadario & Sughrue, 2022; Dadario et al., 2021). While treatments can affect more than one circuit, it is usually the case that we cannot fix everything with one treatment, be that treatment medical, stimulatory, or other. Thus, we have to define goals, and this will tell us where to look.

As stated repeatedly in this book, the fundamental theorem of connectomic medicine posits that relieving a symptom involves improving function in a circuit relevant to that brain function. Thus, abnormal function can result from a damage or loss of part of the structural architecture, from

Connectomic Medicine. https://doi.org/10.1016/B978-0-443-19089-6.00012-4

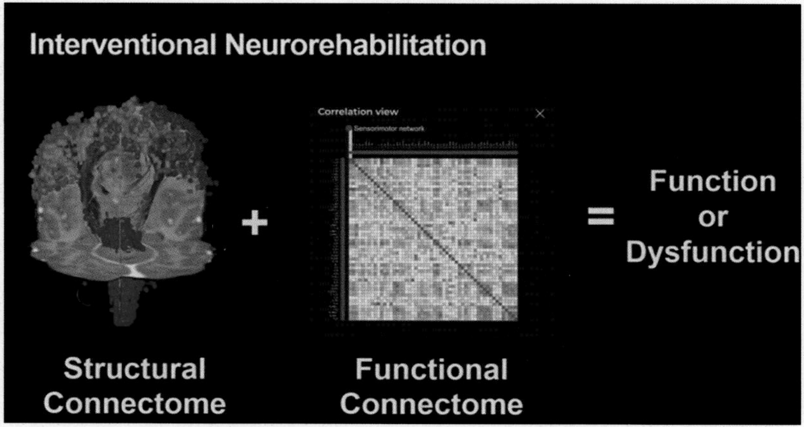

FIGURE 9.1

Evaluating a patients connectome. To appropriately define a patients problem from a connectomic standpoint, both the underlying structural and functional connectomes should be examined. This will allow for an improved understanding of the functional or dysfunctional circuits which may or may not need subsequent rehabilitation or be the primary cause of the patient presentation or pathology.

inappropriate signals running through that structural architecture. Not all diseases cause visible damage to the structure, so sometimes this step can be skipped for diseases like depression, for example. Diseases like stroke, multiple sclerosis, brain tumors, etc., do cause structural damage which can cause changes in functional connectivity, and either in turn can cause abnormal brain circuit functions (Fig. 9.2), and thus both modalities are necessary to understand functional losses in these cases.

Step 1: define the problem

Connectomic tools do not eliminate the need to be a doctor. You still have to take a history, generate a differential diagnosis, and make a treatment plan. The reasons why defining the problem is necessary include neuroscientific (you are unlikely to help a problem if you are treating the wrong part of the brain), statistical (as pointed out in other parts of the book, searching randomly in the connectome makes false discovery nearly certain, and narrowing the search space to look at fewer areas where the pretest probability is already high reduces the chance of treating false positives), and logistical (even with fast protocols like theta burst stimulation (TBS) treating 20 targets is unrealistic, as is trying to do multiple therapies between treatments).

Given the numerous potential symptoms present in a single patient, balancing medical benefits, and patient goals benefits from careful discussion, and usually objective measurement tools to help really assess the severity of various issues. Also, precisely defining issues that often are hard to disentangle are made simpler by working in mechanistic transdiagnostic frameworks like described in previous chapters. This process is also critical for setting realistic expectations, especially for symptoms you do not plan to treat.

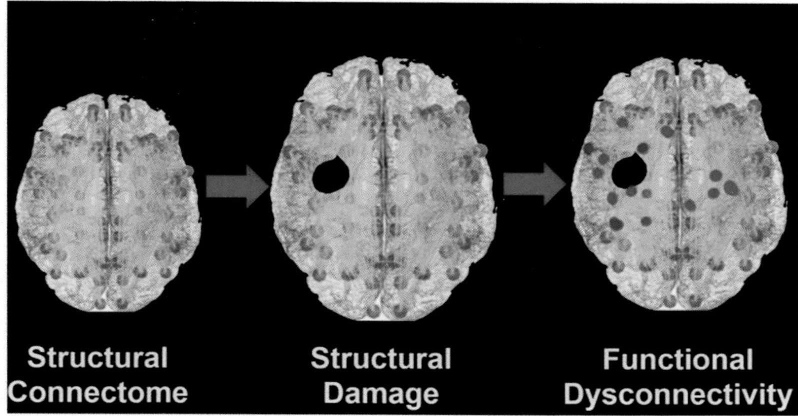

FIGURE 9.2

Structure influences function. Many neurological patients present with pathology that causes disruption in structural connectivity due to underlying white matter damage, such as in stroke or brain tumors. However, functional connectivity is partially defined by its underlying structural connectivity and therefore both of these modalities should be examined together.

Step 2: look at the relevant parts of structural connectome

It is important to recognize that you cannot expect TMS to stimulate a tract which is no longer there. So it is critical in cases where the mechanism might involve brain injury to look at the network or tracts that are involved in performing that function. First, we need to have reasonable expectations. For example, performing motor therapy in a case where there is no corticospinal tract seems unlikely to work.

Additionally, when we think about complex injuries, looking at the structural data can often be necessary to figure what possible targets are still available. Not having the primary pathway might not mean the function cannot recover, but you need to understand what alternate pathway still exists and is available for stimulation.

By defining the problem, we get an idea where to look.

Step 3: look at the relevant parts of the functional connectome

As highlighted by Figs. 9.1 and 9.2, in our goal to find and normalize the affected circuit, we need to identify which parts of the brain are likely referable to the neurologic dysfunction and identify abnormal functional connectivity.

Fig. 9.3 shows what this looks like in practice. In normal individuals, looking at a subset of the connectivity matrix of areas groups by network affiliations, shows mostly red squares indicating high degrees of correlation between areas, as opposed to blue squares which indicate anticorrelated areas, and white squares which indicate unrelated brain regions. Injury to the left arcuate fasciculus in the patient on the right caused not only abnormal anticorrelations, but also abnormal increased

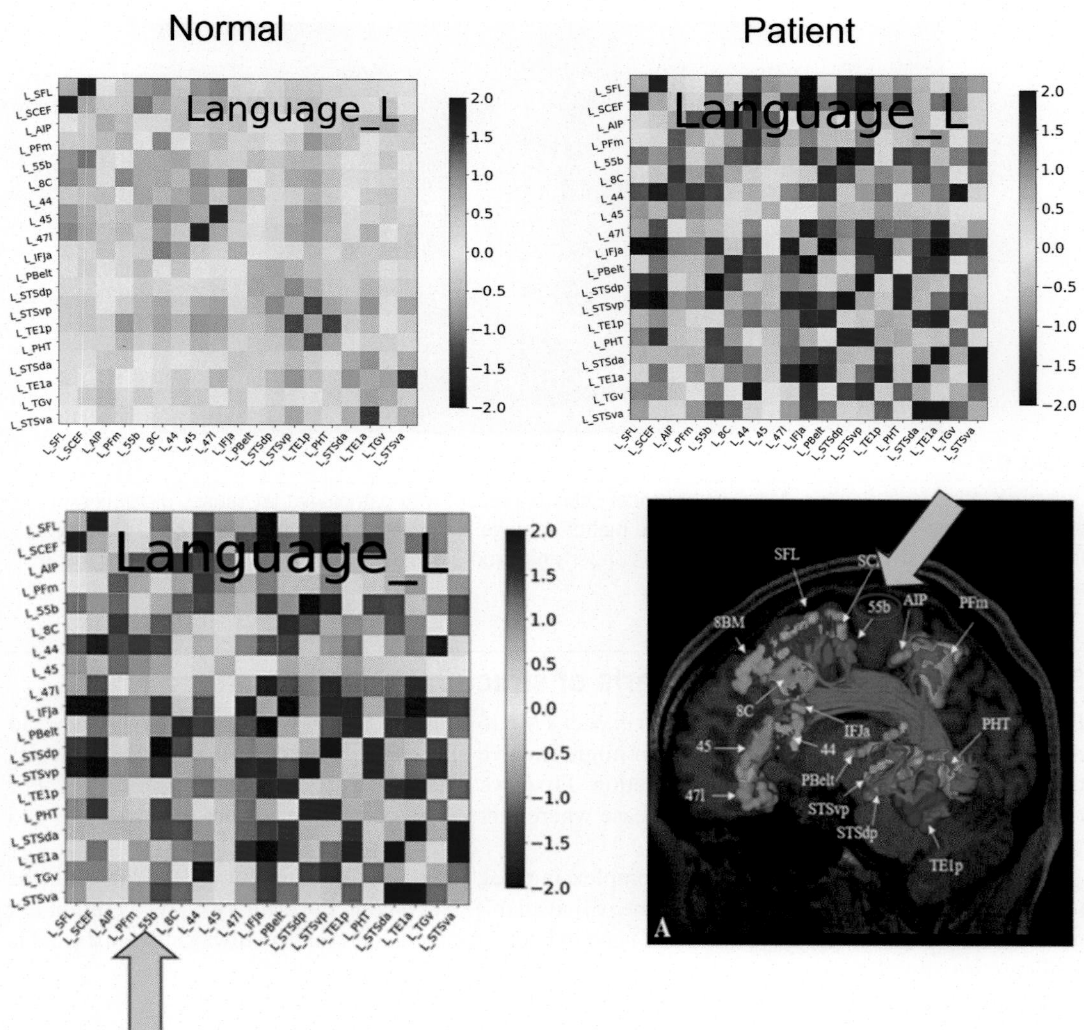

FIGURE 9.3

Examining a patients functional connectivity matrix. This figure compares a healthy subject with a patient with injury to the left arcuate fasciculus on the right (A). As compared to the healthy subject, the patient's functional connectivity matrix demonstrates not only abnormal anticorrelations (blue), but also abnormal increased correlations in other parts of the network (red).

correlations in other parts of the network. Note that in TMS, we only treat one area at a time, and one area corresponds to one column (or row). Expressing functional connectivity on a real brain is shown one column at a time but reflect the same data.

Sometimes, if we have prior information about an approach using raw data, such as the SAINT approach (Cole et al., 2020), certain approaches in stroke therapy, or information from a machine learning approach, can be enough. However, many times we do not know a target a priori, and we just want to target something we can at least conclude is abnormal. Fig. 9.4 shows a view displaying statistical outliers in functional connectivity subsetted by two networks. We read columns with multiple anomalies as more abnormal as this area communicates with other areas in the network abnormally. The basic idea is to treat areas with many anomalies as they are functioning out of the norm, and lie in the circuit of interest.

FIGURE 9.4

Anomaly detection matrices. Example of anomaly detection matrices for the default mode network (DMN) and central executive network (CEN) based on functional connectivity data. Targets were generally represented by columns demonstrating multiple anomalies in functional connectivity. White squares are considered to be functioning normally when compared to a normal dataset, and black squares are areas that are highly variable, and are thus excluded from the analysis. The color scale on the right hand side indicates connectivity, with red representing hyperconnectivity and blue representing hypoconnectivity. The chosen targets for treatment are highlighted by black outline, in this case the selected target is LPGs.

Picking targets

When we look at structural and functional data, usually specific strategies are somewhat evidence from anomalies, raw correlation, and/or the structural data. But a few additional ideas are useful.

(1) *Using standard approaches*. In short, using standard targets like motor cortex or left DLPFC is preferred when they make sense and are abnormal. Sometimes they can be combined with other targets in a multiple target strategy, where the principle target is the heavy lifting for dealing with many of the core symptoms and the additional targets to focus on additional anomalies.

(2) *Anatomical limitations*. Regardless of how outright abnormal the connections stemming from a specific parcellation may be, it is not always amenable to TMS treatment. This is due to the fact that there are anatomical limitations related to this treatment modality. TMS in itself is designed for targeting cortical surfaces. There are cortical surfaces that are beyond the reach of stimulation. Even if we are using a figure 8 coil, the depth of simulation is no more than 3.5 cm. In reality, we have to account for the patient's scalp thickness, thickness of the skull, and the actual brain region. For example, if the anomaly matrix suggests a target within the mesial temporal lobe, it would be difficult for the magnetic field to penetrate deep enough and generate enough of an electric field for effective stimulation. The same is with a target area within the orbitofrontal cortex, where the clinician will have to place the coil directly in front of the forehead. Even if so, domestic field would not be able to penetrate deep enough for effective stimulation. Another anatomical limitation concerns a patient's ability to tolerate the treatment itself. Although TMS is proven to be safe, it can cause scalp discomfort for the patient. Just as the magnetic field can induce an effective electric field to directly act on neurons, areas of the cranium that are covered with thick musculature, such as the temporalis, can be areas of discomfort when the muscles are induced to fire. If a patient cannot tolerate this type of discomfort, then in an area near the temporal lobe might not be feasible due to the intolerability.

Step 4: determine a treatment which addresses these issues and integrates into the overall care plan

Adjunctive therapies during TMS

One of the main goals of connectome-based medicine is to induce long-lasting therapeutic changes to our brains. In the above section we have mentioned about how TMS therapy can induce cortical changes that are long lasting. The overall goal is to have a form of behavioral treatment that also coactivates the target network in order to reinforce the changes we are trying to create.

In diseases like brain tumors and stroke, often the therapy path is well worked out. It can involve physical, occupational, speech, or even cognitive therapy. When it comes to other problems, some flexibility often becomes necessary.

In the treatment of depression, medication is often combined with cognitive behavioral therapy. In the same manner, TMS has been tested together with behavioral therapies for treatment of depression. In particular, numerous benefits have been seen when combining TMS with a type of structured, skill-based therapy for depression called self-system therapy (Neacsiu et al., 2018; Strauman et al., 2006).

Self-system therapy is similar to cognitive behavioral therapy that focuses primarily on self-evaluation as opposed to focus on cognitive distortions. Overall, strong increases in activity between the right VMPFC and hippocampus utilizing this combined treatment has been found. Therapies with different goals, divided between promotion and prevention, induced changes in the VMPFC and the hippocampus, respectively. This is important as prevention failure is associated with depressed patients who have comorbid generalized anxiety disorder and showed increased right prefrontal cortex activation following prevention priming (Eddington et al., 2009). Hyperactivation is associated with vigilance and anxiety. In contrast, a failure in promotion demonstrates decreased activity in left prefrontal cortex activation, which is associated with dysphoria and anhedonia (Rogers et al., 2004). This suggests to us that even small nuances in adjunctive therapy as related to detailed history taking and diagnosis can activate different areas of the brain, reinforcing the power of behavioral therapy in shaping functional connectivity. Furthermore, in the studies discussed above, patients who underwent TMS with adjunctive therapy demonstrated increased frontotemporal connectivity, compared to either therapy alone. We will now expand this concept into the broad field of TMS brain stimulation for neurological and psychocognitive disorders.

Neurological. It is well documented that patients suffering from ischemic stroke demonstrate positive brain connectivity changes with therapies targeted toward their neurological deficit (Chen et al., 2023; Silasi & Murphy, 2014). Even chronic stroke patients can maintain the capacity to enhance neural synchronization between different brain regions. This is not surprising as we now know that stroke patients benefit from therapy and any physical manifestation is likely to originate from changes in brain connectivity (Baldassarre et al., 2016). Keeping in line with this thought, it would make sense to incorporate therapies along with TMS and we are trying to treat neurological conditions, whether it is post-surgical or ischemic in nature. For example, a patient who has had an ischemic stroke can be a candidate for rTMS treatment (Chen et al., 2023; Hsu et al., 2012). It would only make sense that we combine our TMS treatment together with physical therapy that has been shown to be beneficial for stroke rehabilitation. In the treatment of patients with language deficits, it would make sense to incorporate speech therapy that targets specific areas of improvement, whether it is expressive, receptive, or has to do with writing or reading. We concede that at the writing of this book, there is still no gold standard for what kind of therapy to accompany each type of condition. At this time, it would make sense to incorporate a type of adjunctive therapy that is either established in the literature or makes sense.

Psychocognitive. In line with the above philosophy, cognitive behavioral therapy is an important arm of treatment in addition to other interventions, whether it is medical therapy or brain stimulation which we are focusing on. For the treatment of depression, it would be important to incorporate the direction of licensed professional to engage the patient in appropriate cognitive behavioral therapy or self-system therapy. For the treatment of anxiety, it would be appropriate to engage the patients in relaxation methods, such as meditation or exposure to calming media. It is important to note that these types of adjuncts are based on prior research showing that emotional regulation therapy can significantly alter brain connectivity among different networks involved in major depressive disorder and generalized anxiety disorder (Scult et al., 2019). In fact, emotional regulation therapy can demonstrate significant changes in both the DMN and salience network nodes (Scult et al., 2019). Similarly, cognitive training is also an effective intervention that can improve cognitive functioning in neurogenerative diseases and has been shown to alter brain activity and connectivity as well (van Balkom et al., 2020).

It is difficult to list all kinds of adjunctive therapies that should accompany brain stimulation in the treatment of various neurological and psychocognitive disorders. Data is still emerging. In addition, it is still unclear what role various psychoactive medications may play in the treatment of any of the above disorders together with brain stimulation. It is our opinion that it is likely not feasible to conduct iterations of various forms of treatment with varying localization and dosages in combination with many modes of physical therapy, occupational therapy, and various forms of cognitive behavioral therapies. We foresee the future of connectomic medicine to be a highly personalized one—each patient will need personalized stimulation targets based on individual brain mapping, anomaly detection, and adjunctive therapies tailored to each patient's condition and situation.

Conclusion

In this chapter, we introduced a framework for approaching patients with a brain disease when being evaluated for brain stimulation, such as with transcranial magnetic stimulation (TMS). In particular, these steps include: defining the problem, looking at the relevant parts of structural and functional connectomes, and then determining a treatment which addresses these issues and integrates into the overall care plan. In the next chapter, we discuss using the connectome in psychotherapy and other psychiatric therapies.

References

Baldassarre, A., Ramsey, L. E., Siegel, J. S., Shulman, G. L., & Corbetta, M. (2016). Brain connectivity and neurological disorders after stroke. *Current Opinion in Neurology, 29*(6), 706–713. https://doi.org/10.1097/wco.0000000000000396

Chen, R., Dadario, N. B., Cook, B., Sun, L., Wang, X., Li, Y., Hu, X., Zhang, X., & Sughrue, M. E. (2023). Connectomic insight into unique stroke patient recovery after rTMS treatment. *Frontiers in neurology, 14*, 1063408. https://doi.org/10.3389/fneur.2023.1063408

Cole, E. J., Stimpson, K. H., Bentzley, B. S., Gulser, M., Cherian, K., Tischler, C., Nejad, R., Pankow, H., Choi, E., Aaron, H., Espil, F. M., Pannu, J., Xiao, X., Duvio, D., Solvason, H. B., Hawkins, J., Guerra, A., Jo, B., Raj, K. S. ..., & Williams, N. R. (2020). Stanford accelerated intelligent neuromodulation therapy for treatment-resistant depression. *American Journal of Psychiatry, 177*(8), 716–726. https://doi.org/10.1176/appi.ajp.2019.19070720

Dadario, N. B., & Sughrue, M. E. (2022). Should neurosurgeons try to preserve non-traditional brain networks? A systematic review of the neuroscientific evidence. *Journal of Personalized Medicine, 12*(4), 587.

Dadario, N. B., Brahimaj, B., Yeung, J., & Sughrue, M. E. (2021). Reducing the cognitive footprint of brain tumor surgery. *Frontiers in Neurology, 1342*.

Eddington, K. M., Dolcos, F., McLean, A. N., Krishnan, K. R., Cabeza, R., & Strauman, T. J. (2009). Neural correlates of idiographic goal priming in depression: Goal-specific dysfunctions in the orbitofrontal cortex. *Social Cognitive and Affective Neuroscience, 4*(3), 238–246. https://doi.org/10.1093/scan/nsp016

Hsu, W. Y., Cheng, C. H., Liao, K. K., Lee, I. H., & Lin, Y. Y. (2012). Effects of repetitive transcranial magnetic stimulation on motor functions in patients with stroke: A meta-analysis. *Stroke, 43*(7), 1849–1857. https://doi.org/10.1161/strokeaha.111.649756

Neacsiu, A. D., Luber, B. M., Davis, S. W., Bernhardt, E., Strauman, T. J., & Lisanby, S. H. (2018). On the concurrent use of self-system therapy and functional magnetic resonance imaging-guided transcranial magnetic stimulation as treatment for depression. *The Journal of ECT, 34*(4), 266–273. https://doi.org/10.1097/yct.0000000000000545

Rogers, M. A., Kasai, K., Koji, M., Fukuda, R., Iwanami, A., Nakagome, K., Fukuda, M., & Kato, N. (2004). Executive and prefrontal dysfunction in unipolar depression: A review of neuropsychological and imaging evidence. *Neuroscience Research, 50*(1), 1–11. https://doi.org/10.1016/j.neures.2004.05.003

Scult, M. A., Fresco, D. M., Gunning, F. M., Liston, C., Seeley, S. H., García, E., & Mennin, D. S. (2019). Changes in functional connectivity following treatment with emotion regulation therapy. *Frontiers in Behavioral Neuroscience, 13*, 10. https://doi.org/10.3389/fnbeh.2019.00010

Silasi, G., & Murphy, Timothy H. (2014). Stroke and the connectome: How connectivity guides therapeutic intervention. *Neuron, 83*(6), 1354–1368. https://doi.org/10.1016/j.neuron.2014.08.052

Strauman, T. J., Vieth, A. Z., Merrill, K. A., Kolden, G. G., Woods, T. E., Klein, M. H., Papadakis, A. A., Schneider, K. L., & Kwapil, L. (2006). Self-system therapy as an intervention for self-regulatory dysfunction in depression: A randomized comparison with cognitive therapy. *Journal of Consulting and Clinical Psychology, 74*(2), 367–376. https://doi.org/10.1037/0022-006x.74.2.367

van Balkom, T. D., van den Heuvel, O. A., Berendse, H. W., van der Werf, Y. D., & Vriend, C. (2020). The effects of cognitive training on brain network activity and connectivity in aging and neurodegenerative diseases: A systematic review. *Neuropsychology Review, 30*(2), 267–286. https://doi.org/10.1007/s11065-020-09440-w

SECTION

2

Applications

CHAPTER

Using the connectome in psychotherapy and other psychiatric therapies

10

Introduction

Most of this book involves doing anatomically based procedures like transcranial magnetic stimulation (TMS) or brain surgery. This is for the obvious reason that connectomic ideas, image guidance, anatomic target selection, etc., not only are key parts of TMS and brain surgery, and fit these indications quite well, but people are doing many of these techniques in some form presently. Furthermore, we do TMS ourselves every day, and thus, this is what we have the most to explain and the most examples of which we feel are beneficial to those who are new to this topic.

In this chapter, we attempt to do something different. Specifically, we discuss our idea of how the connectome can improve the rest of brain medicine, including psychiatry. A bit of warning that we are not psychiatrists, or experts in psychopharmacology; however, we have spent a great deal of time talking to these kinds of people and listening to their thoughts. Many such people work in our clinic and we talk to them often. As such, we will focus on concepts, and theory, and will leave defining the best practices to experts. We do, however, envision a world where all brain treatments are eventually based on objective tests, and connectomics thus far appears to be the most plausible path to this world we have heard to date.

How could a brain MRI impact psychotherapy?

We believe we have spent more time pondering this idea than any neurosurgeons in history. However, it seems likely that there is a great deal to improve on in this area. To grasp this fully, it is necessary to review a (very) brief history of psychotherapy. It is not comprehensive, nor is it intended to be, but instead is aimed at highlighting an important dichotomy in the field, and where our new abilities to map the connectome lie in this dichotomy.

The historical trajectory of psychotherapy

In our opinion, the field of psychology and associated domains of psychoanalysis and psychotherapy are historically best thought of as a hybrid of ideas from philosophy of mind and social science. In some ways, psychodynamic theories heavily borrow from, and have influenced modern philosophy, and the astute reader of many schools of therapeutic approaches should note substantial similarities between the disciplines. Without such overlap, the idea of existential therapy would not exist.

Connectomic Medicine. https://doi.org/10.1016/B978-0-443-19089-6.00016-1

Freud's principal contribution to psychology was the belief that psychoanalytic techniques could be used to understand root causes of mental illnesses and related life challenges. Contemporary eyes reading the psychoanalytic theories of Freud and Jung, among countless others, are often immediately aware of how the theories are deep, but also cited without data or statistics. Such systematic methods, statistical analytic techniques, and controlled experimentation came decades later; however, the well-trodden approach of people lying on a couch discussing early childhood experiences remained for quite some time. The stated goal of these approaches was and still is to identify the mechanisms of thought and pathologic thought, with the idea that identifying the cause provided a path toward fixing the psychopathology.

The principal critique of the psychoanalysis was that most of its theories were not falsifiable and thus not really testable. This general lack of a firm evidence base behind these techniques and their underlying theories lead many to seek more scientific approaches to understanding and addressing psychopathology. The field of behavioral theory evolved in the 1950s and aimed to determine reproducible and measurable psychologic-physiologic mechanisms in mental illnesses, notably phobias, and to utilize techniques which modulate these interactions. Most of these techniques have a basis in operant conditioning.

Aaron Beck was a psychologist who became aware of the relative lack of scientific basis of psychoanalytic methods, and in response generated the framework of cognitive behavioral therapy (CBT), which in its numerous variations, is the most commonly used form of therapy in the world. It differed from its predecessors in two fundamental ways (Beck & Fleming, 2021). The first is its insistence on measuring outcomes and progress, both inside the therapeutic environment, and in the evaluations of the methods and their effectiveness. CBT is not surprisingly by far the basic type of therapy with the most evidence demonstrating its effectiveness.

More interesting was Beck's insight that negative thoughts and unhelpful thinking often occur in an automatic fashion, predating deep thought about the validity of those thoughts, or even full conscious awareness. Thus, a cornerstone of the CBT approach is to evaluate evidence for or against thoughts, and to develop alternative interpretation to, or healthy responses to, those thoughts. In that way, CBT is essentially a methodology for strengthening the psychodynamic response and reaction to negative, or otherwise counterproductive, psychopathologic thoughts.

Why would a patient benefit by adding an MRI to this process?

Aaron Beck made one fundamental observation which was decades ahead of his time: that mentally ill people often have negative thoughts that enter their conscious thinking, and that it was generally less effective to endlessly try to chase hidden causes of these thoughts, and better to try to build up specific tools for addressing the consequences of these thoughts and to strengthen patient's cognitive responses to promote productive, measurable outcomes. As biological mechanistic psychiatry has increasingly come to dominate the thinking of the field, this is one reason CBT has maintained relevance: it accepts negative or counterproductive thoughts as a given, and starts its work from this premise. It predated deep biologic explanations for these thoughts by many years.

An interesting first question worth asking about CBT's use for specific cases: while it seems obvious how such an approach would be useful in case of a depressed patient thinking their life is worthless, which on its face seems to be a complex mix of biology, psychodynamics, and social dynamics, but how could such an approach be used in a more overtly, even highly genetic disease such as schizophrenia? On the face, it seems a lot less intuitively plausible that obviously biologic problems like delusions or hallucinations can be solved by talking to someone. CBT begins by taking this biology as a given, using common knowledge about patterns of psychopathology in schizophrenic patients to quickly understand the nature of the problem. The therapy is then aimed at helping the patient in reality test their delusions, and to develop mechanisms for addressing future such delusions. Note this does not replace biological therapy, as it is not directly aimed at the root cause of the problem per se, though obviously there is evidence that CBT causes biologic changes in the brain similar to medical therapy (Mason et al., 2017).

Herein lies the vast potential of connectomics to improve this process (Fig. 10.1). Experienced psychotherapists have repeatedly told us that one of the biggest problems of psychotherapy is defining the problem clearly. It is well known from decades of experience that patients can spend years in nonproductive rambling without finding a clear explanation of what is wrong specifically. Transdiagnostic approaches to analyzing the functional connectome, combined with clinical interviewing, can help more rapidly identify the problem, and ensure that this is the problem we are focusing on helping the patient develop appropriate responses to (Dufford et al., 2022). Similar to the schizophrenic patient with delusions of control, understanding that a problem objectively is or is not related to abnormalities of reward delay can focus therapies on addressing automatic thinking related to this objective biologic process. A large number of transdiagnostic circuits have forms of CBT and other treatments which are evidence-based and directed at those specific domains, and can potentially focus

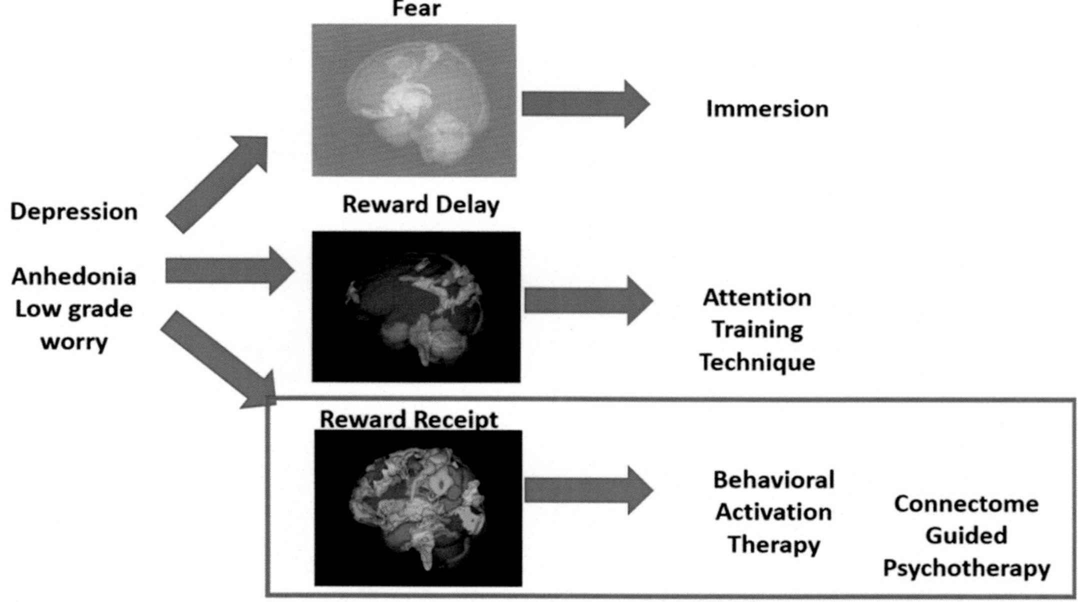

FIGURE 10.1

An example of how connectome based analyses can identify specific therapeutic pathways according to an underlying dysfunctional circuit.

on strengthening coping mechanisms against these circuits. Fig. 10.1 highlights an example of this approach, whereby specific therapeutic pathways can be structured around results of the connectome study.

Additionally, it is not seriously debated that the lack of objective demonstration of the biologic mechanisms behind psychopathology is an impediment to care, largely due to both the difficulty many patients have with accepting the observations of their therapist (rigid thinking can be hard to break) and with lengthening the process of defining the problem clearly and definitively. It seems likely that imaging-based tools can tackle both issues by both focusing the discussion on a biologic process in the brain which the patient experiences and can externalize from their own metacognitive defenses. In other words, it is easier to accept when it is not just faulty thinking, but is biologically linked to how that person's brain works. It also improves confidence in the administered therapies to know they are directed at an objective problem.

How could a brain MRI impact pharmacotherapy?

To start, we should be clear that we are not exactly here yet. However, we are convinced we will be, and it is worth stating clearly what this solution will look like.

Let's begin by outlining some facts which are generally safe to conclude based on the best available evidence on this topic.

(1) Mentally ill patients demonstrate outliers of functional connectivity patterns on resting state fMRI which occur less frequently, or to less extreme degrees in people without mental illness (Menon, 2011).

(2) These connectivity patterns are not unifocal, but they also are not diffuse and brain wide: they are multifocal as some areas are affected and others are normal in almost all patients studied.

(3) Different parts of the brain are affected in different patients with the same diseases and between different diseases, and there is a fair bit of evidence linking abnormalities with different symptoms and response to therapy. Symptoms seem to link.

(4) Pharmacotherapies affect the patterns of brain connectivity (Li et al., 2022). As described in previous chapters, changing the neurotransmitter levels in an area of cortex changes the neurochemical gain between regions, and thus alters functional connectivity in complex ways at a regional and global level.

(5) Neurotransmitter receptors, especially for neurotransmitters targeted by psychotropic medications, are not uniformly expressed in the brain, and thus should reasonably be expected to change physiology and thus connectivity differently in different parts of the brain, even if given systemically.

Taken together, we can formulate what we call the *fundamental theorem of connectomic medicine* (outlined in Fig. 10.2).

In order to alleviate a symptom, a therapy must make a functionally relevant circuit somewhere in the brain fire more normally

Thus, for a psychotropic medicine, brain stimulation, CBT, or whatever other therapy we choose to attempt, to make a symptom disappear, we need to figure out what circuits in the brain perform the function that normally performs the task that is failing—*and then make it work more normal.*

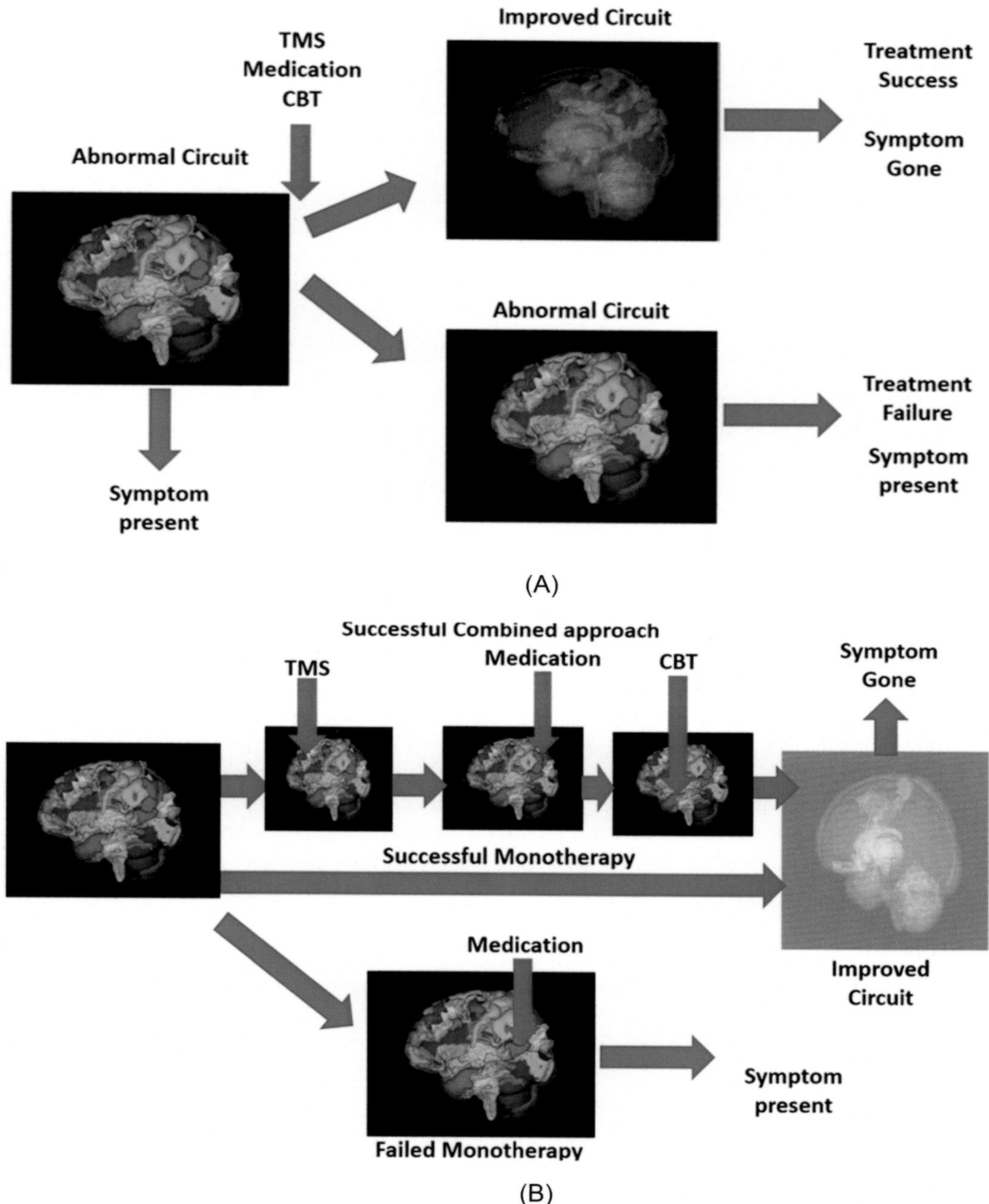

FIGURE 10.2

The fundamental theorem of connectomic medicine. This theory suggests that in order to alleviate a symptom, a therapy must make a functionally relevant circuit somewhere in the brain fire more normally. Specific deficits should be tied to particular underlying network dysfunction or disruption (A). By finding this abnormal circuit, then we may be able to better normalize this circuit, whether through a psychotropic medicine, brain stimulation, CBT or whatever other therapy we choose to attempt (B).

Note this theorem is still valid even if we consider firing events triggered outside the brain. For example, giving local anesthetic to a painful laceration would be expected to make the pain circuits fire less. What it does do is force us to tie our explanations of the mechanisms of our treatments to neurophysiologic events in a rational way, which views the brain as a set of parallel neuroanatomic circuits, and to attempt to tackle the complexity of these problems in steps based ultimately in neuroanatomy.

A few ideas are useful to consider. It may be difficult to totally normalize all aspects of a connectome. For example, genetic diseases like schizophrenia are likely wired to be very prone to lapsing into pathologic rhythms. This makes the need to view these illnesses as a collection of independent symptoms more obvious: we can only fix the symptoms which come from misfiring of a circuit which we are able to normalize.

It is important to consider what an idealized tool would look like which is capable of helping us think like this. Importantly, this tool would be able to estimate a "destination connectome," which is a functional connectome that is more normal than the patient's initial connectome, but importantly may not be entirely normal, but merely may be better able to compensate for biologically abnormal circuits. This tool would also be able to compute the summed effects of a treatment proposed and estimate whether this treatment would be expected to get the patient closer to the destination connectome, and specifically what circuits would it normalize and have a reasonable chance of fixing a symptom from that circuit.

This tool is not available currently, but it is not entirely science fiction (right now). Importantly, it may be the case that machine learning approaches can learn responses to medicines, and estimate the likelihood of a treatment working from previous information and patterns it has learned. Such a tool would be mainly in the prediction type machine learning output, but open box tools can also possibly be useful (Doyen & Dadario, 2022).

Either way, hopefully this discussion has emphasized that mapping the connectome, and using this to guide our thinking and decisions has the potential to drastically change how we practice brain medicine, especially mental illness treatment, both by defining the issue we are treating more clearly, and estimating the likelihood of success before we engage on long and risky trial and error.

Conclusion

In this chapter, we discussed how the connectome can be leveraged to improve brain medicine, specifically for psychiatric illnesses. We introduced the *fundamental theorem of connectomic medicine* as a powerful tool to find a dysfunctional circuit causing a particular symptom and attempt to normalize its firing pattern. In the next chapter, we discuss how these connectomic ideas can be formulated at scale into a large connectomics-driven neuroscience clinic.

References

Beck, J. S., & Fleming, S. (2021). A brief history of Aaron T. Beck, MD, and cognitive behavior therapy. *Clinical Psychology in Europe, 3*(2), 1–7. https://doi.org/10.32872/cpe.6701

Doyen, S., & Dadario, N. B. (2022). 12 plagues of AI in healthcare: A practical guide to current issues with using machine learning in a medical context. *Frontiers in Digital Health, 4*.

Dufford, A., Kimble, V., Tejavibulya, L., Dadashkarimi, J., & Scheinost, D. (2022). Predicting transdiagnostic social impairments in childhood using connectome-based predictive modeling. *Biological Psychiatry, 91*(9), S87. https://doi.org/10.1016/j.biopsych.2022.02.234

Li, W., Lei, D., Tallman, M. J., Ai, Y., Welge, J. A., Blom, T. J., Fleck, D. E., Klein, C. C, Patino, L. R., Strawn, J. R., Gong, Q., Strakowski, S. M., Sweeney, J. A, Adler, C. M, & DelBello, M. P. (2022). Pretreatment alterations and acute medication treatment effects on brain task−related functional connectivity in youth with bipolar disorder: A neuroimaging randomized clinical trial. *Journal of the American Academy of Child & Adolescent Psychiatry, 61*(8), 1023−1033. https://doi.org/10.1016/j.jaac.2021.12.015

Mason, L., Peters, E., Williams, S. C., & Kumari, V. (2017). Brain connectivity changes occurring following cognitive behavioural therapy for psychosis predict long-term recovery. *Translational Psychiatry, 7*(1), e1001. https://doi.org/10.1038/tp.2016.263

Menon, V. (2011). Large-scale brain networks and psychopathology: A unifying triple network model. *Trends in Cognitive Sciences, 15*(10), 483−506. https://doi.org/10.1016/j.tics.2011.08.003

CHAPTER

How to organize a connectomics-driven neuroscience clinic 11

Introduction

TMS is thought to work by potentiating neuroplasticity for a period of around an hour where the stimulated cortex is in an electrically modified state which is thought to be prone to neuroplasticity (Gersner et al., 2011). Thus, this chapter is about how we have organized a practice entirely based around maximizing this window of opportunity.

In this chapter, we share logistical lessons and ideas we have learned building a clinic based in Sydney, Australia (Cingulum Health), which is capable of addressing neurological and psychocognitive disorders in a streamlined fashion. In other words, how do you use TMS and connectomic medicine in novel and agile ways to treat patients with a variety of different neurologic, emotional, or cognitive symptoms across a variety of diagnoses, and to combine patient-specific targeting with patient-specific rehab. We learned a lot of lessons creating this kind of clinic, as we did not have a template for this specific model.

Floor plan of the clinic

The aim of our clinic was to be as little like a medical practice place as possible. In an accelerated theta burst paradigm, patients are with us for a while, and spend time between stimulations undergoing other treatments. Thus, we do not want this to be anxiety provoking, artificial, or awkward.

Fig. 11.1 shows the floor plan for our clinic that is designed to integrate patient intake, consultation, administration, TMS treatment, and adjunctive therapies all under the same roof for seamless, personalized treatment based on connectomic medicine.

Processes to consider for running a TMS clinic

Patient referral. We will first start off with the issue of patient engagement and physician referral. Without patients, there can be no clinic. Depending on the scope of the reader's practice, it is important to advertise the concept of a connectomics-driven clinic to the referring providers. For clinicians who focus on neurological rehabilitation, such as postsurgical deficits or ischemic patients, the clinician should reach out to neurosurgeons and neurologists in the community. Those who are interested in treating psychiatric disorder should reach out to local psychologists and psychiatrists as these patients require the foremost attention of those primary mental health providers.

Connectomic Medicine. https://doi.org/10.1016/B978-0-443-19089-6.00006-9

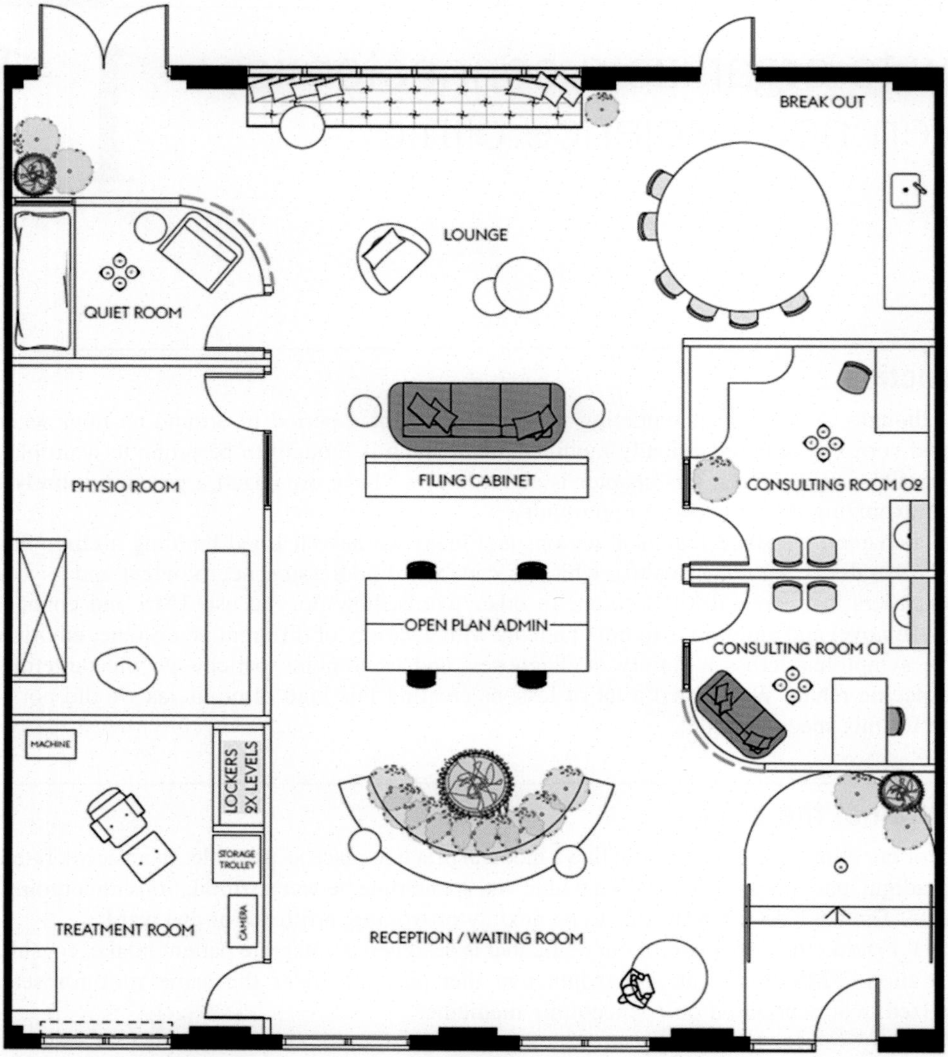

FIGURE 11.1

Floor plan of our connectomics-driven neuroscience clinic.

It is important to remember that TMS or other forms of brain simulation are seen as an adjunct to the gold standard of psychiatric care, which includes medication and cognitive behavioral therapies. Collaboration with these primary referring clinicians will be instrumental in the success of patient treatment and establishing such a practice. One of the hurdles in establishing such a clinic is for people to embrace this new field of medicine. It is a complex concept that is not easily understandable for people without adequate background in the subject matter. This is completely natural as our respective clinical training often only includes knowledge from the past and it is not forward-looking. Medicine is undoubtedly a highly conservative field and for good reasons to safeguard the welfare of the patients. Therefore, when one approaches potential referring providers, it is important to have a firm foundation on this field of medicine. One will begin to question the strength of evidence behind machine learning, industrial grade analytics of structural connectivity, and brain stimulation methods. It is important to emphasize that each of these components is built from solid research. For example, TMS is proven to be safe and efficacious and has the ability to induce lasting changes in connectivity (Einstein et al., 2022b; Hallett et al., 2017). Conversely, it is also at the same time important to concede that this is an emerging field. In the end, the decision to provide any treatment will be based on the risk-to-benefit ratio. With TMS being an extremely safe treatment and with positive evidence, it should not be hard to convince other practitioners in referring patients with various neurological and mental disorders who may have otherwise run out of effective treatment options.

Financial reimbursements. In the United States, TMS is only FDA-approved for the treatments of depression, obsessive compulsive disorder (OCD), and smoking cessation. There are specific insurance requirements prior to insurance approval to reimburse the cost of these treatments. Most guidelines from professional societies indicate TMS is a recognized treatment for patients who have not benefited from one or more antidepressant medications. However, coverage by commercial insurance companies is not uniform. Some companies require a patient to fail four antidepressant medication trials with the addition of psychotherapy in order to qualify reimbursement. Specifically for depression, FDA approval only applies to machines from specific vendors, with left DLPFC targeted at 120% of the motor threshold, with varying dosages, and more recently theta burst protocols have been approved. Also, FDA approvals for OCD and short-term smoking cessation using TMS do not mean that they would be covered by insurance in the United States. Therefore, this will continue to be an emerging topic, especially given the nature of personalized connectomic medicine.

Patient consent. As mentioned above, what we are discussing involves the next step in integrating functional connectomics into clinical practice. Most of the methods discussed here to treat various diseases are forward-looking and will be up to governmental scrutiny in the near future. It remains to be seen how regulatory bodies will react and readjust their policy-making based on personalized medicine. At the current standard, where an indication for TMS treatment is based on a specific brain target, which is unlikely to capture the pathophysiology of all patients, the use of TMS for brain modulation outside of depression, OCD, short-term smoking cessation, will be off-label. In light of this, it is very important to inform the patients of the ***off-label*** use of TMS. Any clinical practice should have a well-detailed consent form that details the potential benefits and risks of the procedure. This should be outlined before the patient visits the clinic and also reinforced during the consultation. It is important to outline the potential risks of seizures to the patients especially in those who have undergone cranial surgeries or have an underlying diagnosis of epilepsy.

Having said this, it is also important to point out the long track record of safety, and the extensive evidence showing utility of many of these off-label treatments (Loo et al., 2008). It is worth noting that being "on-label" means that a company applied to FDA or other regulators for a marketing claim. It does not mean that other treatments with the same device that they have not approached FDA are invalid, dangerous, or even unproven. In some cases, no one has bothered to fill-out the application. We have seen very clear benefits to patients with many off-label therapies (Dadario et al., 2022; Einstein et al., 2022a; Poologaindran et al., 2022; Stephens et al., 2021).

It should also highlight the idea that given the extensive data collection, paperwork, and time required to gain regulatory approval (we have done this many times and in most major jurisdictions on earth, trust us we know), and the rapid progress machine learning type approaches can provide to learning new reasonable variations of this therapy, some openness to new ideas will be necessary to keep pace with the rate of discovery, lest patients suffer unnecessarily while we wait to do the 201st TMS randomized trial before we decide to give an incredibly safe treatment aiming to help a patient who may lack other options for a horrible condition. Medicine is an art, and informed consent in this art involves balancing risks, benefits, and alternatives. In neuroscience, we are not treating a cold, and we need to remember how bad many of these problems are.

Workflow from consultation to treatment

Radiology. Once your practice has received the referral from an outside physician or from direct referral from the patient, it is in our opinion that the patients should arrive with relevant radiology prior to the visit. This includes obtaining an anatomical MRI along with DTI and resting state fMRI (rs-fMRI) imaging that is required, as mentioned in previous chapters. It is important to establish a relationship early on with your local radiology facilities and practitioners because most local community radiology facilities are likely not used to obtaining DTI and rs-fMRI imaging. This is despite the fact that most modern MRI machines are capable of doing so. It requires work and coordination with your local radiologists to coordinate the imaging acquisition protocols in order to obtain proper imaging fit for advanced analytics. This cannot be understated as this is likely to be **the major bottleneck** for establishing a personalized connectomics medicine clinic.

When do you reimage patients after treatment? At minimum, we do this if symptoms return and we are considering repeat treatment. Many patients want to obtain repeat scans after treatment to see if it improved their connectivity. We recommend at earliest waiting 1 month after the end of treatment before doing any repeat imaging. This is because the connectome is often very dynamic earlier and making sense of things is impossible until they hit steady state. Also, TMS response can take this much time to show a full effect also. Additionally, while it is fulfilling to see normalization of the connectomic changes you were treating, this does not always correlate perfectly and about 10% of time (approximately based on our gut feel really) responders show basically the same anomalies we treated. We are not sure what this means, but ultimately treat the *patient* not the *scan*.

Objective assessments. It is our practice to have the patient fill-out a detailed electronic form with inquiries regarding their detailed medical history, including mental health and social history, and also a consent detailing the potential benefits and risks of TMS treatment. At the time of visit, but before the actual consultation with the clinician, the patient is offered a psychological battery test if they are visiting for mental health disorders. Some examples would be the Phq9 or Beck's Depression Inventory for people afflicted with MDD or GAD7 for anxiety disorders. Each practice can select the test based on the clinicians' preferences. In our practice, we utilize the Cambridge Neuroscience Assessments for all of our patients in addition to standard battery tests. The point we are trying to make here is that we should keep track of a patient's progress pre- and post-treatment in an objective fashion. In addition to these specific tests, we also evaluate the patient's general quality of life through EQ-5D metrics.

We aggressively pursue data for all patients. There are four reasons for this. First, we want to know if what we do works. Second, when we find something that does work, we want to show that it does work. Third, and most uniquely, when we treat mentally ill patients, some of them have negative thought patterns, and will tell you they are sick when their data shows they are better, and this is helpful for putting partial responses in context. Finally, TMS can wear off over time, and as such it is useful to base retreatment decisions on objective recurrence of symptoms.

Consultation. This is perhaps the most important part of the workflow when the clinician interviews the patient. If a patient comes with a mental health complaint, it is important to ascertain all related symptoms in the most thorough fashion possible. One should elicit the beginning of symptom onset. The medical interview should be conducted in an open-ended, patient-centric fashion so that the patient may present symptoms or information that might not necessarily be connected to the clinical presentation at first glance, but may have implications in the pathogenesis of the condition. For example, patients with chronic tinnitus are often afflicted in the past with other psychiatric comorbidities such as depression and anxiety (Bhatt et al., 2017). If we only focus on the auditory aspect of the clinical presentation, then we would miss out on other comorbid conditions that might be part of the pathophysiology. In terms of the transdiagnostic framework, we would be missing out on interrogating networks that they may have an interplay with the auditory network as well. For patients with neurological deficits, it is very important to perform a thorough neurological examination that details the nature of the deficits. For example, if a stroke patient is not able to produce speech, one has to figure out whether it is a word finding problem or if it's a pharyngeal motor problem. Once again, this relates to the networks that we will be interrogating as part of the hypothesis generation. We will then integrate our clinical findings with expert opinions from prior psychiatrist, speech pathologist, neurologist, and other specialists to formulate an investigative protocol transdiagnostically using functional and structural connectomics.

Treatment. We perform accelerated theta burst protocols as they increase compliance, work faster, and make this kind of clinic possible (Caeyenberghs et al., 2018). The patient should be made aware of TMS treatment schedule so that they can arrange their lives accordingly. Even in the accelerated TMS protocol, it does require 1 week of uninterrupted commitment from the patient.

Between stimulation sessions, the patient needs to undergo some kind of therapy relevant to the area we are trying to treat (Vedeniapin et al., 2010). For speech or motor problems, this involves physical occupational, and/or speech therapy. For other conditions, therapies have to be more flexible, patient-specific, and often creative. You cannot get a patient to do meditation or use a CBT app if they are against this as a treatment.

Logistically, we find that it is of vital importance to have consistent partnership with established therapists who can also work consistently with the patient during this 1 week treatment. Note that coordinating this treatment to meet the patients' needs for the week requires agility, as we do not generally recommend directly employing therapists full time unless your clinic has a narrow type of patient being treated. A physical therapist will have nothing to do if you have a week of all depression patients. Additionally, we often have therapists who bring their patients to the clinic and participate in the therapies between TMS treatments as they are often impressed with the results we can obtain with difficult patients.

Follow-up and continued assessments. At the end of the treatment, we have patients complete the objective assessments to assess any change compared to pretreatment. We try to reassess the patient in person 1 month after treatment with repeated objective assessments using specific batteries. Following this, we give patients the choice of completing their assessments repeatedly at 3, 6, and 12 months. To maximize adherence to the completion of these assessments, we allow patients to remotely complete these assessments electronically. We take into account the patients' subjective and objective responses in order to determine whether there was treatment success or failure, or if any "top up" treatment is indicated, as in what is commonly seen in TMS treatments for depression (Jelovac et al., 2013; Senova et al., 2019).

Conclusion

In this chapter, we provide an expansive overview of how you can culminate the lessons discussed in previous chapters in order to create and manage a connectomics-driven neuroscience clinic. Many factors go into this process which must be considered, ranging from patient selection process to the radiological workflow to choosing which type of therapist is needed alongside the TMS treatment. In the next chapter, we discuss connectomic approaches for intraaxial brain surgery.

References

Bhatt, J. M., Bhattacharyya, N., & Lin, H. W. (2017). Relationships between tinnitus and the prevalence of anxiety and depression. *The Laryngoscope, 127*(2), 466–469. https://doi.org/10.1002/lary.26107

Caeyenberghs, K., Duprat, R., Leemans, A., Hosseini, H., Wilson, P. H., Klooster, D., & Baeken, C. (2018). Accelerated intermittent theta burst stimulation in major depression induces decreases in modularity: A connectome analysis. *Network Neuroscience, 3*(1), 157–172. https://doi.org/10.1162/netn_a_00060

Dadario, N. B., Young, I. M., Zhang, X., Teo, C., Doyen, S., & Sughrue, M. E. (2022). Prehabilitation and rehabilitation using data-driven, parcel-guided transcranial magnetic stimulation treatment for brain tumor surgery: Proof of concept case report. *Brain Network and Modulation, 1*(1), 48.

Einstein, E. H., Dadario, N. B., Khilji, H., Silverstein, J. W., Sughrue, M. E., & D'Amico, R. S. (2022a). Transcranial magnetic stimulation for post-operative neurorehabilitation in neuro-oncology: A review of the literature and future directions. *Journal of Neuro-Oncology, 157*(3), 445.

Einstein, E. H., Dadario, N. B., Khilji, H., Silverstein, J. W., Sughrue, M. E., & D'Amico, R. S. (2022b). Transcranial magnetic stimulation for post-operative neurorehabilitation in neuro-oncology: A review of the literature and future directions. *Journal of Neuro-Oncology*, 1–9.

Gersner, R., Kravetz, E., Feil, J., Pell, G., & Zangen, A. (2011). Long-term effects of repetitive transcranial magnetic stimulation on markers for neuroplasticity: Differential outcomes in anesthetized and awake animals. *Journal of Neuroscience, 31*(20), 7521–7526. https://doi.org/10.1523/jneurosci.6751-10.2011

Hallett, M., Di Iorio, R., Rossini, P. M., Park, J. E., Chen, R., Celnik, P., Strafella, A. P., Matsumoto, H., & Ugawa, Y. (2017). Contribution of transcranial magnetic stimulation to assessment of brain connectivity and networks. *Clinical Neurophysiology, 128*(11), 2125–2139. https://doi.org/10.1016/j.clinph.2017.08.007

Jelovac, A., Kolshus, E., & McLoughlin, D. M. (2013). Relapse following successful electroconvulsive therapy for major depression: A meta-analysis. *Neuropsychopharmacology, 38*(12), 2467–2474. https://doi.org/10.1038/npp.2013.149

Loo, C. K., McFarquhar, T. F., & Mitchell, P. B. (2008). A review of the safety of repetitive transcranial magnetic stimulation as a clinical treatment for depression. *International Journal of Neuropsychopharmacology, 11*(1), 131–147. https://doi.org/10.1017/S1461145707007717

Poologaindran, A., Profyris, C., Young, I. M., Dadario, N. B., Ahsan, S. A., Chendeb, K., Briggs, R. G., Teo, C., Romero-Garcia, R., Suckling, J., & Sughrue, M. E. (2022). Interventional neurorehabilitation for promoting functional recovery post-craniotomy: A proof-of-concept. *Scientific Reports, 12*(1), 3039. https://doi.org/10.1038/s41598-022-06766-8

Senova, S., Cotovio, G., Pascual-Leone, A., & Oliveira-Maia, A. J. (2019). Durability of antidepressant response to repetitive transcranial magnetic stimulation: Systematic review and meta-analysis. *Brain Stimulation, 12*(1), 119–128. https://doi.org/10.1016/j.brs.2018.10.001

Stephens, T. M., Young, I. M., O'Neal, C. M., Dadario, N. B., Briggs, R. G., Teo, C., & Sughrue, M. E. (2021). Akinetic mutism reversed by inferior parietal lobule repetitive theta burst stimulation: Can we restore default mode network function for therapeutic benefit? *Brain and Behavior, 11*(8), e02180. https://doi.org/10.1002/brb3.2180

Vedeniapin, A., Cheng, L., & George, M. S. (2010). Feasibility of simultaneous cognitive behavioral therapy and left prefrontal rTMS for treatment resistant depression. *Brain Stimulation, 3*(4), 207–210. https://doi.org/10.1016/j.brs.2010.03.005

CHAPTER

Connectomic approaches to neurosurgical planning

12

Introduction

When we operate in the cerebral cortex for some reason, we are hampered by the near total lack of anatomic landmarks visible to the naked eye. Thus, we are forced to try to define anatomy, and remain safe within those confines. Dr. Sughrue wrote an entire textbook previously on this topic (Baker, Burks, Briggs, Conner, et al., 2018), so we will not be able to exhaustively cover all of these issues, but this chapter highlights some of the key issues, and makes it easier to get the basic idea of how connectomics help with this goal.

Intracranial neurosurgery carries a significant rate of causing new neuropsychocognitive deficits (Drewes et al., 2018; Rijnen et al., 2019). This mainly has to do with violation of brain matter during the surgical approach. This is not intentional, but usually occurs in large part because neurosurgeons lack an understanding of how the cognitive and emotional systems of the brain are organized and thus lack a mental idea on how to avoid these problems (Dadario et al., 2021).

In the past two decades, neurosurgeons have utilized diffusion tractography imaging and task-based functional MRI for surgical planning in order to avoid transgressing important and eloquent brain areas (Fig. 12.1). However, the interpretation of diffusion tractography is usually limited to that of corticospinal tracts. Task-based functional MRIs are in the same manner limited to the interpretation of sensorimotor areas and language areas. They are not as useful for identifying areas of higher cognitive function, which are generally more complex networks that can be done using manual segmentation.

It is clear from the literature that basically all brain tumor patients benefit from having as much of the tumor removed as possible; however, sometimes this means a cure and other times a very short extension of life. However, we know the transgression of certain brain structures will lead to suboptimal neurocognitive outcomes. Therefore, the modern philosophy for brain tumor surgery must be to achieve onco-functional balance. What that means is to find the perfect balance of achieving maximal safe resection while keeping in mind that a patient will need to be fully functional in order to continue to have an acceptable quality of life.

In the past, the term "*eloquence*" was the dividing line for this decision-making process (Kahn et al., 2017). This has been with us for sometime, but was strengthened into dogma by a landmark paper by Dr. Robert Spetzler that stratifies the preoperative risks of arteriovenous malformation resection (Spetzler & Martin, 1986). In general, that paper identified eloquent brain areas as those involved in motor, language, and deep brain structures, such as the thalamus. While easy to learn, this concept neglects other areas of the brain that might be responsible for subtle yet more complex

Connectomic Medicine. https://doi.org/10.1016/B978-0-443-19089-6.00011-2

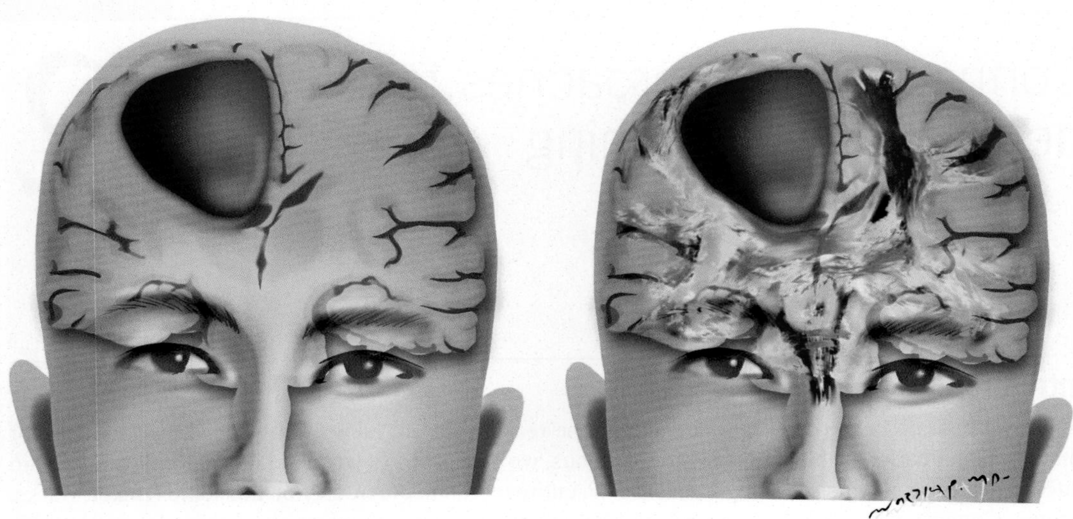

FIGURE 12.1

This figure illustrates basic DTI imaging with whole-brain tractography surrounding a right-hemispheric tumor.

neurological function, which may have complex interactions, and which may individually be responsible for specific degrees of redundancy and tolerance.

Overall, the field has been fairly successful in avoiding obvious neurological deficits including language and motor skills by utilizing our anatomical knowledge of specific cortical structures. In addition, when appropriate, we can perform awake surgery to monitor obvious neurocognitive functions. The question with awake surgery is how do we monitor highly complex functions, such as our perception of the outside world, internal processing of our emotions, or simply how we deal with other humans. It is almost inconceivable to test for these functions in a reliable and meaningful way during an already taxing procedure for the patient who is under minor sedation. These methods are time-consuming, increase surgical time, and are not applicable to most facilities outside of academic centers. Even more so, many brain tumor patients may be unable to actually complete the task.

Thus, current advancements in clinical neurosurgery are not enough to prevent subtle deficits seen in patients with higher order cognitive functions, nor does it explain why we can perform the same surgery in traditionally "non-eloquent" tissue 100 times, but there will always be a few patients who end up with different outcomes. How then, in the era of personalized connectomic medicine, do we determine our surgical approach based on functional connectivity and their inherent importance?

We now have computational advancements involving machine learning and artificial intelligence for us to appreciate highly dimensional neural imaging data so that we can start interpreting the vast amount of raw data in a clinically meaningful manner (Yeung, Taylor, Nicholas et al., 2021). In order for us to integrate connectomic imaging into planning brain surgery, we have the following considerations.

(1) Thorough history taking. When we encounter a patient afflicted with an intraaxial brain condition, such as a brain tumor, it is important to perform thorough history taking in order to identify deficits incurred by the tumor itself. Most of the time, neurosurgeons tend to focus solely on the physical examination, trying to identify neurological deficits. However, it is more seldom that neurosurgeons ask about changes in a patient's quality of life and also whether a patient is still able to perform daily activities of living that require higher cognitive functions, such as the ability to organize payments and paying attention to schedules. We cannot learn from our mistakes if we are unaware of them.

(2) Fund of anatomical knowledge. Before integrating connectivity and network anatomy into neurosurgical planning, we ought to have anatomical familiarity surrounding the tumor location. It is essential to study the basic structural MRI information in order to understand whether a tumor is invasive or well circumscribed as that would influence the method of surgical approach.

(3) Obtaining connectomic data. The step of obtaining DTI or functional MRI imaging might seem trivial, given almost all of these patients get image guided MRIs anyway for surgery, but we have to assess whether a patient is able to remain still in the MRI scanner in order to participate in these studies. Furthermore, it is important to work with your local radiology center to make sure that they are capable of executing the protocols needed in the above image and the postprocessing. Once we have obtained DTI imaging, we can use advanced analytics and visualization to strategize our surgical approach to an intracerebral lesion (Dadario & Sughrue, 2022a).

Preoperative planning. We can now overlay cerebral networks onto individual brains, taking into accounts of cerebral edema and mass effect, using techniques detailed in earlier chapters (Yeung, Taylor, Nicholas et al., 2021). The next step that we should undertake is to identify networks that are immediately adjacent to the lesion. Part of the beauty of integrating connectomics into presurgical discussion is the ability to better inform the patient of potential neurocognitive deficits that may arise from the surgical approach (Gao et al., 2022). Before, using canonical anatomical knowledge, we may be able to tell people whether they're going to have a language deficit or not. However, the integration of connectomics into surgical planning allows us to consider the potential ramifications of surgery involving higher cognitive function, such as our ability to pay attention to our surroundings or our ability to process external and internal information.

In the field of neurosurgical oncology, the extent of resection must be balanced with the functional outcome (Tang et al., 2022). Beyond just quality of life, functional status can affect eligibility for chemoradiation or clinical trials. From a complication standpoint, whether from a medicolegal or emotional acceptance standpoint, an unexpected complication is vastly different than a predicted risk based on functional connectomics that has been accepted by the patient before surgery.

Operative strategy. Most neurosurgeons are trained to debulk a tumor and resect outwards toward the tumor-brain interface. This strategy can be combined with subcortical stimulation and surgery to identify important fiber tracks, such as the corticospinal tract. We will not belabor the limitations of this approach, but rather discuss the concept of "**disconnection surgery.**" The philosophy of disconnection surgery is founded on the principle of determining whether a network needs to be preserved or could be sacrificed. So how exactly do we determine whether a part of a brain network can be cut out without enduring the neurocognitive sequelae? This is the point where we introduced the concepts of ***essentiality*** versus ***redundancy*** (Hennig et al., 2018). We know from the accumulated wisdom of past neurosurgeons that we can successfully remove parts of the brain without incurring permanent neurological deficit.

Such an example would be the supplementary motor area, which is commonly involved in low grade gliomas. The complete resection of this area usually results in SMA syndrome, where the patient has trouble initiating action, language, or even thoughts (Palmisciano et al., 2022). This is due to the fact that the SMA area is part of the prefrontal cognition initiation axis (Briggs et al., 2021). From retrospective studies and experience passed on from our predecessors, we know that most people will recover from SMA syndrome as long as parts of the bridging frontal aslant tract connecting to the contralateral SMA area is preserved, allowing the contralateral SMA area to compensate (Baker, Burks, Briggs, Smitherman, et al., 2018; Palmisciano et al., 2022). This is a great example of redundancy in the cerebral networks.

Redundancy indicates that the loss of some areas can be compensated for, often over time. One example is explained by the complete destruction of the primary motor cortex. Unless an individual is of young age with notable brain plasticity, it is unlikely that the patient's contralateral motor function will ever recover. This is an example of essentiality. Admittedly, the field of connectomic medicine is constantly evolving. We do not yet have enough information to completely list which parcellation within a network is essential or redundant. However, it is usually a good assumption that any central hub of a network, such as the splenial portion of the default mode network, is vital without redundancy. We will discuss later on about the current concepts and the future directions in determining the essentiality or eloquence of a specific brain network parcellation (Tanglay, Young, et al., 2022).

Once we are able to identify the networks bordering the neurosurgical lesion, we then determine whether the network can be safely sacrificed or not (see subsequent section). We can plan the circumference of the craniotomy to encompass specific cuts that we plan to perform on the periphery of the tumor such that we can disconnect the tumor from the surrounding networks. This operative strategy is termed "disconnection surgery." Using the example from Fig. 12.2, the tumor encompasses the left insula (Fig. 12.2A) and the anterior portion of the temporal lobe (Fig. 12.2B). In order to excise the tumor and to gain unhindered access to the insula, we can first disconnect the anterior temporal portion of the tumor in front of the presumed Wernicke portion of the language network, thinking access to the insular portion of the tumor. In another example of the right frontal tumor (Fig. 12.3), we visualize the tumor being surrounded laterally by the central executive network, posteriorly by the sensory motor network, medial and deep by the singular hub of the default mode network. In this specific scenario, we can plan a craniotomy to make specific linear cuts to isolate the tumor from the surrounding networks. Attention would then need to be paid to include SMA area if it is involved, but spare the crossing fiber of the frontal aslant tract, which would be deep and medial to the tumor. This type of disconnection surgical strategy enables precise and efficient supramaximal resection of the tumor while taking into considerations of the surrounding networks.

Explanations of noncanonical postoperative deficits

Neuropsychocognitive dysfunction after intracerebral surgery is a well-documented phenomenon, but frequently left unaddressed. These deficits incur significant socioeconomic costs as patients cannot fully participate in normal daily activities of living and they cannot contribute effectively to society. At this time, the neurological community does not have the tools to readily predict or explain atypical postoperative deficits that are outside of obvious motor and language functions. How then can we leverage connectomic medicine to better account and predict these unwanted outcomes?

(A)

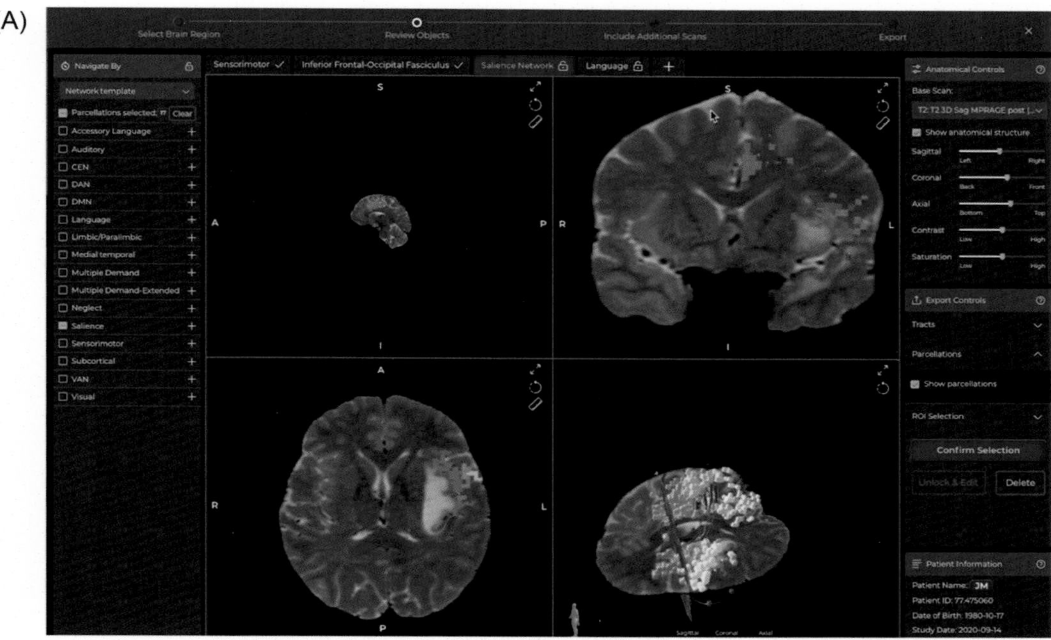

(B)

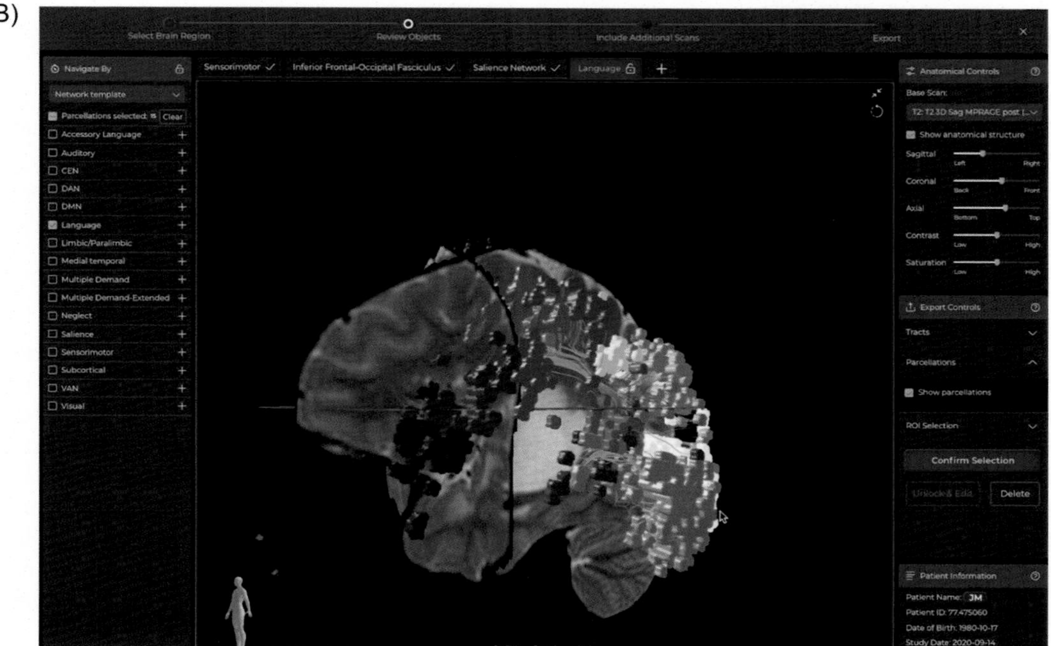

FIGURE 12.2

A case example of a tumor compressing relevant network anatomy in the Quicktome software. This case highlights a tumor compressing relevant connectomic structures along the left insula (A) and anterior portion of the temporal lobe (B).

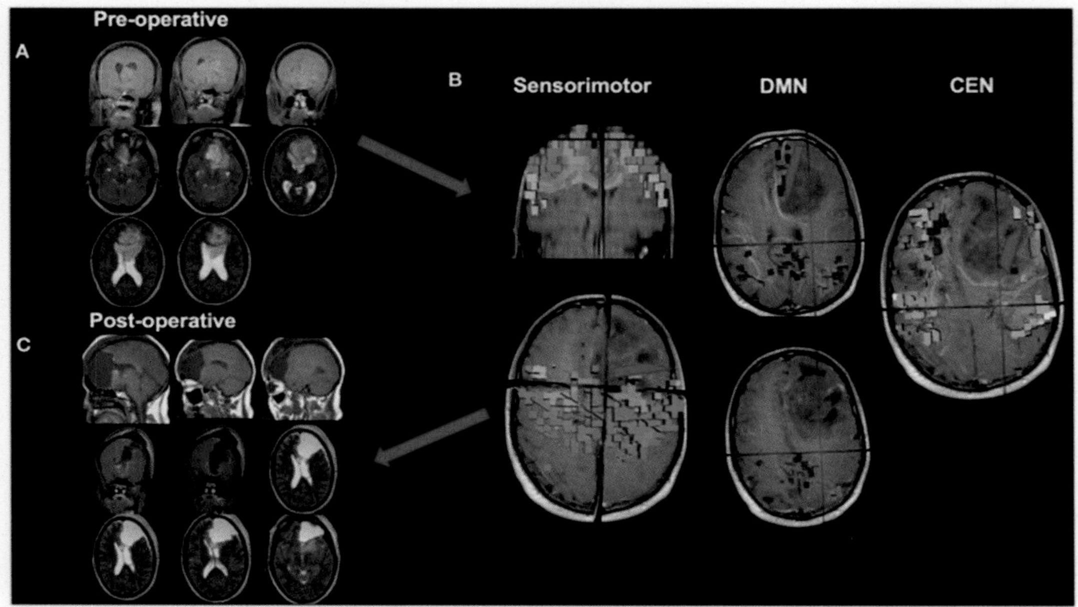

FIGURE 12.3

Disconnection surgery for a right frontal tumor. (A) This seemingly complex case can be reduced to a series of disconnections that can be defined against three known brain networks. From a network perspective, the boundaries are as follows: the DMN on medial boundary, the CEN on the lateral boundary, and the sensorimotor network on the posterior boundary. (B) The neurosurgeon can make more informed decisions during surgery by understanding information on the spatial relationship of the tumor to relevant white matter tracts and major networks, such as where or how far to disconnect normal tissue (blue lines) infiltrated by a tumor up until certain key fibers or nodes are met. This decision should be made based on patient predefined goals and patient prognosis among other factors, and further work will hopefully clarify how much of specific networks can be safely disconnected without compromising certain functions (C).

Adapted from Dadario et al. (2021).

Pay attention to what others have discovered. Neurosurgeons must acknowledge that traditional neurosurgical training does not encompass the findings by modern neuroscientists. This includes the study of brain networks involved in higher complex cognitive functions (Dadario & Sughrue, 2022b; Morell et al., 2022; Wu et al., 2022). In the patient above (Fig. 12.2), the patient had preoperative depression that was severely exacerbated after resection of her left insular temporal tumor. This can be attributed to the initial compression of the anterior insular hub of the salience network and further exacerbation from tumor resection. We know from recent studies that the salience network is involved in the processing of external and internal stimuli (Briggs et al., 2022). This includes the processing of one's own emotions and one's reactions to outside cues. Once we acknowledge the involvement of these networks in higher cognitive functions (see Chapter 3), we have found that many of these problems stop being unsolved mysteries: *we know why they happened.*

Not everyone is wired the same. An inherent limitation in the current field in functional connectomics is the assumption that everyone is wired the same. We know that this is not true because our connectome is likely the culmination of genetics and environmental factors during our upbringing, sculpted by our cultural differences as well. It would therefore make sense that we should take caution in assuming that everyone's functional networks, broken down into parcellations, are comparable to one another.

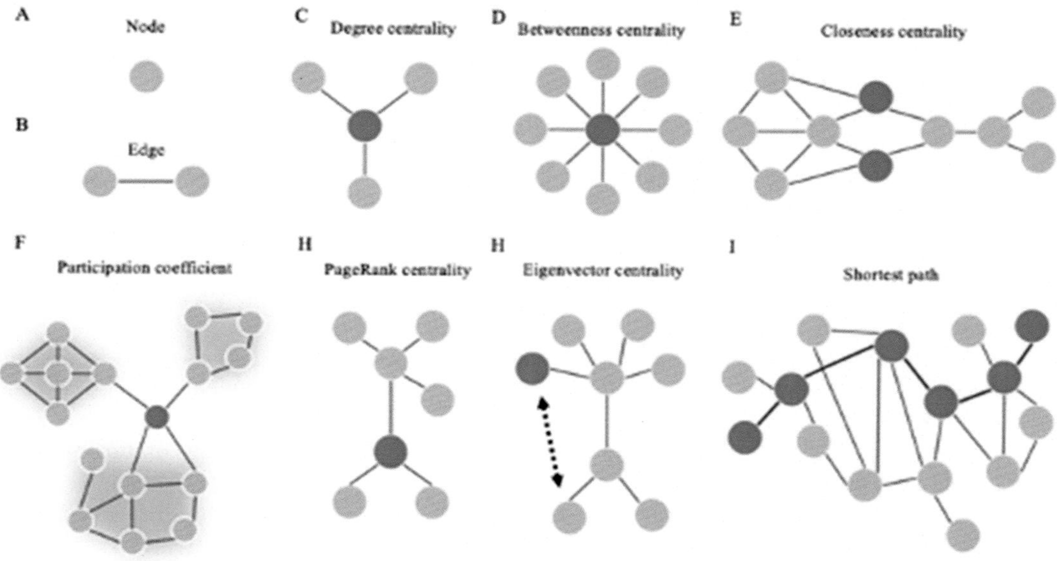

FIGURE 12.4

Graph theoretical measures of network centrality. In graphs of the brain, nodes (A) correspond to parts of the brain, while edges (B) signify either structural or functional connections. The nodes in purple display the highest corresponding centrality. (C) The degree centrality is the number of direct neighbors a node has, such that the degree of the purple node is 3. (D) The betweenness centrality measures the extent to which a node acts as a bridge between two other nodes. (E) Closeness centrality measures how fast a given node can access every other node in the graph. (F) Participation coefficient is a measure of the distribution of a node's connections among the modules in a network, with each module represented in orange. (G) PageRank centrality scales the influence of a node's neighbors by its degree. (H) Eigenvector centrality considers the quality of a node's neighbors when quantifying its centrality. The purple node has a higher eigenvector centrality than the gray node (pointed to by the dashed arrow), despite their degrees being the same. (I) Path length is a measure used to calculate the efficiency of information flow in a given network.

Adapted from Tanglay, Dadario, et al. (2022), Tanglay, Young, et al. (2022).

With advancements in machine learning and artificial intelligence, we are able to compare individual connectome to a "standard" connectome derived from a population average. This is one way of how we can identify whether one's connectome is vastly different than what would be expected at a population level. We will discuss this approach in later chapters when we discuss how we select targets for personalized brain stimulation. However, this approach does not provide mechanistic insight into how we are wired, differently, or whether there are specific parcellations that we should avoid transgressing during surgery.

One promising method to potentially tackle this problem is to employ graph theory to understand how important each parcellation is within the framework of brain connectivity (Hart et al., 2016). If we treat each parcellation as a node within a web of connections, then we can calculate the importance of each node based on their connections and their secondary and even tertiary connections, and so forth (Fig. 12.4). It turns out that if we employ graph theory to analyse parcellations of all the brain networks, we can recreate high-ranking parcellations that resemble brain areas that are traditionally considered as eloquent (Ahsan et al., 2020). Taking this concept one step further, we have identified that in a small percentage of the population, otherwise low-ranking parcellations at a population level can be high-ranking and "unexpected hubs" for some individuals (Yeung, Taylor, Young et al., 2021). These findings shed light on why there could be unexpected outcomes in a small percentage of patients from surgeries that are routinely performed in the same fashion and in the same brain regions. Using insight from novel metrics, such as those derived from graph theory, in addition to traditional anatomical models can provide a more inisghtful and individualized approach to network-based neurosurgery (Hormovas et al., 2023).

In summary, neurosurgeons are forced to do dangerous procedures in a setting of inadequate information about how the brain works. This is true for most of brain medicine, including neurology and psychiatry: *we are often forced to act with what information we have*. Having anatomic representation of the brain networks linked to the image guidance, they can incorporate the concepts of disconnection surgery, and make safer and more deliberative cuts into the brain to separate it from the tumor tissue. The understanding of the surrounding networks will enable neurosurgeons to better inform the patients preoperatively with regards to the risks of surgery and possible noncanonical neurological deficits that may occur post-surgery (Dadario & Sughrue, 2022b; Morell et al., 2022). Furthermore, the incorporation of machine learning and graph theory will enable neurosurgeons to avoid the transgressions of areas that are not traditionally considered as eloquent, but may embody unexpected importance at the individual level.

Conclusions

In this chapter, we discussed how connectomics can provide an improved understanding of the structural and functional anatomy of the human brain for brain tumor surgery. In disconnection surgery, a series of surgical cuts can be planned and defined against the known network architecture so as to maximize the onco-functional after surgery. Furthermore, novel ways of examining this data such as with graph theory allows us to better consider important and eloquent anatomy on an individualized basis. In the next chapter, we discuss how this information can be leveraged to guide neurocognitive rehabilitation after surgery.

References

Ahsan, S. A., Chendeb, K., Briggs, R. G., Fletcher, L. R., Jones, R. G., Chakraborty, A. R., Nix, C. E., Jacobs, C. C., Lack, A. M., Griffin, D. T., Teo, C., & Sughrue, M. E. (2020). Beyond eloquence and onto centrality: a new paradigm in planning supratentorial neurosurgery. *Journal of Neuro-Oncology, 146*(2), 229–238. https://doi.org/10.1007/s11060-019-03327-4

Baker, C. M., Burks, J. D., Briggs, R. G., Conner, A. K., Glenn, C. A., Sali, G., McCoy, T. M., Battiste, J. D., O'Donoghue, D. L., & Sughrue, M. E. (2018). A connectomic atlas of the human cerebrum-chapter 1: Introduction, methods, and significance. *Operative Neurosurgery (Hagerstown, Md.), 15*(Suppl_1), S1–S9. https://doi.org/10.1093/ons/opy253

Baker, C. M., Burks, J. D., Briggs, R. G., Smitherman, A. D., Glenn, C. A., Conner, A. K., Wu, D. H., & Sughrue, M. E. (2018). The crossed frontal aslant tract: A possible pathway involved in the recovery of supplementary motor area syndrome. *Brain and Behavior, 8*(3), e00926. https://doi.org/10.1002/brb3.926

Briggs, R. G., Allan, P. G., Poologaindran, A., Dadario, N. B., Young, I. M., Ahsan, S. A., Teo, C., & Sughrue, M. E. (2021). The frontal aslant tract and supplementary motor area syndrome: Moving towards a connectomic initiation axis. *Cancers, 13*(5), 1116. https://doi.org/10.3390/cancers13051116

Briggs, R. G., Young, I. M., Dadario, N. B., Fonseka, R. D., Hormovas, J., Allan, P., Larsen, M. L., Lin, Y. H., Tanglay, O., Maxwell, B. D., Conner, A. K., Stafford, J. F., Glenn, C. A., Teo, C., & Sughrue, M. E. (2022). Parcellation-based tractographic modeling of the salience network through meta-analysis. *Brain and Behavior, 12*(7), e2646. https://doi.org/10.1002/brb3.2646

Dadario, N. B., & Sughrue, M. E. (2022a). Advanced neuroimaging of the subcortical space: Connectomics in brain surgery. In *Subcortical neurosurgery* (pp. 29–47). Cham: Springer.

Dadario, N. B., & Sughrue, M. E. (2022b). Should neurosurgeons try to preserve non-traditional brain networks? A systematic review of the neuroscientific evidence. *Journal of Personalized Medicine, 12*(4), 587.

Dadario, N. B., Brahimaj, B., Yeung, J., & Sughrue, M. E. (2021). Reducing the cognitive footprint of brain tumor surgery. *Frontiers in Neurology, 1342*.

Drewes, C., Sagberg, L. M., Jakola, A. S., & Solheim, O. (2018). Perioperative and postoperative quality of life in patients with glioma—A longitudinal cohort study. *World Neurosurgery, 117*, e465–e474. https://doi.org/10.1016/j.wneu.2018.06.052

Gao, M., Lam, C. L. M., Lui, W. M., Lau, K. K., & Lee, T. M. C. (2022). Preoperative brain connectome predicts postoperative changes in processing speed in moyamoya disease. *Brain Communications, 4*(5). https://doi.org/10.1093/braincomms/fcac213. fcac213.

Hart, M. G., Ypma, R. J., Romero-Garcia, R., Price, S. J., & Suckling, J. (2016). Graph theory analysis of complex brain networks: New concepts in brain mapping applied to neurosurgery. *Journal of Neurosurgery, 124*(6), 1665–1678. https://doi.org/10.3171/2015.4.Jns142683

Hennig, J. A., Golub, M. D., Lund, P. J., Sadtler, P. T., Oby, E. R., Quick, K. M., Ryu, S. I., Tyler-Kabara, E. C., Batista, A. P., Yu, B. M., & Chase, S. M. (2018). Constraints on neural redundancy. *Elife, 7*. https://doi.org/10.7554/eLife.36774

Hormovas, J., Dadario, N. B., Tang, S. J., Nicholas, P., Dhanaraj, V., Young, I., Doyen, S., & Sughrue, M. E. (2023). Parcellation-based connectivity model of the judgement core. *Journal of Personalized Medicine, 13*(9), 1384. https://doi.org/10.3390/jpm13091384

Kahn, E., Lane, M., & Sagher, O. (2017). Eloquent: History of a word's adoption into the neurosurgical lexicon. *Journal of Neurosurgery, 127*(6), 1461–1466. https://doi.org/10.3171/2017.3.JNS17659

Morell, A. A., Eichberg, D. G., Shah, A. H., Luther, E., Lu, V. M., Kader, M., Higgins, D. M. O., Merenzon, M., Patel, N. V., Komotar, R. J., & Ivan, M. E. (2022). Using machine learning to evaluate large-scale brain networks in patients with brain tumors: Traditional and non-traditional eloquent areas. *Neuro-oncology Advances, 4*(1), vdac142. https://doi.org/10.1093/noajnl/vdac142

Palmisciano, P., Haider, A. S., Balasubramanian, K., Dadario, N. B., Robertson, F. C., Silverstein, J. W., & D'Amico, R. S. (2022). Supplementary motor area syndrome after brain tumor surgery: A systematic review. *World Neurosurgery, 165*, 160.e2-171.e2. https://doi.org/10.1016/j.wneu.2022.06.080

Rijnen, S. J. M., Kaya, G., Gehring, K., Verheul, J. B., Wallis, O. C., Sitskoorn, M. M., & Rutten, G. M. (2019). Cognitive functioning in patients with low-grade glioma: effects of hemispheric tumor location and surgical procedure. *Journal of Neurosurgery, 133*(6), 1671–1682. https://doi.org/10.3171/2019.8.JNS191667

Spetzler, R. F., & Martin, N. A. (1986). A proposed grading system for arteriovenous malformations. *Journal of Neurosurgery, 65*(4), 476–483. https://doi.org/10.3171/jns.1986.65.4.0476

Tang, S. J., Holle, J., Lesslar, O., Teo, C., Sughrue, M., & Yeung, J. (2022). Improving quality of life post-tumor craniotomy using personalized, parcel-guided TMS: safety and proof of concept. *Journal of Neuro-Oncology, 160*(2), 413–422. https://doi.org/10.1007/s11060-022-04160-y

Tanglay, O., Dadario, N., Chong, E., Tang, S., Young, I. M., & Sughrue, M. E. (2022). Graph theory measures and their application to neurosurgical eloquence. *Cancers*.

Tanglay, O., Young, I. M., Dadario, N. B., Taylor, H. M., Nicholas, P. J., Doyen, S., & Sughrue, M. E. (2022). Eigenvector PageRank difference as a measure to reveal topological characteristics of the brain connectome for neurosurgery. *Journal of Neuro-oncology, 157*(1), 49–61. https://doi.org/10.1007/s11060-021-03935-z

Wu, Z., Hu, G., Cao, B., Liu, X., Zhang, Z., Dadario, N. B., Shi, Q., Fan, X., Tang, Y., Cheng, Z., Wang, X., Zhang, X., Hu, X., Zhang, J., & You, Y. (2023). Non-traditional cognitive brain network involvement in insulo-Sylvian gliomas: a case series study and clinical experience using Quicktome. *Chinese Neurosurgical Journal, 9*(1), 16. https://doi.org/10.1186/s41016-023-00325-4

Yeung, J. T., Taylor, H. M., Nicholas, P. J., Young, I. M., Jiang, I., Doyen, S., Sughrue, M. E., & Teo, C. (2021). Using quicktome for intracerebral surgery: Early retrospective study and proof of concept. *World Neurosurgery, 154*, e734–e742. https://doi.org/10.1016/j.wneu.2021.07.127

Yeung, J., Taylor, H., Young, I., Nicholas, P., Doyen, S., & Sughrue, M. (2021). Unexpected hubness: A proof-of-concept study of the human connectome using pagerank centrality and implications for intracerebral neurosurgery. *Journal of Neuro-Oncology, 151*, 1–8. https://doi.org/10.1007/s11060-020-03659-6

CHAPTER

Connectomic strategies for post-neurosurgical applications

13

Introduction

Brain surgery is difficult.

It is no secret that brain surgery is a high-risk endeavor. Specifically, surgical approaches that involve violation of the brain parenchyma are associated with not just neurological deficits, but also psychocognitive ones (Dadario & Sughrue, 2022; Morell et al., 2022; Rijnen et al., 2019). The Glioma Outcomes Project, a concerted effort in the early 2000s that included the tracking of 800 patients who had undergone excision of gliomas in the United states, has taught us that up to 32%, 35%, and 16% of patients had motor deficits, memory loss, and altered levels of consciousness, respectively. What is most fascinating is that over 90% of patients reported depression like symptoms compared to a baseline 15% of patients before surgery.

It should be no surprise to neurosurgeons that violation of the brain would incur alterations in structural and functional connectivity. The combination of potential neurological and psychocognitive deficits can hinder a patient from returning to normal life. What is also of clinical importance is these limitations may preclude patients from further adjunctive chemotherapy and radiation treatments, thus resulting in shorter survival. We also have to ask the question—*in patients with high grade malignancies, how can we best improve the quality of life in these patients with limited life span?* Even if we employ the strategies outlined in Chapter 12 to minimize surgical insults to the brain, what if complications were to occur? What if we wanted to intentionally disrupt the system for the sake of maximizing resection? Is there potentially a way to salvage function based on redundancy in the networks?

Importantly, as discussed in previous chapters, brain activity can be modulated post-craniotomy in a very safe, feasible, and effective way with transcranial magnetic stimulation (TMS) with the goal of functional recovery (Einstein et al., 2022). Below, we discuss how to facilitate neurocognitive rehabilitation for a variety of postoperative deficits which may be encountered after brain surgery. We define how to consider the problem, the relevant network anatomy involved, and how to begin addressing this anatomy.

Connectomic Medicine. https://doi.org/10.1016/B978-0-443-19089-6.00005-7

TMS principles for post-neurosurgical rehab: language disturbances

Step 1: Define the problem

Speech and language disturbances can be a complex combination of phonemic, motor, initiation, verbal memory, or semantic disturbances, and this needs to be clarified as different problems are caused by different systems.

Step 2: Look at the relevant parts of structural connectome

Language network
Accessory Language system
Face motor
Contralateral SMA

Step 3: Look at the relevant parts of the functional connectome

Specifically, looking for either anomalies or lack of correlation in parts of the networks:

Language network
Accessory language system
Face motor
Contralateral SMA
Basal ganglia

Step 4: Determine a treatment which addresses these issues and integrates into the overall care plan

This is usually speech and language rehab modified to the specific situation.

TMS principles for post-neurosurgical rehab: sensorimotor problems

Step 1: Define the problem

Motor problems can be weakness, incoordination, lack of eccentric control, or rigidity. It is obviously important to distinguish arm, face, or leg weakness as these involve different portions of the motor system.

Step 2: Look at the relevant parts of structural connectome

Bilateral motor cortices
Primary sensory cortex
SMA
Corticospinal tract
Dorsal and ventral attention network
Basa ganglia

Step 3: Look at the relevant parts of the functional connectome

Looking for contralateral hyperconnectivity or ipsilateral hypoconnectivity, or anomalies in the following:

Sensorimotor network
Basal ganglia

Step 4: Determine a treatment which addresses these issues and integrates into the overall care plan

This is usually combined with physical therapy aimed at the affected limb.

TMS principles for post-neurosurgical rehab: alertness and initiation problems

Step 1: Define the problem

Patients who wake up with the inability to meaningfully participate in therapy can develop this issue from a variety of causes. Famously, the most serious injuries occur from reticular activating system and thalamic injuries, but these need to be differentiated from SMA syndrome, akinetic mutism, initiation problems, and forms of abulia, which can look very severe also.

Step 2: Look at the relevant parts of structural connectome

DMN
Salience
SMA
Basal ganglia and thalamus
Brainstem, especially reticular activating system

Step 3: Look at the relevant parts of the functional connectome

Looking for hypoconnectivity, or anomalies in the following:

DMN
Salience
SMA
Basal ganglia and thalamus
Brainstem

Step 4: Determine a treatment which addresses these issues and integrates into the overall care plan

This is usually with coma stimulation like therapeutic maneuvers.

Personalized, multinetwork rTMS treatment for postsurgical neurorehabilitation

The transdiagnostic approach goes beyond our foundational, yet oversimplified, understanding of brain functions. It is increasingly gaining traction in the mental health community for understanding psychiatric disorders with myriad overlapping symptoms. In the setting of patients who had undergone cranial neurosurgery and are afflicted with physical and mental ailments, the transdiagnostic approach allows us to breakdown the patient symptoms by networks, namely identifying aberrations in brain circuitry. Especially given the fact that tumor localization and symptom manifestations can be variable, in addition to variable patient demographics, a personalized, targeted approach is needed to enhance postsurgical neurorehabilitation outside of what we already offer in the form of supportive physical, occupational, speech therapies, and psychoactive medications (Poologaindran et al., 2022).

Our ability to target multiple symptoms in various regions of the brain stems from our enhanced ability to understand functional connectivity at a personalized, semi-quantitative manner. In our experience, patients oftentimes can present with more than just neurological deficits. This goes back to our earlier points about thorough history taking. This is evidence-based in the sense that data has shown patients are more likely to have postsurgical mental disorders (D'Angelo et al., 2008). All we have to do is ask. The goal for all is to not just address the physical and mental disabilities after neurosurgery, but to also enhance the quality of life outcomes in these patients.

There is an established literature on the role of TMS in motor and language recovery in stroke patients. We derive our understanding for neurohabilitation for postsurgical patients using the fundamental knowledge derived from prior experiences (He et al., 2020). It is established that TMS is useful in poststroke hemiplegia and cortical reorganization can take place between primary and secondary motor cortices (Sharma et al., 2009). Similar strategies can theoretically be used to improve language functions as well (Kawashima et al., 2013). Clearly, TMS does not create de novo connections that never existed, but it may allow the reorganization of existing circuitry to compensate for new deficits and recruits what clinicians usually notes as there is "neurological reserve." Some examples are provided below.

Case study #1 (Fig. 13.1)

We have a 63-year-old right-handed gentleman who presented with right-sided paralysis following resection of a grade 3 glioma located in the left posterior thalamus with extension into the midbrain period (Fig. 13.1A). The surgery itself was uneventful, but post-surgically, the patient was rendered paralyzed on the left side. He had no function in the right lower extremity and trace movement only in the right upper extremity.

This possible postsurgical sequelae was thoroughly discussed preoperatively with the patient who agreed with radical resection of the tumor. The tumor was located more posteriorly in the thalamus and the motor deficit was out of proportion to the radiographic resection. Upon review of the sensory motor tractography, it was noted that although slightly diminished in numbers compared to the right side, the corticospinal tracts remain robust (Fig. 13.1B). However, now that this has occurred, how do we identify the issue and try to augment a patient physiological reserve to regain any of his movements? We turned attention to the patient's anomaly matrix focusing on the sensory motor network and their connections with subcortical structures (Fig. 13.1C). We identify robust, inconsistent negative correlations of the left ventral diencephalon (L_ventralDC) with other subcortical structures. We further noted consistent abnormalities in area 6 bilaterally (L_6v and R_6v). Brodmann area 6 is located ventral to the primary motor cortex; and it is responsible for motor planning. As both of these areas demonstrate significantly positive correlations (hyperactive) with other parcellations of the sensory motor network, a decision was made to target bilateral area 6 with continuous TBS (cTBS) for theoretical inhibition. The patient was treated within accelerated protocol with five sessions daily for 5 days at 80% of the MTS, determine on the unaffected side. During the simulation sessions, the patient was engaged in intense physical therapy to address the side of paralysis. After 5 days of treatment, the patient had notable improvement in both the upper extremity and lower extremity functional scores (Figs. 13.1D). At 4 months after treatment, the patient has sustained improvement.

Our ability to target more than one brain area utilizes the idea that networks are sensitive to influences from remote cortical areas. Even if deep structures are affected by surgery, as long as the intentional circuitry still remain intact, such as the sensory motor fibers in the above case, we can potentially target other areas in the circuit to augment the disabled function. In the above case, if corticospinal tracts were severed, then we will be limited in what we can achieve with new rehabilitation.

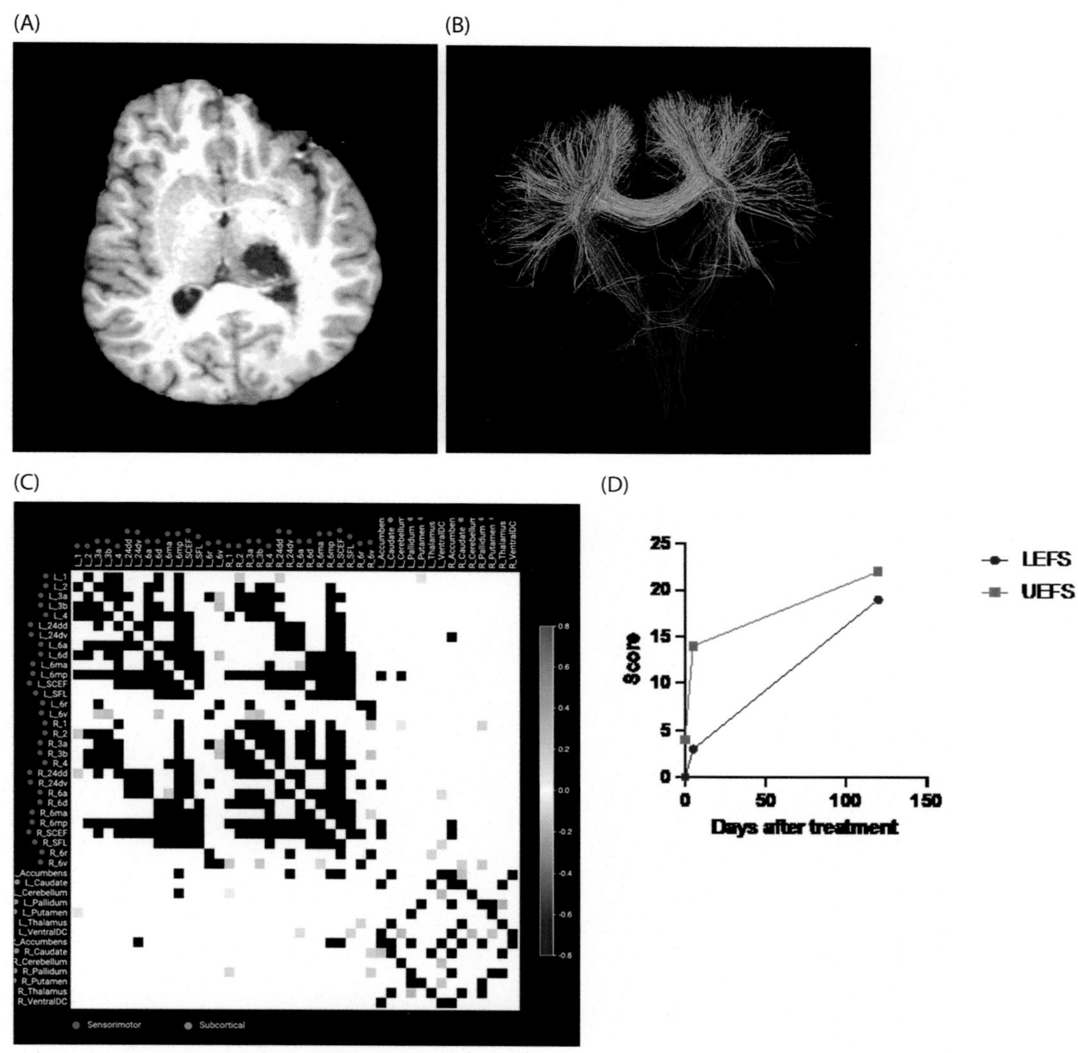

FIGURE 13.1

Case example following glioma resection. The case highlights a right-sided paralysis following resection of a grade 3 glioma located in the left posterior thalamus with extension into the midbrain period (A). After confirming a relatively robust CST (B), attention was turned to the patient's anomaly matrices concerning the sensorimotor network and subcortical structures (C). Significant improvements following continuous TBS (cTBS) treatment was noted in motor scores (D).

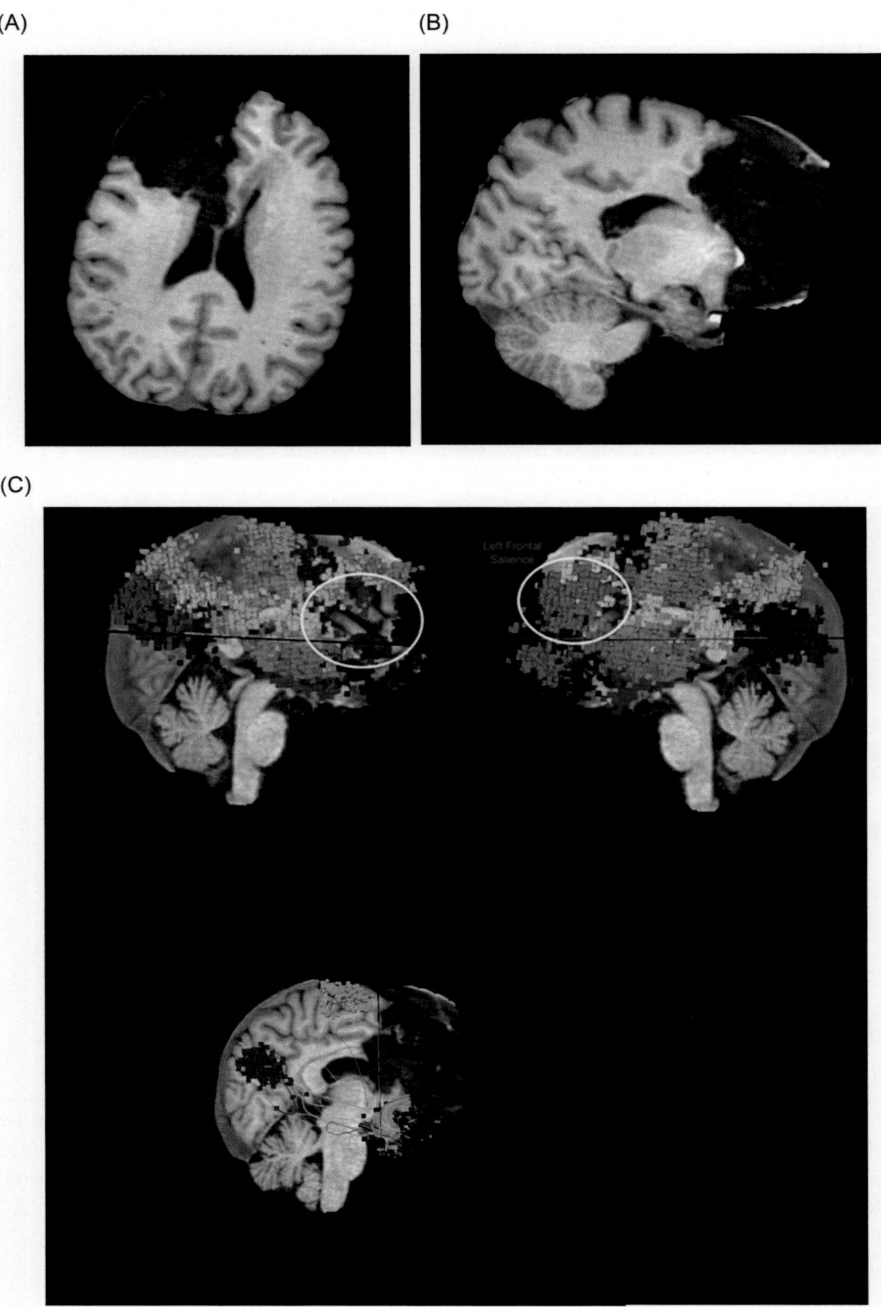

FIGURE 13.2

Case study 2 following glioma resection. The case highlights a case of severe hypobulia following a right frontal craniotomy for resection of a grade 3 oligodendroglioma (A and B). Disruption was noted in the patients default mode network, salience network, and a supplementary motor area, all of which comprise the prefrontal cognitive initiation axis (C).

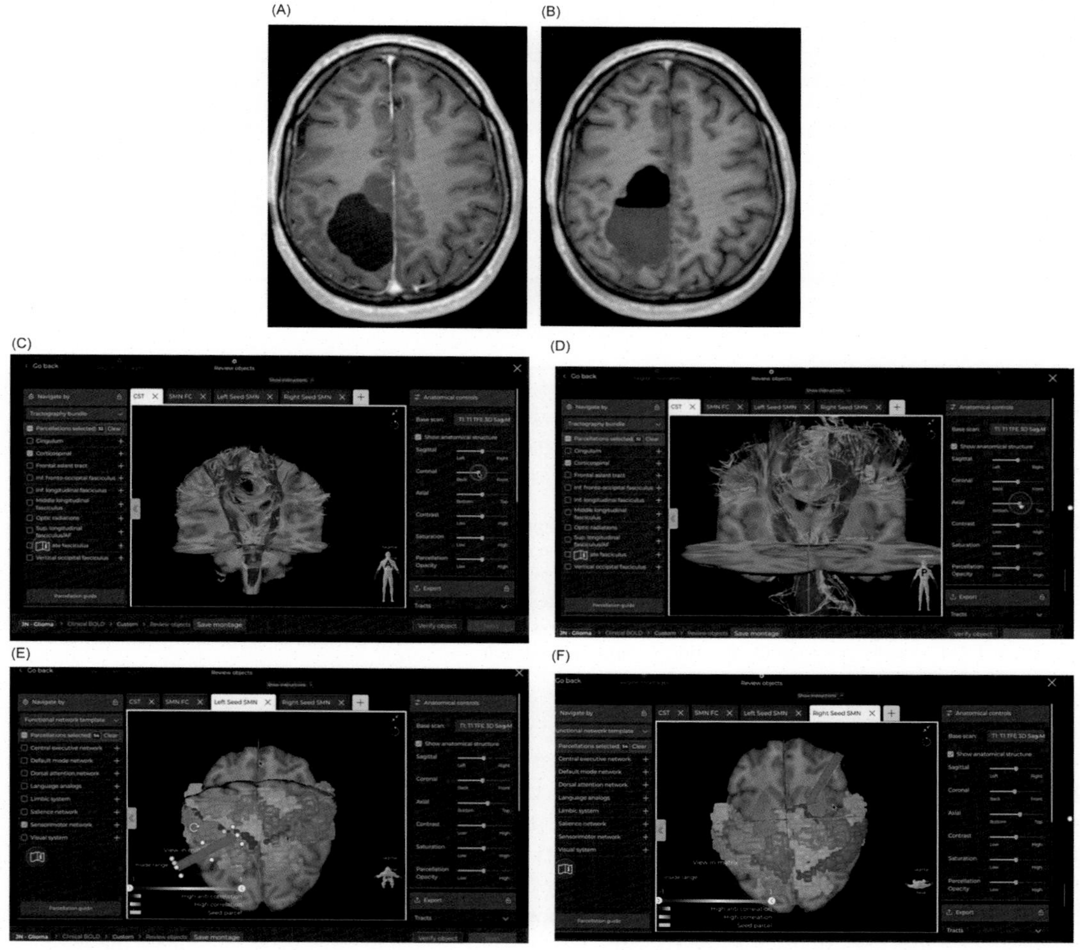

FIGURE 13.3

Case study 3 following glioma resection. The case highlights a case of a significant hemiplegia following a right parietal craniotomy for resection of a grade 3 glioma (A and B). The CST and motor networks were intact according to tractographic analyses (C and D). Functional connectivity data highlight normal left (E) and affected right (F) motor cortices, and demonstrate strong hyperconnectivity in the left sensory cortex and hypoconnectivity in the right SMA, which is common with right motor region injury.

Case study #2 (Fig. 13.2)

We have a 49-year-old right-handed male who underwent right frontal craniotomy for resection of a grade 3 oligodendroglioma (Fig. 13.2A and B). Postoperatively, the patient developed severe hypobulia. He was not interested in engaging in conversation or activities. He was otherwise motor intact. In this case, we suspected that there were aberrations in the patients default mode network, salience network, and a supplementary motor area, all of which comprise the prefrontal cognitive initiation axis (Fig. 13.2C). Upon review of the anomaly matrix focusing on these networks, we identified R6ma (cTBS), RPGi (iTBS), and RTGd (iTBS) as rTMS targets 2 days after surgery. He was engaged with cognitive training in between simulations. His hypobulia was completely resolved. His quality of life metric (EQ-5D) significantly improved from −0.74 to 0.796 five days after treatment and to 1 (perfect score) at 2 months follow-up.

Case study #3 (Fig. 13.3)

We have a 49-year-old right-handed male who underwent right parietal craniotomy for resection of a grade 3 glioma (Fig. 13.3A and B). Postoperatively, the patient developed significant hemiplegia. Left leg was 0/5 strength with 2/5 right arm strength.

The corticospinal tracts were intact bilaterally and the motor cortex was also intact (Figs. 13.3C and D). Analysis of the functional connectivity data, using seed-based methods in the normal left (E) and affected right (F) motor cortices, demonstrated strong hyperconnectivity in the left sensory cortex and hypoconnectivity in the right SMA, which is common with right motor region injury. There were our two targets, treated with cTBS on the left and iTBS on the right. Recovery was immediate and dramatic, returning to near normal in 2 days. This is likely due to the fact that the corticospinal tract was intact, and TMS rapidly corrected the functional connectivity deficit.

In our experience in treating patients who had postsurgical deficits, every patient had TMS targets outside of the immediate resection sites that were identified using our proposed method. Even in patients who had infratentorial brain tumor resections, by targeting susceptible parcellations in supratentorial regions, we were able to improve the patient's symptoms and their quality-of-life metrics. To date, we have not had any seizure events in patients during TMS treatment even in patients who had brain tumors and prior seizures. This may be due to the fact that our strategy is based on the entire connectome and modulating circuitry as opposed to stimulating near the sites of resection, where seizure foci are usually located. Despite this, we encourage practitioners to educate the patients of these theoretical risks and have appropriate medical equipment to handle these situations should they arise.

Conclusions

In this chapter, we discussed the concept of connectome-based, TMS-guided neurorehabilitation post-craniotomy. Importantly, this is a safe and feasible approach to promote functional recovery in brain tumor patients after surgery who often times will face severe morbidity due to both the difficulty and primary goals of the resection. In the next chapter, we switch gears and discuss how connectomic information can be leveraged in the setting of stroke patients.

References

Dadario, N. B., & Sughrue, M. E. (2022). Should neurosurgeons try to preserve non-traditional brain networks? A systematic review of the neuroscientific evidence. *Journal of Personalized Medicine, 12*(4), 587.

D'Angelo, C., Mirijello, A., Leggio, L., Ferrulli, A., Carotenuto, V., Icolaro, N., Miceli, A., D'Angelo, V., Gasbarrini, G., & Addolorato, G. (2008). State and trait anxiety and depression in patients with primary brain tumors before and after surgery: 1-year longitudinal study. *Journal of Neurosurgery, 108*(2), 281–286. https://doi.org/10.3171/jns/2008/108/2/0281

Einstein, E. H., Dadario, N. B., Khilji, H., Silverstein, J. W., Sughrue, M. E., & D'Amico, R. S. (2022). Transcranial magnetic stimulation for post-operative neurorehabilitation in neuro-oncology: A review of the literature and future directions. *Journal of Neuro-Oncology*, 1–9.

He, Y., Li, K., Chen, Q., Yin, J., & Bai, D. (2020). Repetitive transcranial magnetic stimulation on motor recovery for patients with stroke: A PRISMA compliant systematic review and meta-analysis. *American Journal of Physical Medicine and Rehabilitation, 99*(2), 99–108. https://doi.org/10.1097/phm.0000000000001277

Kawashima, A., Krieg, S. M., Faust, K., Schneider, H., Vajkoczy, P., & Picht, T. (2013). Plastic reshaping of cortical language areas evaluated by navigated transcranial magnetic stimulation in a surgical case of glioblastoma multiforme. *Clinical Neurology and Neurosurgery, 115*(10), 2226–2229. https://doi.org/10.1016/j.clineuro.2013.07.012

Morell, A. A., Eichberg, D. G., Shah, A. H., Luther, E., Lu, V. M., Kader, M., Higgins, D. M. O., Merenzon, M., Patel, N. V., Komotar, R. J., & Ivan, M. E. (2022). Using machine learning to evaluate large-scale brain networks in patients with brain tumors: Traditional and non-traditional eloquent areas. *Neuro-Oncology Advances, 4*(1), vdac142. https://doi.org/10.1093/noajnl/vdac142

Poologaindran, A., Profyris, C., Young, I. M., Dadario, N. B., Ahsan, S. A., Chendeb, K., Briggs, R. G., Teo, C., Romero-Garcia, R., Suckling, J., & Sughrue, M. E. (2022). Interventional neurorehabilitation for promoting functional recovery post-craniotomy: A proof-of-concept. *Scientific Reports, 12*(1), 3039. https://doi.org/10.1038/s41598-022-06766-8

Rijnen, S. J. M., Kaya, G., Gehring, K., Verheul, J. B., Wallis, O. C., Sitskoorn, M. M., & Rutten, G. M. (2019). Cognitive functioning in patients with low-grade glioma: Effects of hemispheric tumor location and surgical procedure. *Journal of Neurosurgery, 133*(6), 1671–1682. https://doi.org/10.3171/2019.8.Jns191667

Sharma, N., Baron, J. C., & Rowe, J. B. (2009). Motor imagery after stroke: Relating outcome to motor network connectivity. *Annals of Neurology, 66*(5), 604–616. https://doi.org/10.1002/ana.21810

CHAPTER

Connectomic strategies for stroke patients

14

Introduction

Ischemic strokes are a leading cause of disability in the world. It carries tremendously heavy socio-economic burden as patients do not necessarily expire after an ischemic stroke, but they are rendered unable to perform daily activities of living, including employment activities that can contribute to societal welfare (Kelly-Hayes et al., 2003). Therefore, the neurological and psychocognitive disabilities resulting from ischemic strokes can accumulate rolling personal and societal burden. Over two-thirds of patients develop severe upper limb impairment 6 months after the initial stroke event (Ackerley et al., 2016; Tosun et al., 2017). The addition of Tissue Plasminogen Activator (TPA) and mechanical thrombectomy have revolutionized the treatment for ischemic strokes. However, patients who are not eligible for these therapies and are left with severe deficits and have limited options aside from physical, occupational, in cognitive therapies. There are currently no established interventional rehabilitation brain modulation strategies for these patients.

Repetitive transcranial magnetic stimulation (rTMS) has been investigated for its potential in hastening neurorehabilitation. It has been shown repeatedly to be a safe intervention with the most common side effects as minor dizziness, minor superficial site discomfort, and headaches (Caulfield et al., 2022; Loo et al., 2008; Perera et al., 2016). It has the ability to rewire neural networks and potentiate reorganizations of neural circuitry, as long as essential tracts are intact. It is established in the literature that patients with chronic stroke can derive improvements in motor performance with noninvasive brain stimulation (O'Brien et al., 2018). Specifically, rTMS demonstrates encouraging data in promoting motor learning and rehabilitation (Hoyer & Celnik, 2011). The theory behind using rTMS for stroke rehabilitation is based on the concept of interhemispheric inhibition, where the ischemic hemisphere experiences increased inhibition from the contralateral hemisphere. This is due to the fact that at baseline, contralateral motor cortex exhibits constant inhibitory reflex to the ipsilateral cortex to maintain homeostasis. Based on this, the concept of previous clinical trials has been to administer rTMS to the contralateral side to decrease inhibition or to the ischemic side to stimulate and increase excitability. We also have to take into account that the effects of rTMS is likely dependent on stimulation frequency as well. Low frequency stimulation is likely to result in transient reduction in cortical excitability, as opposed to high frequency stimulation that increases excitability (Lefaucheur, 2019; Wang et al., 2012). The issue with this approach is that it overly simplifies our functional connectome and it's likely only applicable to motor function. It would not apply to the language network, for instance, where it is usually a unilateral phenomenon. Also, if we consider that there are at least 180 functional parcellations per hemisphere, the above approach would be an oversimplification

Connectomic Medicine. https://doi.org/10.1016/B978-0-443-19089-6.00003-3

to tackle the many types of neurocognitive and psychiatric deficits brought about by ischemic strokes. This is where personalized connectomic medicine will play a great role in reshaping this field (Chen et al., 2023).

Aside from a plethora of randomized control trials using sham treatment in patients who had suffered motor deficits after ischemic strokes, how do we know that there are underlying changes to functional connectivity? It is well established that neuroplasticity is closely related to stroke recovery and changes in tractography have been previously demonstrated in poststroke patients (Kierońska et al., 2021). In fact, the strucutral and functional connectivity patterns at stroke presentation can shed insight on the ultimate recovery tracjectories of individual patients after rTMS treatment (Chen et al., 2023). We have also previously demonstrated that even patients who are considered to have "chronic" stroke can benefit from rTMS treatments (Yeung et al., 2021). Accompanying objective measurable improvements in motor function, we also notice significant recruitment of parcellations based on rs-fMRI and increase in crossing fibers across sensorimotor networks. Below, we provide a case example.

Case study #1 (Figs. 14.1 and 14.2)

The patient was a 19-year-old right-handed female who had suffered a left superior MCA trunk ischemic stroke resulting in deficits in right upper extremity function and phonological text alexia, reading. The patient presented 20 months after standard intense physical, occupational, and speech therapy after which she has plateaued in terms of her functional recovery.

Analysis of structural connectivity demonstrated near absence of the left arcuate and SLF, and minimal presence of Broca's area, from the stroke (Fig. 14.1A) raising the question of how she was able to speak at all. We felt that looking at connections from the opposite side SMA, this seemed like the only plausible connection to what was left of the frontal language areas (Fig. 14.1B). Looking at functional connectivity, we noted that when selecting a seed in the remaining Broca's area, neither SMA showed strong functional connectivity to the language areas so it made sense to stimulate the contralateral SMA to try to drive improvement of function (Fig. 14.1C). It would be hard to identify this target without connectomics.

We identified three targets for rTMS including (1) left area 1 (L_1, primary motor area) to improve her hand function, (2) left area 45 (L_45), which is one of the only remaining areas of her language network that was not damaged by the stroke, and (3) right area SFL (R_SFL), which was thought to be a tract serving to support her remaining language network. In between stimulation, division underwent intense physical therapy focusing on her right upper extremity and speech therapy as well encompassing speech production and reading comprehension. At 5 months follow-up, her language and right upper extremity movements had significantly improved.

There was a noticeable increase in crossing fiber bundle (Fig. 14.2A and B) involved in the left sensorimotor network before (Fig. 14.2A) and after (Fig. 14.2B) rTMS treatment. There was an increase in the frontal cortical representations (red dotted circles) detected in the language before (Fig. 14.2C) and after (Fig. 14.2D) rTMS treatment (Yeung et al., 2021). The patient demonstrated notable improvements in right upper extremity function when comparing pre and post TMS cores (Fig. 14.2E).

In order to conceive a treatment strategy for a patient with ischemic stroke that may display a variety of neurological and psychocognitive deficits, we must try to break down the patient's symptoms and deficits in a very systematic fashion. Since there is no cookbook recipe for using personalized rTMS in stroke recovery, it is encouraged that we apply what we know as evidence-based knowledge and have to think outside of the box when we are dealing with noncanonical deficits.

Thorough history taking. The initial patient intake is an incredibly important part of treatment planning. We always encourage the patient to be accompanied by a family member or a person who knows the patient well before, and after, the stroke has occurred. The reason is so that we can have a

A

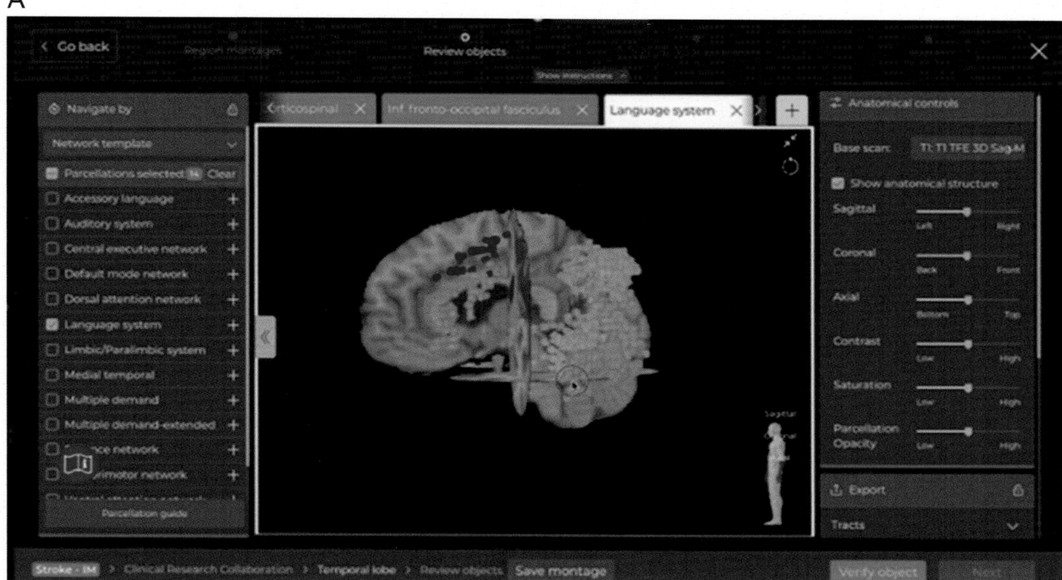

B

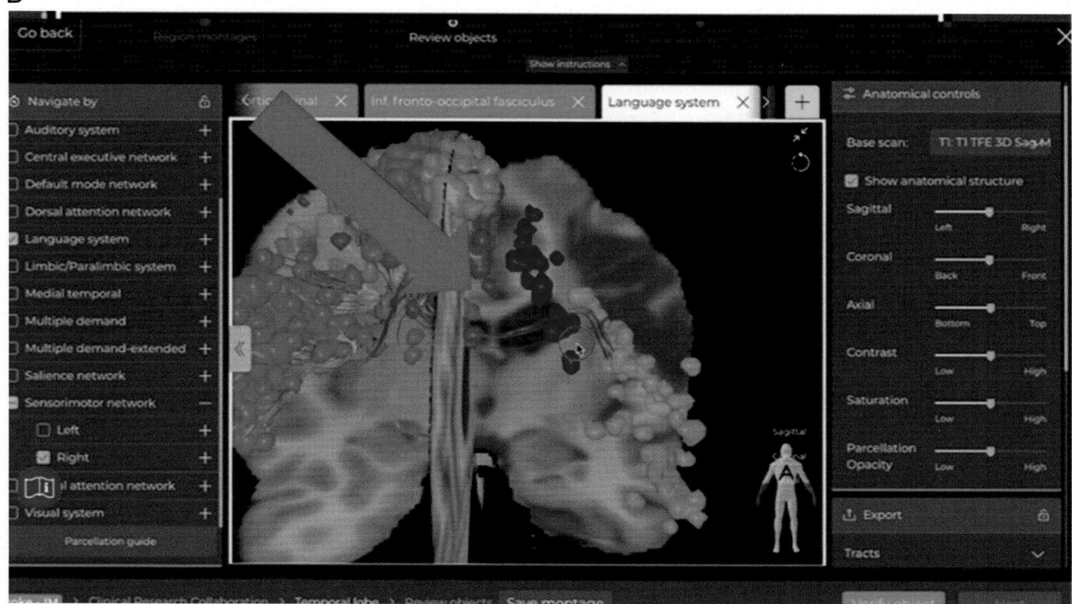

FIGURE 14.1

Case study example 1 following stroke. Structural connectivity analysis on the patient's language system suggests near absence of the left arcuate and superior longitudinal fasciculus fibers (SLF) as well as Broca's area (A). It was believed the contralateral supplementary motor area (SMA) connections may be the primary connections left with the frontal language areas and therefore the contralateral SMA was stimulated (B, C).

C

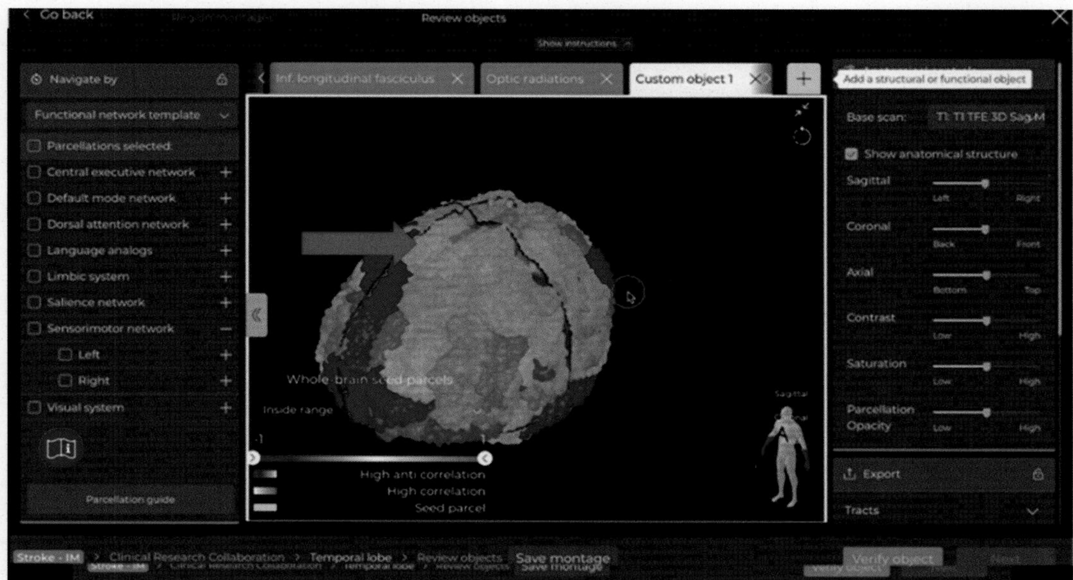

FIGURE 14.1 Cont'd

secondary account of how the patient was functionally before and how they were covering despite conventional therapies. This will provide us a baseline for which the patient might try to aim. It is also important to understand whether the patient has plateaued with traditional therapies and set an expectation for our rTMS treatment. This is especially important in patients who have cognitive limitations or if they cannot express their thoughts effectively. What is also important is the review of objective assessments by previous therapists and to use them as a resource for gaining objective insight into the patients' deficits. At this point in time, we encouraged asking unconventional questions concerning daily activities of living to gain better understanding of how a patient functionality may improve with treatment and thereby increase the quality of life. It is important to discern whether the patient has any poststroke mood disorders that may be underrecognized and undiagnosed. Remember—the spirit of connectomics medicine allows us to appreciate various networks in the brain that we may not usually consider in these patients.

Goal setting. It is essential to identify specific goals for rTMS together with patients and their families. Especially for rTMS, it is unclear what the results could be if more than one brain area is stimulated. In our experience, we have had success in treating multiple networks, but no more than three targets at a time. We have to understand that a limitation of brain stimulation is that we do not know the cumulative effects of targeting multiple brain areas. With this limitation, it is important for us to set very defined goals with the patients. This will also help us in the selection of adjunctive therapies that would take place in between stimulation sessions. We advice focusing on no more than two functional recovery goals, such as motor/language, language/depression, etc.

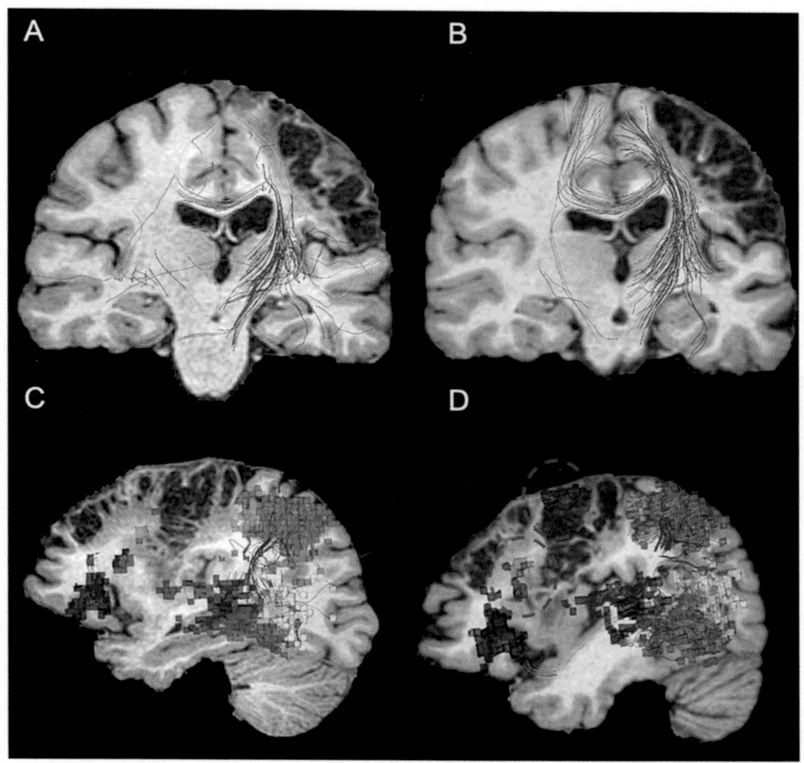

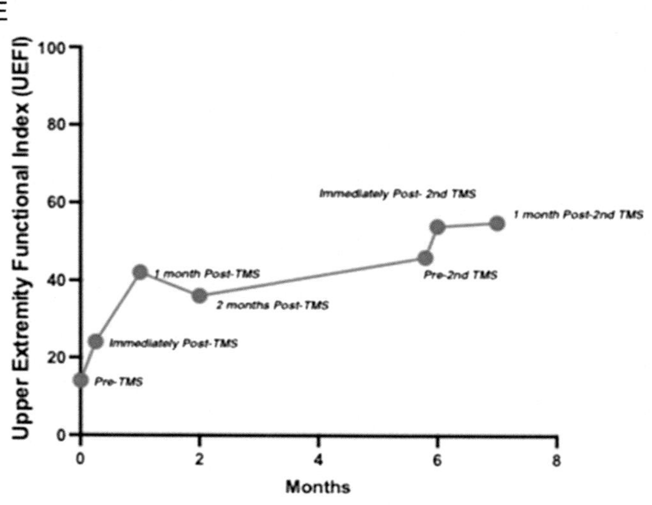

FIGURE 14.2

Pre- and post-rTMS treatment for case 1. After stimulation, there was a noticeable increase in the left sensorimotor network before (A) and after (B) rTMS treatment. There was an increase in the frontal cortical representations (*red dotted circles*) detected in the language before (C) and after (D) rTMS treatment. (E) The patient's upper extremity functional index scores pre- and post-TMS treatment.

Adapted from Yeung et al. (2021).

Basic principles. The clinician should analyse the patient's deficits and utilize the transdiagnostic approach to dissect the symptoms and associate them with the proper networks (Barron et al., 2021). In analysing connectivity during anomaly detection, we have to be able to identify the misconnections and try to make sense out of them. After misconnections or anomalies are identified using no subjective pretense, we have to look at these anomalies to see whether they make sense or pass the "sniff test." We can employ some basic principles derived from what has been established in clinical trials, such as inhibition of the contralateral motor cortex to facilitate and promote the function of the affected hemisphere. It is our experience that our unbiased, objective approach to anomaly detection oftentimes coincides with established principles of cortical promotion or inhibition in a treatment of motor and language-related strokes. Overall, we urge clinicians to not only focus on the obvious corticospinal pathways connecting the primary motor cortex and down, but also pay attention to premotor planning and supplemental motor areas. Many times these can be affected such that there is poor activation and coordination of motor function, resulting in deficits. When it comes to other noncanonical deficits, such as poor memory or attention, we would oftentimes be able to identify miscommunications related to temporal lobe parcellations (especially as part of the default mode network) and parcellations in dorsal/ventral attention networks. An extremely important principle lastly to remember is that brain modulation does not bring dead brain back to life, but the goal is to enhance existing, likely compensatory, pathways.

Below, we outline a few principles for poststroke rehabilitation using TMS for a variety of deficits including how to define the problem and the relevant network anatomy to address.

TMS principles for poststroke rehab: language disturbances

Step 1: Define the problem

Speech and language disturbances can be a complex combination of phonemic, motor, initiation, verbal memory, or semantic disturbances, and this needs to be clarified as different problems are caused by different systems.

Step 2: Look at the relevant parts of structural connectome

- Language network
- Accessory language system
- Face motor
- Contralateral SMA

Step 3: Look at the relevant parts of the functional connectome

Specifically looking for either anomalies or lack of correlation in parts of the networks:

- Language network
- Accessory language system
- Face motor
- Contralateral SMA
- Basal ganglia

Step 4: Determine a treatment which addresses these issues and integrates into the overall care plan

This is usually speech and language rehab modified to the specific situation.

TMS principles for poststroke rehab: sensorimotor problems

Step 1: Define the problem

Motor problems can be weakness, incoordination, lack of eccentric control, or rigidity. It is obviously important to distinguish arm, face, or leg weakness as these involve different portions of the motor system.

Step 2: Look at the relevant parts of structural connectome

Bilateral motor cortices
Primary sensory cortex
SMA
Corticospinal tract
Dorsal and ventral attention network
Basal ganglia

Step 3: Look at the relevant parts of the functional connectome

Looking for contralateral hyperconnectivity or ipsilateral hypoconnectivity, or anomalies in the following:
Sensorimotor network
Basal ganglia

Step 4: Determine a treatment which addresses these issues and integrates into the overall care plan

This is usually combined with physical therapy aimed at the affected limb.

Adjunctive therapies. Interstimulation therapies are as important as the actual brain stimulation (Yang et al., 2020). It is theorized that any sort of adjunctive therapies, including physical therapy and occupational therapy, during stroke rehabilitation is essential for helping to reinforce the results of brain stimulation. It is important to remember that rTMS merely opens the lock, or otherwise can be thought as lowering the threshold, and the adjunctive therapy will help to reinforce those pathways for motor relearning. During this time, it is important for the clinician to communicate with the therapist about this specific form of therapy administered, and that they should be focused on retraining the nervous system as opposed to a focal strengthening. The matter is less definite when it comes to other forms of cognitive training, and at times the clinician will have to be creative in instituting different forms of adjunctive therapies. Patients with memory deficits can undergo cognitive exercises for reinforcing short-term memory training (Zhou et al., 2021). Patients with language deficits should work with a speech therapist to address all regards of language functions, not limited to just speech production but comprehension, reading, or writing. Below we provide a brief case example.

Case study #2 (Fig. 14.3)

This was a 72-year-old female with a history of open heart surgery for aortic valve replacement and sustained a postoperative embolic stroke 3 years ago. An immediate angiogram showed that there were no emboli or clot to retrieve, but she certainly had a complete stroke of her right posterior internal capsule. She was left with speech disturbance and a right-sided hemiparesis which has improved somewhat, but certainly not back to a functional state. She walked with a limp and the right arm is essentially without any functionality. Her speech is also slightly dysphasic from a motor viewpoint with maintained comprehension and her dysphasia was barely discernible.

During the consultation with the patient, we determined treatment goals (goal setting) in a patient-centered fashion. The patient indicated that her language function was only minimally affected. She wanted to focus on her right upper extremity function as that is the main hindrance to our quality of life. In light of this goal, we focused on her sensory motor network and the communication among each other. We examined the patient tractography which demonstrated grossly intact corticospinal tract. We identified L_6V (left ventral area 6) and R_6ma (right medial anterior area 6) as potential brain modulation targets (Fig. 14.3A). Brodmann area 6 is situated right in front of the primary motor cortex and is composed of

Case study #2 (Fig. 14.3)—cont'd

the premotor cortex and the supplementary motor area. Seeing as how both of these areas demonstrated patterns of hyperconnectivity, we elected to stimulate both of these areas with cTBS at 80% MTS, in hopes of modulating this hyperconnectivity. The patient underwent intense motor relearning physical therapy between stimulations.

Immediately after treatment, the patient's right upper extremity functional score more than doubled from 23 to 49. Similar improvement was observed with the lower extremities. At 3 months follow-up, this type of functionality was maintained. The patient's quality of life metrics also improved along with the motor scores (Fig. 14.3B and C).

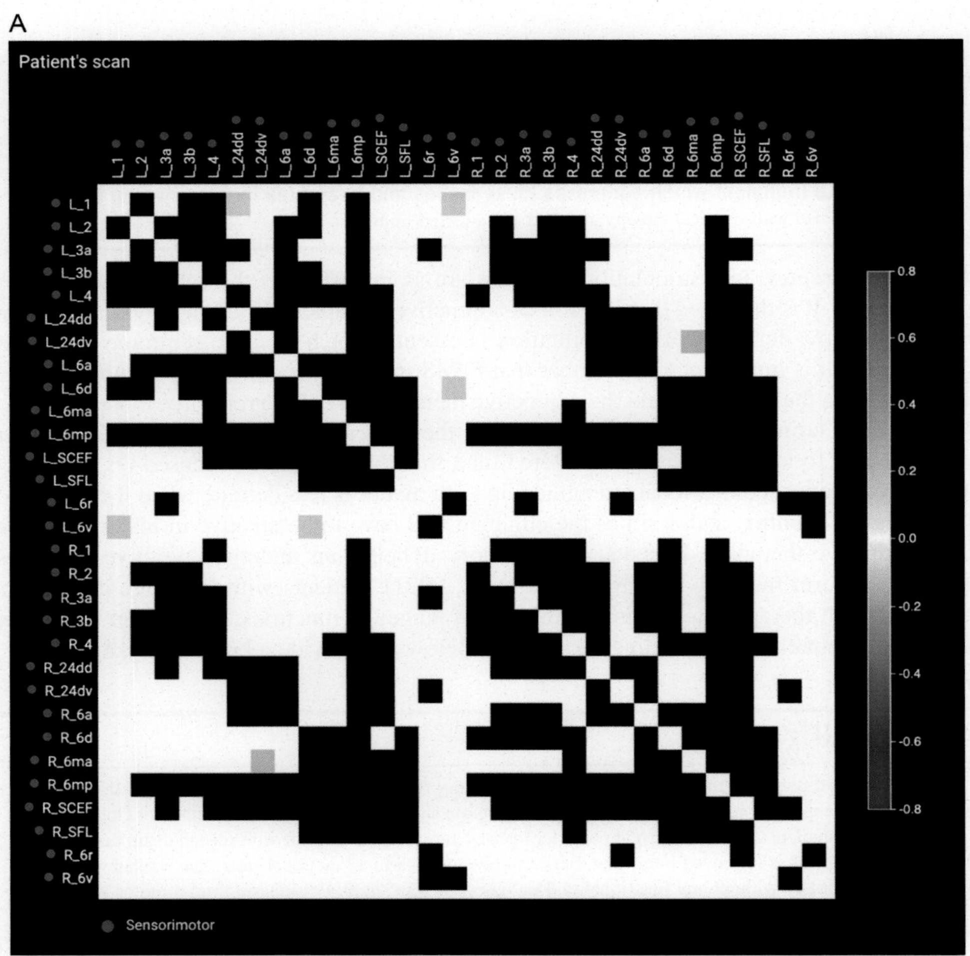

FIGURE 14.3

Case study example 2 following stroke. Functional connectivity anomalies were identified in the sensorimotor network for possible targets (A). Significant post-TMS improvements were noted in motor functioning and quality of life scores (B and C).

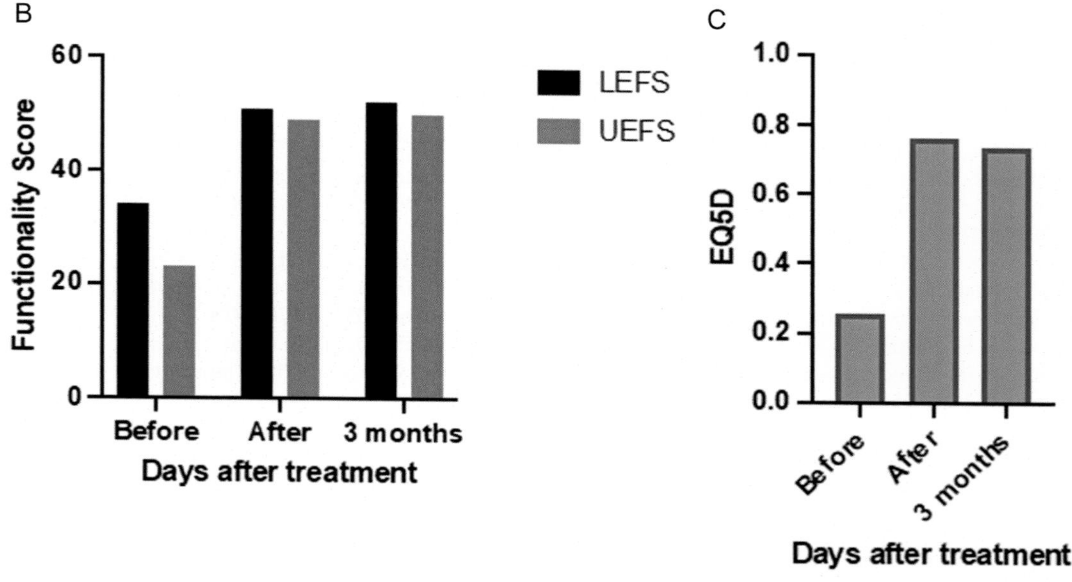

FIGURE 14.3 Cont'd

Conclusions

In this chapter we discussed the potential applications and benefits of using connectivity-guided TMS for stroke rehabilitation. In this context, connectomic information is helpful to both (1) better understand and define the patients' deficits according to lesion-induced network disruption or dysfunction and then (2) outline a feasible plan to restore network synchronicity for functional recovery or compensation. In the next chapter, we discuss connectomic strategies for the treatment of depression and anxiety.

References

Ackerley, S. J., Byblow, W. D., Barber, P. A., MacDonald, H., McIntyre-Robinson, A., & Stinear, C. M. (2016). Primed physical therapy enhances recovery of upper limb function in chronic stroke patients. *Neurorehabilitation and Neural Repair, 30*(4), 339–348. https://doi.org/10.1177/1545968315595285

Barron, D. S., Gao, S., Dadashkarimi, J., Greene, A. S., Spann, M. N., Noble, S., Lake, E. M. R., Krystal, J. H., Constable, R. T., & Scheinost, D. (2021). Transdiagnostic, Connectome-Based Prediction of Memory Constructs Across Psychiatric Disorders. *Cerebral Cortex (New York, N.Y. : 1991), 31*(5), 2523–2533. https://doi.org/10.1093/cercor/bhaa371

Caulfield, K. A., Fleischmann, H. H., George, M. S., & McTeague, L. M. (2022). A transdiagnostic review of safety, efficacy, and parameter space in accelerated transcranial magnetic stimulation. *Journal of Psychiatric Research, 152*, 384–396. https://doi.org/10.1016/j.jpsychires.2022.06.038

Chen, R., Dadario, N. B., Cook, B., Sun, L., Wang, X., Li, Y., Hu, X., Zhang, X., & Sughrue, M. E. (2023). Connectomic insight into unique stroke patient recovery after rTMS treatment. *Frontiers in Neurology, 14*, 1063408. https://doi.org/10.3389/fneur.2023.1063408

Hoyer, E. H., & Celnik, P. A. (2011). Understanding and enhancing motor recovery after stroke using transcranial magnetic stimulation. *Restorative Neurology and Neuroscience, 29*(6), 395–409. https://doi.org/10.3233/rnn-2011-0611

Kelly-Hayes, M., Beiser, A., Kase, C. S., Scaramucci, A., D'Agostino, R. B., & Wolf, P. A. (2003). The influence of gender and age on disability following ischemic stroke: The framingham study. *Journal of Stroke and Cerebrovascular Diseases, 12*(3), 119–126. https://doi.org/10.1016/S1052-3057(03)00042-9

Kierońska, S., Świtońska, M., Meder, G., Piotrowska, M., & Sokal, P. (2021). Tractography alterations in the arcuate and uncinate fasciculi in post-stroke aphasia. *Brain Sciences, 11*(1). https://doi.org/10.3390/brainsci11010053

Lefaucheur, J.-P. (2019). Chapter 37 - transcranial magnetic stimulation. In K. H. Levin, & P. Chauvel (Eds.), *Handbook of clinical neurology* (pp. 559–580). Elsevier.

Loo, C. K., McFarquhar, T. F., & Mitchell, P. B. (2008). A review of the safety of repetitive transcranial magnetic stimulation as a clinical treatment for depression. *International Journal of Neuropsychopharmacology, 11*(1), 131–147. https://doi.org/10.1017/S1461145707007717

O'Brien, A. T., Bertolucci, F., Torrealba-Acosta, G., Huerta, R., Fregni, F., & Thibaut, A. (2018). Non-invasive brain stimulation for fine motor improvement after stroke: A meta-analysis. *European Journal of Neurology, 25*(8), 1017–1026. https://doi.org/10.1111/ene.13643

Perera, T., George, M. S., Grammer, G., Janicak, P. G., Pascual-Leone, A., & Wirecki, T. S. (2016). The clinical TMS society consensus review and treatment recommendations for TMS therapy for major depressive disorder. *Brain Stimulation, 9*(3), 336–346. https://doi.org/10.1016/j.brs.2016.03.010

Tosun, A., Türe, S., Askin, A., Yardimci, E. U., Demirdal, S. U., Kurt Incesu, T., Tosun, O., Kocyigit, H., Akhan, G., & Gelal, F. M. (2017). Effects of low-frequency repetitive transcranial magnetic stimulation and neuromuscular electrical stimulation on upper extremity motor recovery in the early period after stroke: A preliminary study. *Topics in Stroke Rehabilitation, 24*(5), 361–367. https://doi.org/10.1080/10749357.2017.1305644

Wang, R. Y., Tseng, H. Y., Liao, K. K., Wang, C. J., Lai, K. L., & Yang, Y. R. (2012). rTMS combined with task-oriented training to improve symmetry of interhemispheric corticomotor excitability and gait performance after stroke: a randomized trial. *Neurorehabilitation and Neural Repair, 26*(3), 222–230. https://doi.org/10.1177/1545968311423265

Yang, Y. W., Pan, W. X., & Xie, Q. (2020). Combined effect of repetitive transcranial magnetic stimulation and physical exercise on cortical plasticity. *Neural Regeneration Research, 15*(11), 1986–1994. https://doi.org/10.4103/1673-5374.282239

Yeung, J. T., Young, I. M., Doyen, S., Teo, C., & Sughrue, M. E. (2021). Changes in the brain connectome following repetitive transcranial magnetic stimulation for stroke rehabilitation. *Cureus, 13*(10), e19105. https://doi.org/10.7759/cureus.19105

Zhou, L., Huang, X., Li, H., Guo, R., Wang, J., Zhang, Y., & Lu, Z. (2021). Rehabilitation effect of rTMS combined with cognitive training on cognitive impairment after traumatic brain injury. *American Journal of Translational Research, 13*(10), 11711–11717.

CHAPTER

15 Connectomic strategies for depression and anxiety

Introduction

TMS treatment for major depression is well established and backed by numerous randomized controlled trials for the treatment of treatment-resistant major depression (TRD) (Garnaat et al., 2018). So why change what works?

Well, a few key points should cause some reflection that we might be able to do a lot better than we have been doing:

(1) The response rate in TRD in most trials with traditional rTMS paradigms hovers around 40% (Fitzgerald et al., 2016). This is better than a third or fourth medicine, but it still leaves a bit under two-third failure rate, for patients who lack other good options (and no, electroconvulsive therapy (ECT) is not a good option, it is a bailout option until it makes some progress) … with a potentially fatal illness.

(2) Even good responding patients do not usually improve to symptom free. This argues that even when it works, it is missing much of the pathology. A patient whose depression subsides who still has crushing anxiety is still miserable, regardless of what their scores say.

(3) Despite being a treatment with class 1 evidence, being covered by payers, and having class 1 evidence, of the nearly five million TRD patients in the United States, it is estimated that about 30,000 patients receive TMS, which is less than 1% of eligible people. This is not due to a lack of data, or fantastic alternative options for these patients. It is due in large part to inconsistent results, and impractical demands of the procedure.

Moving forward with a better approach

So while we do not have the answers for all aspects of depression and anxiety, and there are plenty of unsolved questions in the field, we think a few core elements of this procedure seem to be reliably useful in our experience and our interpretation of the literature and discussions with others, and it is safe to say that a good approach involves the following basic principles.

(1) If you do not hit the same area repeatedly, it is probably less effective.

(2) If you do not use image guidance, basically no matter how experienced you are, you are not hitting the same area repeatedly (Rosen et al., 2021). We have published results demonstrating

Connectomic Medicine. https://doi.org/10.1016/B978-0-443-19089-6.00001-X

this, but also it has been known in neurosurgery that image guidance is the standard of care for performing procedures on the cerebrum.

(3) More stimulation seems to work better than less. This could involve more sessions, more pulses, and/or more targets. People have noted success with all of these approaches at different times.
(4) You need to hit the right part of the dlPFC or it is less effective (Moreno-Ortega et al., 2020).
(5) Depression and anxiety are complex diseases which affect many different parts of the brain, which in turn causes the heterogeneous complex of symptoms in the disease (Young et al., 2023).
(6) Different patients have different network problems. The Venn diagram of these patients has some reliable areas but a lot of variability (Drysdale et al., 2017).
(7) It seems less likely that treating a normally functioning brain area in a patient is likely to help, simply because other patients have a similar symptom set and have responses at this area.
(8) Depression and anxiety symptoms cooccur so often that they likely result from similar and overlapping mechanisms.

A personalized approach to target the heterogeneity of MDD

The personalized approach to rTMS treatment of major depressive disorder (MDD) has been previously published in an open-label proof-of-concept study in which rTMS targets were determined by fMRI and then coupled with a type of cognitive behavioral therapy (CBT) known as self-system therapy (SST) (Neacsiu et al., 2018; Sathappan et al., 2019). The results from that study showed evidence of efficacy in all five participants, thereby encouraging the use of personalized rTMS. Importantly, they found a region in the right dlPFC that was discriminately more active for patients with MDD and generalized anxiety comorbidities than patients with MDD without anxiety, and previous studies from this group have found success in targeting this area in patients with comorbid anxiety and depression (Mantovani et al., 2013; Neacsiu et al., 2018). Given how symptoms and connectivity of MDD vastly affects treatment outcome, it is difficult for one method of dlPFC-based TMS to be a catch-all for patients with depression.

Our proposed solution is to create a personalized rTMS treatment based on the patients' specific functional connectivity. The personalized agile method rTMS could benefit patients who were previously excluded from clinical trials of rTMS. The first clinical trial of rTMS (O'Reardon et al., 2007) excluded patients with a failed ECT, thereby overlooking a significant population of patients with depression who do not respond well to treatment and would most benefit from the development of new therapies (O'Reardon et al., 2007). This initial clinical trial study reported a 23% reduction of depressive symptoms with rTMS treatment based on the Hamilton Depression Rating Scale-17 (HAMD-17) (O'Reardon et al., 2007).

Below, we describe how to approach TMS treatment for depression and anxiety, including how to define the problem and approach the relevant network anatomy.

TMS principles for depression/anxiety

Step 1: Define the problem

We think of these patients along three main axes

(1) Depression general
(2) Anxiety general
(3) Additional symptoms

By this, we mean that given we cannot realistically treat every single abnormal circuit in the brains of these patients, we need to first focus on using specific targets to do the heavy lifting, i.e., treat areas that improve several symptoms at once, and then follow this with additional targets aimed at specific, especially bad symptoms. This can be focused on dangerous symptoms (notably suicidal ideations), or symptoms particularly bothersome for the patient.

To this end, we perform several surveys, before and after treatment. Standard depression and anxiety scales are useful for identifying what core targets are needed (a dlPFC depression, an anxiety core target, or both). They can also help identify other symptoms worth treating if one looks at individual items in these scales. Transdiagnostic type surveys are also very useful in defining new targets.

We usually aim to find three targets.

Step 2: Look at the relevant parts of structural connectome

The structural connectome in these patients is usually grossly normal. While perhaps in the future, a valuable structural biomarker may arise, at present, structural information is only useful in cases where you are dealing with secondary depression, such as that from brain surgery, stroke or other diseases.

Step 3: Look at the relevant parts of the functional connectome

Depression core

We look at anticorrelation with area 25 in the left dlPFC (Fig. 15.1).

Anxiety core

We look for anomalies in a montage comparing DMN and CEN. The most common thing we see are anomalies in areas Left 8AV, L_PGs, and/or either TE1M (Fig. 15.2). It is important to also look at fear-related transdiagnostic montages if phobia or panic type symptoms are present.

Symptoms-specific targets

These can be selected using transdiagnostic montages relevant to the patient's domain of problem (Fig. 15.3). We prioritize suicide-specific targets over others if suicidal ideation is a problem and base this on machine learning models. Initial response to reward gives an insight into anhedonia. Reward probability can be useful for understanding negativity. Self-knowledge is often abnormal in patients with rumination.

Alternatively, studying montages of the DMN, CEN, and salience can be useful using an anomaly detector, if a good target is not obvious.

Step 4: Determine a treatment which addresses these issues and integrates into the overall care plan

A number of options are available in our practice. Not all patients are open to all strategies, and these need to be fit to their symptoms, and preferences. Here are some common tools we use, but we have used many additional approaches.

Depression core: CBT, related therapy apps

Anxiety core: Guided meditation, massage, CBT

Individual symptoms: As described in previous chapters, CBT can be aimed at various transdiagnostic domains. Additionally, other strategies can be used.

Fig. 15.4 demonstrates a basic algorithm we have used previously with some success for these patients.

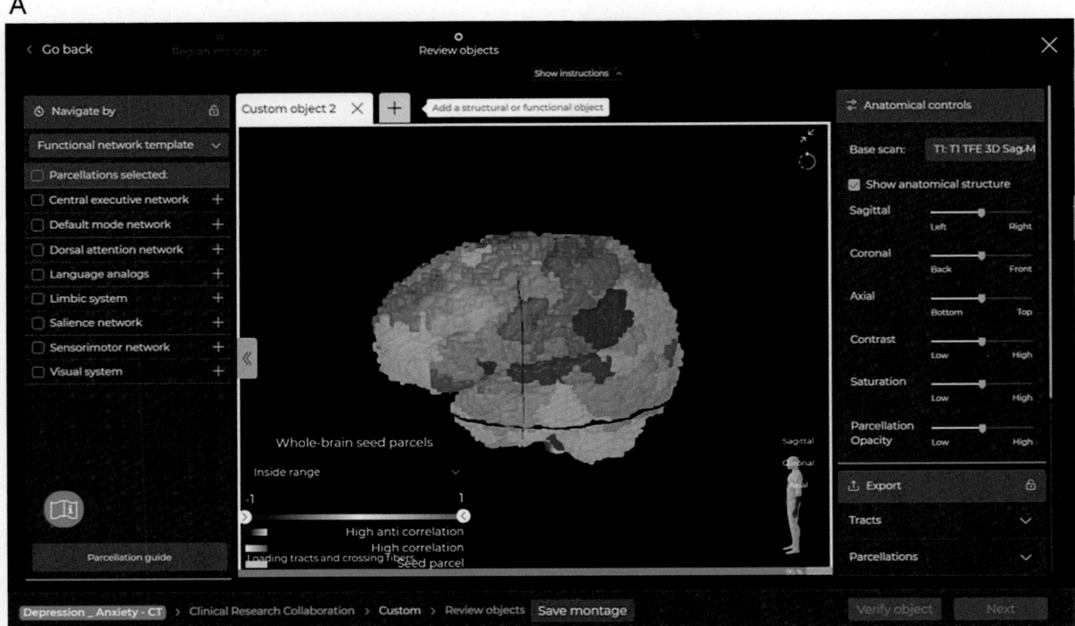

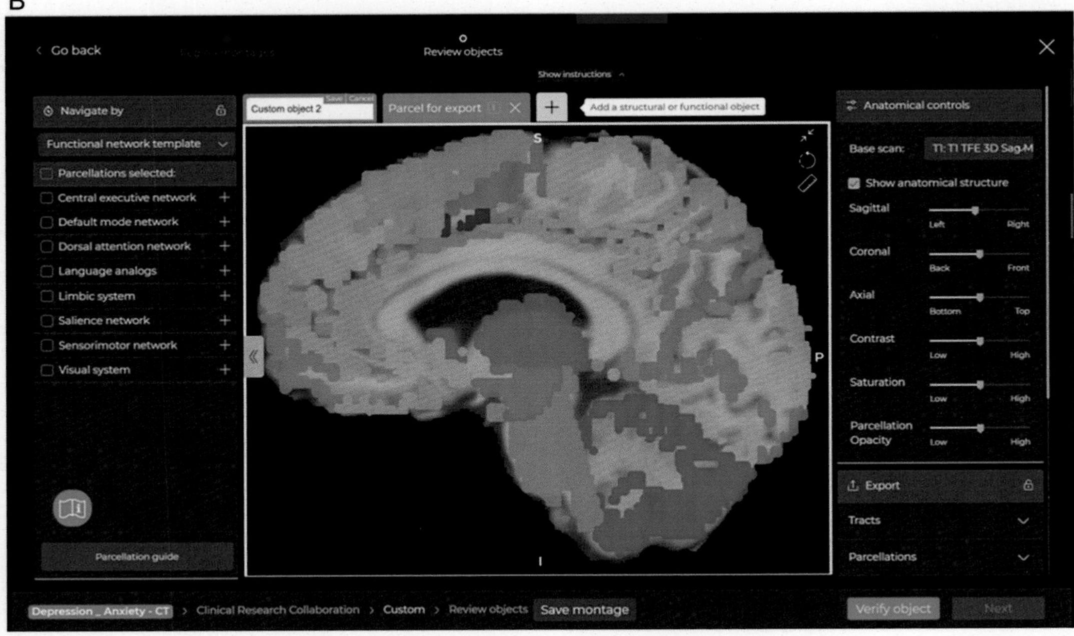

FIGURE 15.1

This is a step-by-step method for finding anticorrelation with subgenual cingulate in the Quicktome software. (A) Here you see the entire functional connectivity atlas. (B) A seed is placed in the subgenual cingulate cortex. (C) Whole-brain functional connectivity is analysed using a standard atlas. (D) Using the sliderbar you can examine the most anticorrelated areas with the dlPFC to look for a specific target. (E) A parcel in the dlPFC is then found which is selected as a target (F) for treatment.

C

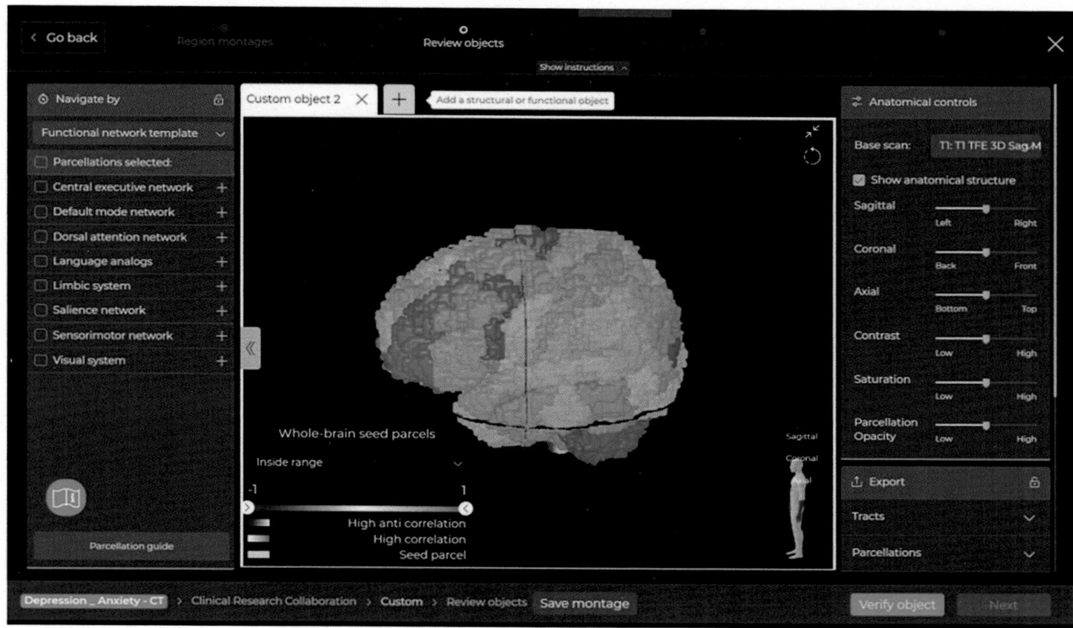

D

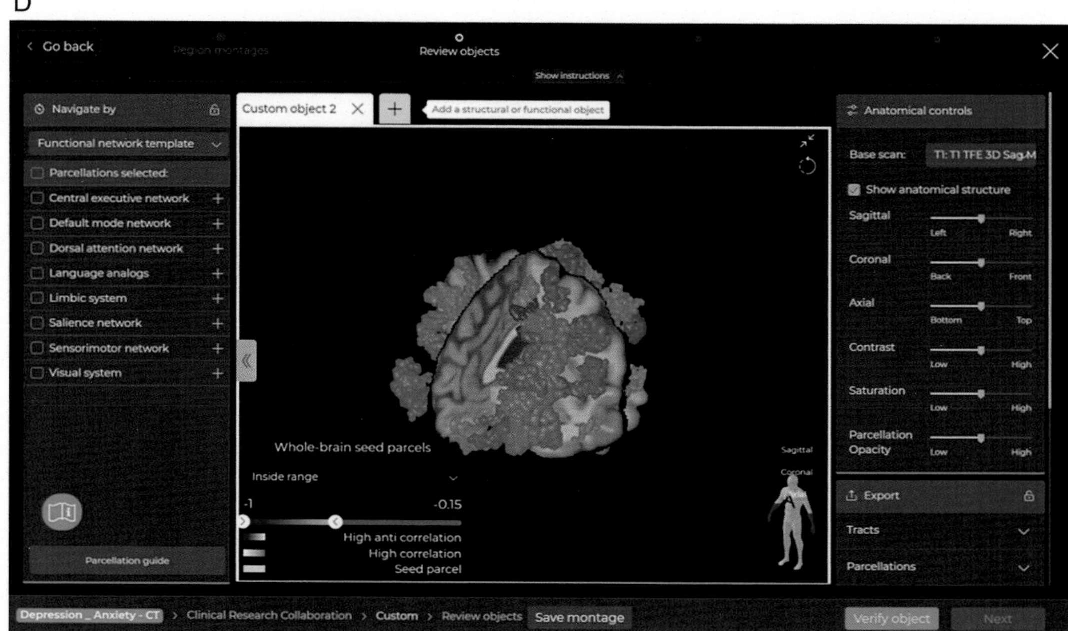

FIGURE 15.1 Cont'd

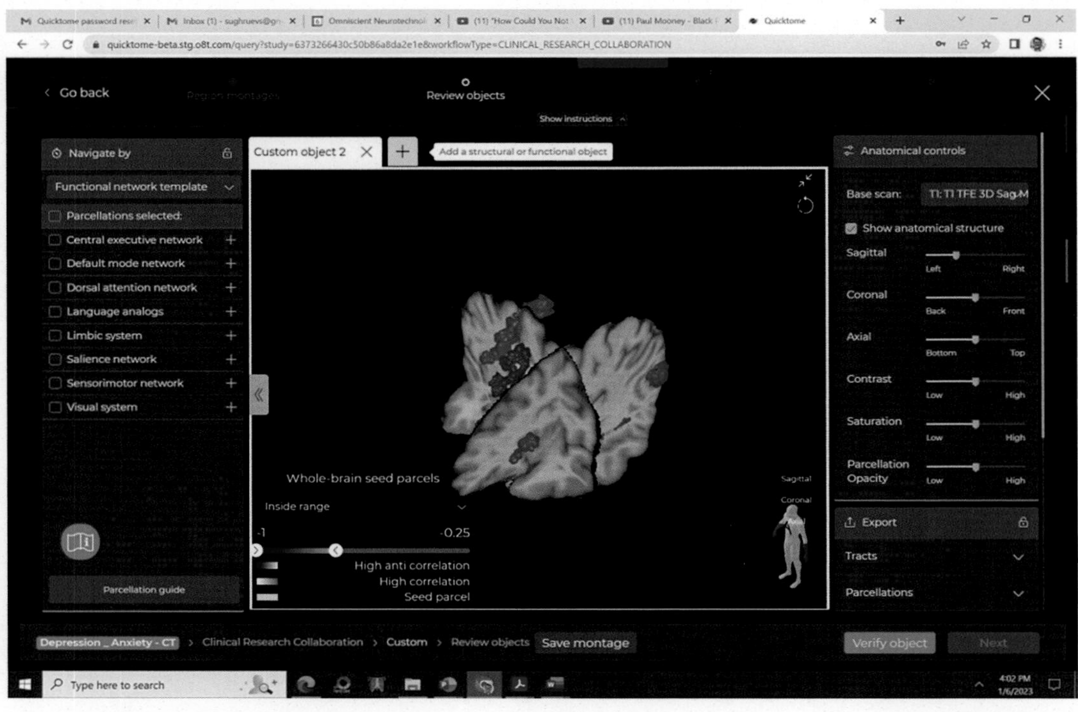

F

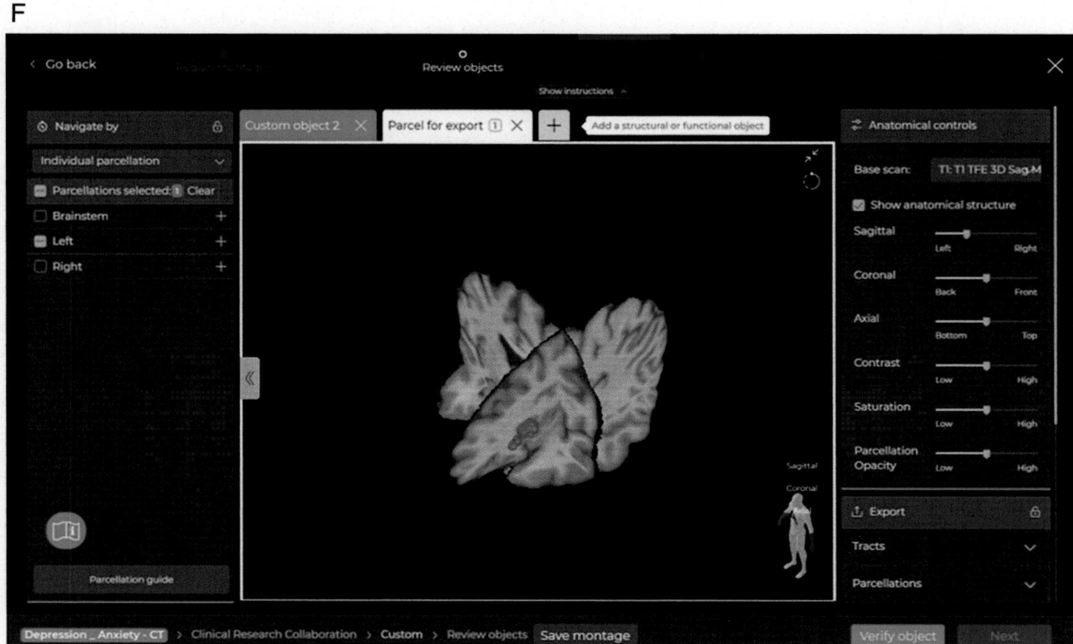

FIGURE 15.1 Cont'd

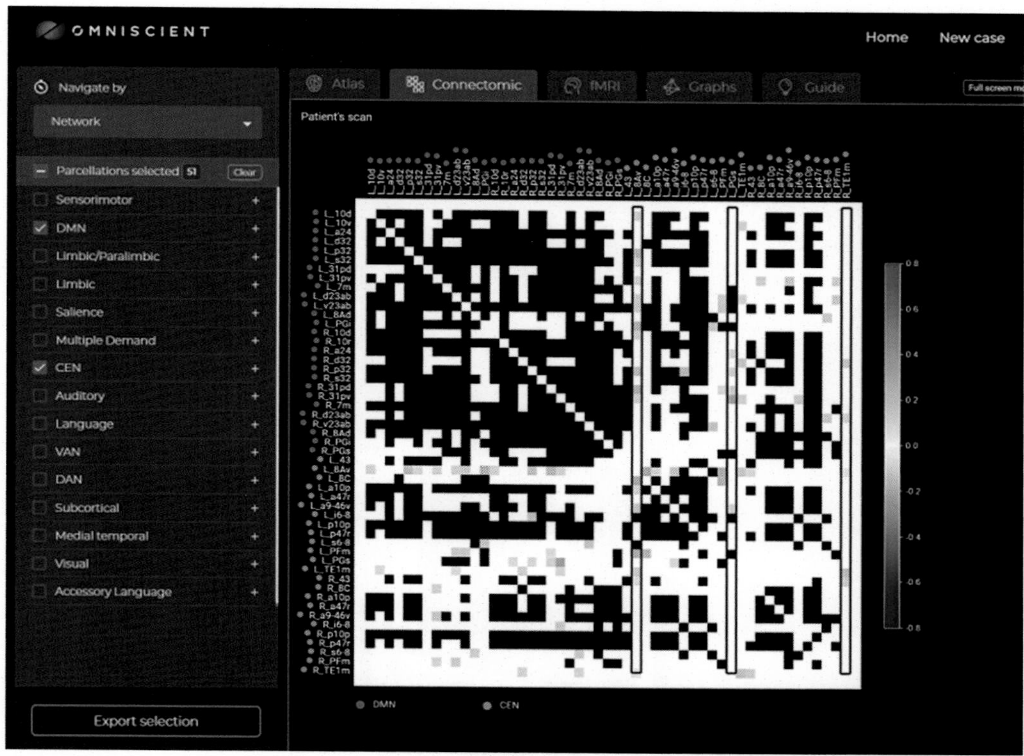

FIGURE 15.2

Functional connectivity anomalies are highlighted in areas L_8AV, L_PGs, and TE1M when comparing the DMN and CEN for the anxiety core.

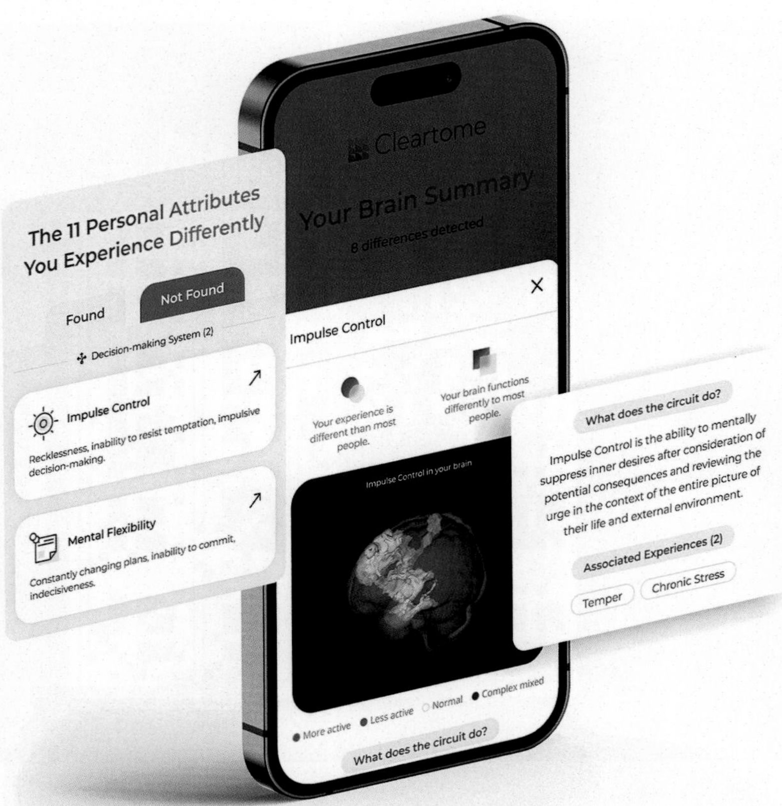

FIGURE 15.3

This figure presents a methodology for picking symptom-specific targets. (A) This is a symptom-specific montage from the Cleartome software. Anomalies are shown on the phone screen. You can see this patient has significantly abnormal impulse control circuit with several anomalies. (B) This is a similar montage in Quicktome (for reward anticipation) demonstrating how you can then pick symptom-specific targets from that dataset.

B

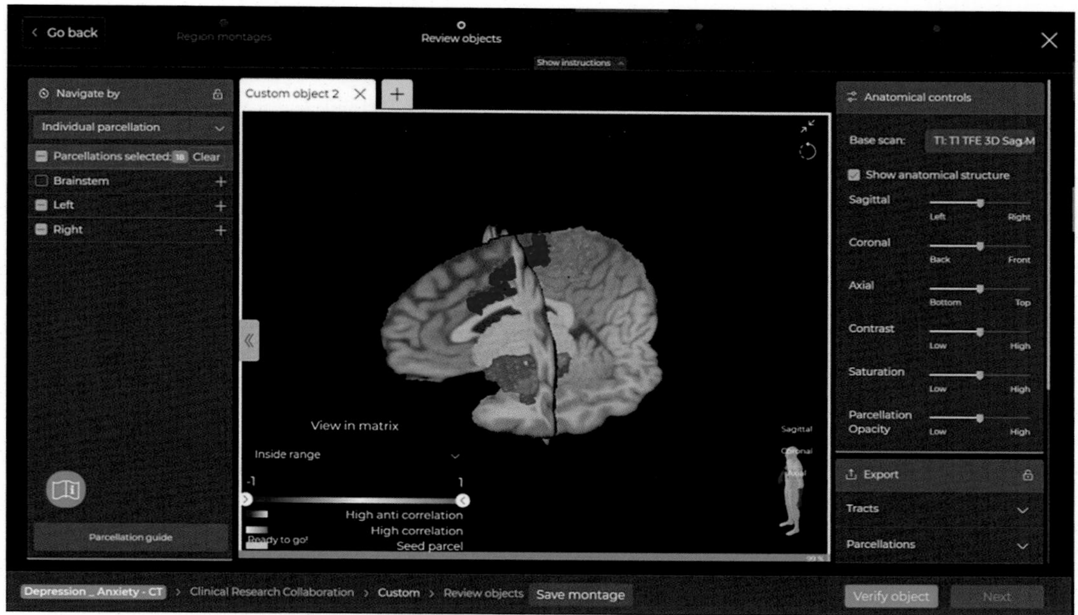

FIGURE 15.3 Cont'd

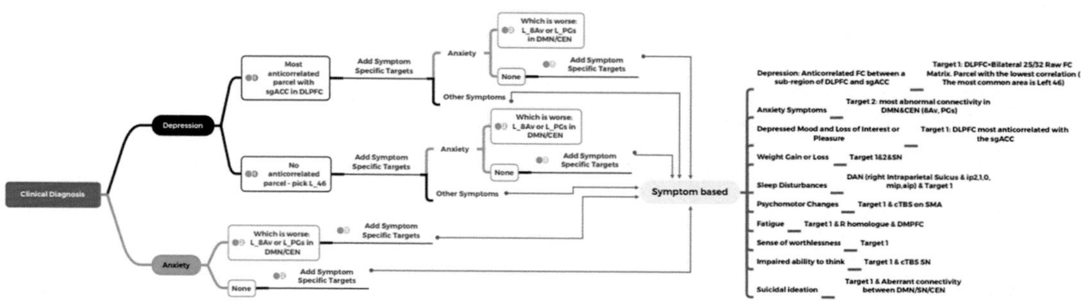

FIGURE 15.4

Basic algorithm to pick targets for treating depression and anxiety with higher success.

Conclusion

In this chapter, we provide a detailed overview of the application of connectivity-guided TMS for the treatment of depression and anxiety. Furthermore, we outline a basic algorithm which can help approach and treat these patients with greater success. In the next chapter, we discuss all of the other less common, but important treatments connectivity-guided TMS can be leveraged toward.

References

Drysdale, A. T., Grosenick, L., Downar, J., Dunlop, K., Mansouri, F., Meng, Y., Fetcho, R. N., Zebley, B., Oathes, D. J., Etkin, A., Schatzberg, A. F., Sudheimer, K., Keller, J., Mayberg, H. S., Gunning, F. M., Alexopoulos, G. S., Fox, M. D., Pascual-Leone, A., Voss, H. U., Casey, B. J., … Liston, C. (2017). Resting-state connectivity biomarkers define neurophysiological subtypes of depression. *Nature Medicine, 23*(1), 28–38. https://doi.org/10.1038/nm.4246

Fitzgerald, P. B., Hoy, K. E., Anderson, R. J., & Daskalakis, Z. J. (2016). A study of the pattern of response to rTMS treatment in depression. *Depression and Anxiety, 33*(8), 746–753. https://doi.org/10.1002/da.22503

Garnaat, S. L., Yuan, S., Wang, H., Philip, N. S., & Carpenter, L. L. (2018). Updates on transcranial magnetic stimulation therapy for major depressive disorder. *Psychiatric Clinics of North America, 41*(3), 419–431. https://doi.org/10.1016/j.psc.2018.04.006

Mantovani, A., Aly, M., Dagan, Y., Allart, A., & Lisanby, S. H. (2013). Randomized sham controlled trial of repetitive transcranial magnetic stimulation to the dorsolateral prefrontal cortex for the treatment of panic disorder with comorbid major depression. *Journal of Affective Disorders, 144*(1–2), 153–159. https://doi.org/10.1016/j.jad.2012.05.038

Moreno-Ortega, M., Kangarlu, A., Lee, S., Perera, T., Kangarlu, J., Palomo, T., Glasser, M. F., & Javitt, D. C. (2020). Parcel-guided rTMS for depression. *Translational Psychiatry, 10*(1), 283. https://doi.org/10.1038/s41398-020-00970-8

Neacsiu, A. D., Luber, B. M., Davis, S. W., Bernhardt, E., Strauman, T. J., & Lisanby, S. H. (2018). On the concurrent use of self-system therapy and functional magnetic resonance imaging–guided transcranial magnetic stimulation as treatment for depression. *The Journal of ECT, 34*(4).

O'Reardon, J. P., Solvason, H. B., Janicak, P. G., Sampson, S., Isenberg, K. E., Nahas, Z., McDonald, W. M., Avery, D., Fitzgerald, P. B., Loo, C., & Demitrack, M. A. (2007). Efficacy and safety of transcranial magnetic stimulation in the acute treatment of major depression: A multisite randomized controlled trial. *Biological Psychiatry, 62*(11), 1208–1216. https://doi.org/10.1016/j.biopsych.2007.01.018

Rosen, A. C., Bhat, J. V., Cardenas, V. A., Ehrlich, T. J., Horwege, A. M., Mathalon, D. H., Roach, B. J., Glover, G. H., Badran, B. W., Forman, S. D., George, M. S., Thase, M. E., Yurgelun-Todd, D., Sughrue, M. E., Doyen, S. P., Nicholas, P. J., Scott, J. C., Tian, L., & Yesavage, J. A. (2021). Targeting location relates to treatment response in active but not sham rTMS stimulation. *Brain Stimulation, 14*(3), 703–709. https://doi.org/10.1016/j.brs.2021.04.010

Sathappan, A. V., Luber, B. M., & Lisanby, S. H. (2019). The Dynamic Duo: Combining noninvasive brain stimulation with cognitive interventions. *Progress in Neuro-Psychopharmacology and Biological Psychiatry, 89*, 347–360. https://doi.org/10.1016/j.pnpbp.2018.10.006

Young, I. M., Dadario, N. B., Tanglay, O., Chen, E., Cook, B., Taylor, H. M., Crawford, L., Yeung, J., Nicholas, P. J., Doyen, S., & Sughrue, M. E. (2023). Connectivity model of the anatomic substrates and network abnormalities in major depressive disorder: A coordinate meta-analysis of resting-state functional connectivity. *Journal of Affective Disorders Reports, 11*, 100478. https://doi.org/10.1016/j.jadr.2023.100478

CHAPTER

Connectomic strategy for the treatment of postconcussive syndrome

16

Introduction

Traumatic brain injury (TBI) is a leading cause of death and disability, affecting 2.5 to 6.5 million people in the United States and 55.5 million people globally (Abdelmalik et al., 2019; GBD 2016 Traumatic Brain Injury and Spinal Cord Injury Collaborators, 2019). Long-term disabilities resulting from the TBI are estimated to affect 3.2 to 5.3 million people (Coronado et al., 2011). Postconcussive symptoms resulting from mild/moderate TBI can be psychological (including depression, anxiety, and irritability), cognitive (memory issues, brain fog, problems with concentration), and sensory in nature (light and noise sensitivity and headaches) (Ryan & Warden, 2003). Symptoms can be divided into early and late onset symptoms: headaches, dizziness, and nausea are early onset PCS and irritability, concentration issues, problems with memory, noise sensitivity, depression, and anxiety are late onset symptoms (Eyres et al., 2005). Current treatment of post-concussive syndrome (PCS) following TBI is often variable and largely aims to treat individual symptoms as separate entities. These drugs may include CNS stimulants, antidepressants, and cholinesterase inhibitors (Crupi et al., 2020).

Our current understanding of the pathophysiology of TBI is explained by the presence of inflammation and edema after the activation of cellular repair, lasting a few weeks after injury (Villamar et al., 2012). In the months that follow, cellular plasticity ensues to promote recovery through long-term potentiation (LTP) and long-term depression (LTD) (Villamar et al., 2012). Neuroimaging studies have shown that TBI leads to changes in connectivity between neural regions in addition to altered resting state networks (Eierud et al., 2014). Resting state fMRI (rsfMRI) studies on patients with mild TBI found an increase in spatial coactivity within regions of the default mode network (DMN) (Nathan et al., 2015). However, it is difficult to obtain an objective measure to characterize radiographically TBI patients from normal patients due to the heterogenous nature of TBI injuries (Nathan et al., 2015). This is in part due to the variability in injury severity and location. As a result of the varied nature of TBI injury, there is a myriad of postconcussive symptoms developed by in patients following TBI.

Transcranial magnetic stimulation (TMS) is a promising noninvasion tool that can induce LTP and LTD, thereby modulating the mechanisms of recovery from brain injury. It is approved by the FDA as a safe, well-tolerated, and noninvasive technology for the treatment of major depressive disorder (MDD), obsessive compulsive disorder (OCD), migraines, smoking cessation, and MDD with anxiety comorbidity (Cohen et al., 2022).

There have been several randomized control trials of repetitive TMS (rTMS) for the treatment of PCS with the dorsolateral prefrontal cortex (dlPFC) as the primary target (Hoy et al., 2019; Koski et al., 2015; Tallus et al., 2013). Patients report a decrease of early reversible symptoms of PCS such as headache, fatigue, and trouble sleeping after rTMS treatment, but these improvements are limited in

Connectomic Medicine. https://doi.org/10.1016/B978-0-443-19089-6.00019-7

duration with symptoms returning after 3 months (Koski et al., 2015). Another approach has been to target specific symptoms of PCS with rTMS instead. Four randomized control trials (RCT) for rTMS in patients with postconcussive headaches found improvements in headache severity and chronicity (Mollica et al., 2022). For later onset symptoms such as depression, four randomized control trials (RCTs) conducted on the effect of TMS in patients with postconcussive depression found mixed results on the improvement of depressive symptoms (Mollica et al., 2022). One study found that the improvements in depressive symptoms with rTMS are no different than with sham treatment (Hoy et al., 2019). One possible interpretation of these results is that targeting only the dlPFC does not effectively treat and account for the myriad of symptoms associated with PCS.

One potential reason is the unreliability of targeting due variability in structure and connectivity at the individual level. To address this, one study has tried to use rTMS targets from individualized dorsal attention network (DAN) and DMN to treat postconcussive depression which resulted in a 29% improvement compared to sham (Siddiqi et al., 2019). More recently, individual resting-state network mapping has been used in conjunction with rTMS to treat postconcussive depression by targeting the subgenual anterior cingulate cortex (sgACC) (Siddiqi et al., 2023). This personalized method of rTMS targeting has shown to improve clinical outcomes (Siddiqi et al., 2023).

Throughout this book, we advocate for the analysis of the brain connectome using machine learning to achieve understanding of structural and functional connectivity at the individual level. Hereafter, we discuss the use of machine learning and parcel-guided rTMS to identify and treat multiple regions of hypo- and hyperconnectivity in the brain following brain injury. We the authors strongly believe that this personalized, parcel-guided rTMS treatment of postconcussive syndrome, known as the Agile Method (Young et al., 2023) is safe and may lead to therapeutic response for patients with TBI.

TMS principles for chronic tinnitus

Step 1: define the problem

Patients with history of TBI are often underdiagnosed for symptoms of postconcussive syndrome. These symptoms may include, but are not limited to, headache, fatigue, vision changes, disturbances in balance, confusion, dizziness, insomnia, and difficulty concentrating. During the initial consultation, it is crucial to conduct a comprehensive history taking to understand the full spectrum of neuro-psycho-cognitive symptoms. It is important to understand whether the symptoms precede the initial head trauma or whether the symptoms truly correlate with the inciting event.

Step 2: look at the relevant parts of structural connectome

This is highly dependent on the mechanisms of trauma as head injuries may induce coup and contra-coup injuries. Longer fiber tracts, such as superior/inferior longitudinal fasciculi and inferior frontooccipital fasciculus (IFOF) are especially susceptible to rotational and linear acceleration-deceleration forces.

Step 3: look at the relevant parts of the functional connectome

Specifically looking for either anomalies or lack of correlation in parts of the networks:

Auditory network
DMN
CEN
Salience
DAN
VAN

*Given the spectrum of possible symptoms, we advise that focus should be paid to networks that correlate to the patient's specific symptoms.

Step 4: determine a treatment which addresses these issues and integrates into the overall care plan

This is usually behavioral and cognitive adjunctive therapies tailored to the patient's situation in between treatments.

In our experience, all patients had rTMS targets outside of the dlPFC, which is not surprising given the wide-spread effect of TBI and myriad of potential symptoms. We find that patient often have significant improvement in their quality of life index (EQ-5D). Similarly, patients can derive improvement in their comorbid anxiety and depression symptoms too. A recent literature review has found a seizure risk for rTMS in healthy adults of 1 in 100,000 and a slightly increased risk for TBI patients, among other neurological conditions, of 67 in 100,000 (Rossi et al., 2021). We recognize that seizures are an inherent but extremely low risk of rTMS but we have not had any seizure-related adverse events with stimulation up to 80% of motor thresholds.

Case study

This was a 57-year-old male patient who, following a TBI in 2017, presented with symptoms consistent with PCS, such as poor concentration, headaches, emotional lability, poor short-term memory. Based on his connectivity matrix, we targeted L8Av (cTBS), LPGs (cTBS), and RTe1m (iTBS) (Fig. 16.1). He had significant improvements in his anxiety, depression, and EQ-5D inventories 6-weeks post-treatment.

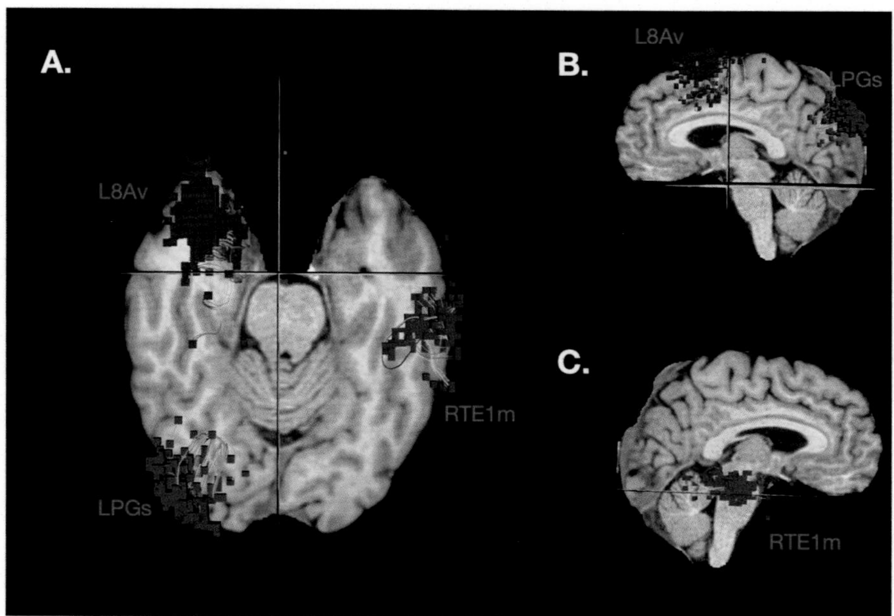

FIGURE 16.1

Anatomical locations of rTMS targets for Patient 1. Above are plane images from the patient's T1-weighted MRI. Axial (A), left sagittal (B), and right sagittal (C).

Importance of the personalized, multisystem rTMS targeting

Results from randomized controlled trials in literature suggest that standard methods of treating neuropsychiatric disorders may not be effective in treating the same symptoms of PCS. For example, the current FDA-approved approach of treating MDD with rTMS by targeting the left dlPFC has a 29.3% response rate compared to 10.4% in the sham group (Horvath et al., 2010). In contrast, a randomized controlled trial of bilateral dlPFC rTMS (which was found in other studies to have superior response rates compared to left dlPFC) of postconcussive depression in TBI patients found improvements that were no different than sham treatment (Hoy et al., 2019). This suggests that the pathophysiology behind a brain injury—induced depression may be different than a non-TBI-related depression. Therefore, it necessitates a difference in treatment for two individuals who may present with similar symptoms of depression (or any other symptom of PCS) from diverse etiologies.

Moreover, the challenge with developing therapies for PCS is its myriad symptoms. While some symptoms have shared network abnormalities, there are also important regions to target that are not shared by these symptoms. For example, the shared networks implicated in both tinnitus and depression include the DMN, Central Executive Network (CEN), and anterior cingulate (Chen et al., 2017; Dutta et al., 2014; Kaiser et al., 2015). At the same time, a treatment for post-concussion chronic tinnitus must also target the primary auditory cortex (Schlee et al., 2008). The heterogeneity of injury, location, and severity makes a one-size-fits-all rTMS protocol difficult. Additionally, there is no approved rTMS treatments currently available that addresses the less common and more subjective symptoms of PCS, let alone a treatment that can target multiple symptoms at once.

With these two points taken together, there is demand for a personalized approach to rTMS treatment for PCS. The use of individualized neuroimaging to treat PCS has been explored by other groups, particularly in treating post-TBI depression through resting-state network mapping (Siddiqi et al., 2019, 2023). One of these studies found improvements of 56% in depressive symptoms with rTMS treatment (Siddiqi et al., 2019). With the addition of machine learning and goal of targeting multiple cortical regions, the Agile Method can improve on efficacy on not only postconcussive depression but other PCS symptoms as well. Functional networks act in global interconnections. Functional connectivity analysis has found that networks are sensitive to influences from remote cortical areas (Adachi et al., 2012). Therefore, it is not just the region and its neighbors that are important for brain function, but the connections between networks and regions as well. Targeting solely one region for rTMS in pursuit of treatment response is therefore inconsistent with our understanding of brain function and connectivity.

Conclusion

The individualized Agile Method of rTMS is safe and can be used in treating the various symptoms of post-concussive syndrome. This method may enable reliable and personalized rTMS targeting, inviting future studies on its efficacy with larger cohorts.

References

Abdelmalik, P. A., Draghic, N., & Ling, G. S. F. (2019). Management of moderate and severe traumatic brain injury. *Transfusion, 59*(S2), 1529–1538.

Adachi, Y., Osada, T., Sporns, O., Watanabe, T., Matsui, T., Miyamoto, K., & Miyashita, Y. (2012). Functional connectivity between anatomically unconnected areas is shaped by collective network-level effects in the macaque cortex. *Cerebral Cortex (New York, N.Y.: 1991), 22*(7), 1586–1592. https://doi.org/10.1093/cercor/bhr234

Chen, Y. C., Bo, F., Xia, W., Liu, S., Wang, P., Su, W., Xu, J. J., Xiong, Z., & Yin, X. (2017). Amygdala functional disconnection with the prefrontal-cingulate-temporal circuit in chronic tinnitus patients with depressive mood. *Progress in Neuro-psychopharmacology & Biological Psychiatry, 79*(Pt B), 249–257. https://doi.org/10.1016/j.pnpbp.2017.07.001

Cohen, S. L., Bikson, M., Badran, B. W., & George, M. S. (2022). A visual and narrative timeline of US FDA milestones for Transcranial Magnetic Stimulation (TMS) devices. *Brain Stimulation, 15*(1), 73–75. https://doi.org/10.1016/j.brs.2021.11.010

Coronado, V. G., Xu, L., Basavaraju, S. V., McGuire, L. C., Wald, M. M., Faul, M. D., Guzman, B. R., Hemphill, J. D., & Centers for Disease Control and Prevention (CDC). (2011). Surveillance for traumatic brain injury-related deaths–United States, 1997–2007. Morbidity and Mortality Weekly Report. *Surveillance Summaries (Washington, D.C.: 2002), 60*(5), 1–32.

Crupi, R., Cordaro, M., Cuzzocrea, S., & Impellizzeri, D. (2020). Management of traumatic brain injury: From present to future. *Antioxidants (Basel, Switzerland), 9*(4), 297. https://doi.org/10.3390/antiox9040297

Dutta, A., McKie, S., & Deakin, J. F. W. (2014). Resting state networks in major depressive disorder. *Psychiatry Research: Neuroimaging, 224*(3), 139–151.

Eierud, C., Craddock, R. C., Fletcher, S., Aulakh, M., King-Casas, B., Kuehl, D., & LaConte, S. M. (2014). Neuroimaging after mild traumatic brain injury: Review and meta-analysis. *NeuroImage: Clinical, 4*, 283–294. https://doi.org/10.1016/j.nicl.2013.12.009

Eyres, S., Carey, A., Gilworth, G., Neumann, V., & Tennant, A. (2005). Construct validity and reliability of the Rivermead Post-Concussion Symptoms Questionnaire. *Clinical Rehabilitation, 19*(8), 878–887. https://doi.org/10.1191/0269215505cr905oa

GBD 2016 Traumatic Brain Injury and Spinal Cord Injury Collaborators. (2019). Global, regional, and national burden of traumatic brain injury and spinal cord injury, 1990–2016: A systematic analysis for the global burden of disease study 2016. *The Lancet Neurology, 18*(1), 56–87.

Horvath, J. C., Mathews, J., Demitrack, M. A., & Pascual-Leone, A. (2010). The NeuroStar TMS device: Conducting the FDA approved protocol for treatment of depression. *Journal of Visualized Experiments: JoVE, 45*, 2345. https://doi.org/10.3791/2345

Hoy, K. E., McQueen, S., Elliot, D., Herring, S. E., Maller, J. J., & Fitzgerald, P. B. (2019). A pilot investigation of repetitive transcranial magnetic stimulation for post-traumatic brain injury depression: Safety, tolerability, and efficacy. *Journal of Neurotrauma, 36*(13), 2092–2098. https://doi.org/10.1089/neu.2018.6097

Kaiser, R. H., Andrews-Hanna, J. R., Wager, T. D., & Pizzagalli, D. A. (2015). Large-scale network dysfunction in major depressive disorder: A meta-analysis of resting-state functional connectivity. *JAMA Psychiatry, 72*(6), 603–611. https://doi.org/10.1001/jamapsychiatry.2015.0071

Koski, L., Kolivakis, T., Yu, C., Chen, J. K., Delaney, S., & Ptito, A. (2015). Noninvasive brain stimulation for persistent postconcussion symptoms in mild traumatic brain injury. *Journal of Neurotrauma, 32*(1), 38–44. https://doi.org/10.1089/neu.2014.3449

Mollica, A., Greben, R., Oriuwa, C., Siddiqi, S. H., & Burke, M. J. (2022). Neuromodulation treatments for mild traumatic brain injury and post-concussive symptoms. *Current Neurology and Neuroscience Reports, 22*(3), 171–181. https://doi.org/10.1007/s11910-022-01183-w

Nathan, D. E., Oakes, T. R., Yeh, P. H., French, L. M., Harper, J. F., Liu, W., Wolfowitz, R. D., Wang, B. Q., Graner, J. L., & Riedy, G. (2015). Exploring variations in functional connectivity of the resting state default mode network in mild traumatic brain injury. *Brain Connectivity, 5*(2), 102–114. https://doi.org/10.1089/brain.2014.0273

Rossi, S., Antal, A., Bestmann, S., Bikson, M., Brewer, C., Brockmöller, J., Carpenter, L. L., Cincotta, M., Chen, R., Daskalakis, J. D., Di Lazzaro, V., Fox, M. D., George, M. S., Gilbert, D., Kimiskidis, V. K., Koch, G., Ilmoniemi, R. J., Lefaucheur, J. P., Leocani, L., Lisanby, S. H., … basis of this article began with a Consensus Statement from the IFCN Workshop on "Present, Future of TMS: Safety, Ethical Guidelines", Siena, October 17–20, 2018, updating through April 2020. (2021). Safety and recommendations for TMS use in healthy subjects and patient populations, with updates on training, ethical and regulatory issues: Expert guidelines. *Clinical Neurophysiology: Official Journal of the International Federation of Clinical Neurophysiology, 132*(1), 269–306. https://doi.org/10.1016/j.clinph.2020.10.003

Ryan, L. M., & Warden, D. L. (2003). Post concussion syndrome. *International Review of Psychiatry, 15*(4), 310–316.

Schlee, W., Weisz, N., Bertrand, O., Hartmann, T., & Elbert, T. (2008). Using auditory steady state responses to outline the functional connectivity in the tinnitus brain. *PLoS One, 3*(11), e3720. https://doi.org/10.1371/journal.pone.0003720

Siddiqi, S. H., Trapp, N. T., Hacker, C. D., Laumann, T. O., Kandala, S., Hong, X., Trillo, L., Shahim, P., Leuthardt, E. C., Carter, A. R., & Brody, D. L. (2019). Repetitive transcranial magnetic stimulation with resting-state network targeting for treatment-resistant depression in traumatic brain injury: A randomized, controlled, double-blinded pilot study. *Journal of Neurotrauma, 36*(8), 1361–1374. https://doi.org/10.1089/neu.2018.5889

Siddiqi, S. H., Kandala, S., Hacker, C. D., Trapp, N. T., Leuthardt, E. C., Carter, A. R., & Brody, D. L. (2023). Individualized precision targeting of dorsal attention and default mode networks with rTMS in traumatic brain injury-associated depression. *Scientific Reports, 13*(1), 4052. https://doi.org/10.1038/s41598-022-21905-x

Tallus, J., Lioumis, P., Hämäläinen, H., Kähkönen, S., & Tenovuo, O. (2013). Transcranial magnetic stimulation-electroencephalography responses in recovered and symptomatic mild traumatic brain injury. *Journal of Neurotrauma, 30*(14), 1270–1277. https://doi.org/10.1089/neu.2012.2760

Villamar, M. F., Santos Portilla, A., Fregni, F., & Zafonte, R. (2012). Noninvasive brain stimulation to modulate neuroplasticity in traumatic brain injury. *Neuromodulation: Journal of the International Neuromodulation Society, 15*(4), 326–338. https://doi.org/10.1111/j.1525-1403.2012.00474.x

Young, I. M., Taylor, H. M., Nicholas, P. J., Mackenzie, A., Tanglay, O., Dadario, N. B., Osipowicz, K., Davis, E., Doyen, S., Teo, C., & Sughrue, M. E. (2023). An agile, data-driven approach for target selection in rTMS therapy for anxiety symptoms: Proof of concept and preliminary data for two novel targets. *Brain and Behavior, 13*(5), e2914. https://doi.org/10.1002/brb3.2914

CHAPTER

17 Connectomic strategies for treating chronic tinnitus associated with psychiatric disorders

Introduction

Tinnitus is characterized by the perception of ringing not resulting from an external acoustic source (Lanting et al., 2009; Mühlau et al., 2006). It affects 10%−15% of the population, with prevalence increasing with age (Kim et al., 2015; Leaver et al., 2016; Rauschecker et al., 2010). Research on tinnitus traditionally focuses on the central auditory pathway. However, a growing body of evidence suggests that brain areas outside of the auditory pathway may be involved in the development of tinnitus (Schlee et al., 2008).

Various brain structures have been studied in relation to the pathophysiology of tinnitus. The most accepted areas are the auditory pathways and the anterior cingulate cortex (ACC) which may also be active, along with other limbic structures, secondary to the emotional distress induced by tinnitus (De Ridder et al., 2016). Both intensity of the tinnitus and distress and mood changes from tinnitus can be attributable to the dorsal anterior cingulate cortex (dACC) and the insula, which are both components of the salience network (SN). De Ridder et al. attempted to permanently suppress dACC hyperactivity in two tinnitus patients who had undergone transcranial magnetic stimulation (TMS) by implanting an electrode on the dACC (De Ridder et al., 2016). Although only one out of the two patients responded to treatment, the authors postulated that increased functional connectivity among the parahippocampal gyrus, subgenual ACC, and insula may result in clinical improvement (De Ridder et al., 2016). In support of that, Chen et al. found abnormal connections between the ACC, auditory cortex, prefrontal cortex, visual cortex, and the default mode network (DMN), suggesting that the tinnitus network is a complex interaction of multiple areas of the brain (Chen et al., 2017). Specifically, it was found that rTMS of the right DLPFC transiently reduced the perception of the loudness of the tinnitus (Vanneste & De Ridder, 2013). Vanneste and De Ridder also found that 1−3 Hz TMS to the prefrontal cortex reduced tinnitus intensity and distress, supporting the hypothesis that nonauditory areas are involved in mediating tinnitus (Vanneste & De Ridder, 2013).

Based on evidence from prior literature suggesting the involvement of networks other than the auditory network in the pathophysiology of tinnitus, we hypothesized that CEN, DMN, and SN could be targeted using rTMS. These networks are often targeted in the treatment of depression and anxiety which are common comorbid psychiatric conditions in patients with chronic tinnitus

Connectomic Medicine. https://doi.org/10.1016/B978-0-443-19089-6.00020-3

(Bhatt et al., 2017; Sullivan et al., 1988; Zoger et al., 2006). Williams showed that perturbances in these major functional networks resulted in depression and anxiety (Williams, 2016). Until recently, the field lacked a readily available tool to visualize and quantify the functional connectome on a personalized level. Throughout this book, we discussed the current advances that allow us to quantify abnormal connections among parcels in the CEN, DMN, and SN that could serve as targets for rTMS modulation to reduce tinnitus intensity by perhaps modulating the perception of the distress. In this chapter, we discuss the use of this agile approach to treat patients with chronic tinnitus and share our insights from using this connectomic-based strategy.

TMS treatment

The treatment paradigm used was an accelerated theta burst stimulation (aTBS) protocol, consisting of five image-guided TBS treatment sessions per day for 5 days with 1-hour gaps between sessions (Sonmez et al., 2019). We used the anomaly detection algorithm to guide the selection of intermittent theta burst stimulation (iTBS) or continuous theta burst stimulation (cTBS) protocols. We used cTBS to induce cortical depression if an area showed mostly hyperconnectivity with other areas (Huang et al., 2005). We used iTBS for areas that were hypoconnected to other parts of these networks (Huang et al., 2005). The targets were consecutively stimulated in each treatment session at 80% of the resting motor threshold. In between stimulations, patients participated in individualized behavioral therapies, such as relaxation therapy for those with comorbid anxiety (Tsagaris et al., 2016).

TMS principles for chronic tinnitus

Step 1: define the problem

Patients with chronic tinnitus ought to undergo evaluation by an otolaryngologist and/or had undergone audiology evaluation prior to being referred for rTMS treatment. The clinician needs to rule out structural causes to the patient's chronic tinnitus. Chronic noise exposure should be inquired. During the initial consultation, it is crucial to conduct a comprehensive history taking as well to understand whether the patient has any comorbid mood disturbances or prior history of post-concussive symptoms.

Step 2: look at the relevant parts of structural connectome

primary auditory cortex

posterior supertemporal gyrus

Step 3: look at the relevant parts of the functional connectome

Specifically looking for either anomalies or lack of correlation in parts of the networks:

Auditory network

DMN

CEN

Salience

Step 4: determine a treatment which addresses these issues and integrates into the overall care plan

These are usually behavioral and cognitive adjunctive therapies tailored to the patient's situation in between treatments.

Case study #1

Patient 1 was a 49-year-old-female patient who presented with symptoms of tinnitus, which emerged in adolescence. She had a history of hospitalization for depression at age 18, with current comorbidities of depression, anxiety, Hashimoto's disease, Raynaud's Syndrome and early menopause. She experienced occasional tinnitus prior to treatment, with a pretreatment Tinnitus Reaction Questionnaire (TRQ) rating of 16. Twenty-five sessions of neuro-navigated continuous theta burst stimulation (cTBS) at 80% of resting motor threshold (RMT) was administered targeting LPFm and L46 (Fig. 17.1). Due to her comorbid anxiety, she underwent meditation between the cTBS sessions. Following treatment, Patient 1 reported alleviation of tinnitus symptoms, experiencing no tinnitus with a TRQ score of 0. Her 3-month follow-up TRQ rating indicated only slight tinnitus symptoms.

Case study #2

Patient 2 was a 48-year-old-female patient who presented with symptoms consistent with postconcussion syndrome resulting from a high velocity motor vehicle accident 3 years prior to the current study. The accident resulted in loss of consciousness, multiple rib fractures and left eye and face lacerations. Although Patient 2 did not experience retrograde amnesia, she experienced other symptoms of post-concussion syndrome including altered taste, difficulties with speech, fatigue, visual disturbances, constant headaches, light and noise sensitivities, and visuospatial disorientation. Patient 2 reported poor sleep and mental blocks particularly during social interactions. After acute treatment of her injuries, she began neurorehabilitation and sought treatment from neuropsychologists.

Tinnitus scales of Patient 2 were not assessed at the time of treatment as her primary diagnosis was postconcussion syndrome. However, her initial TRQ scales were obtained retroactively to represent her perceived symptoms prior to

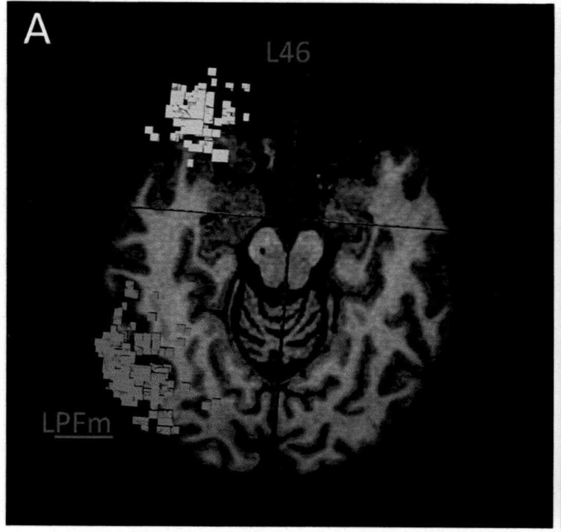

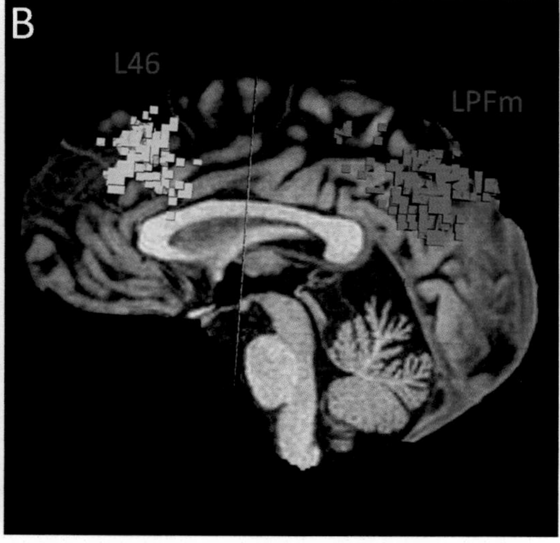

FIGURE 17.1

Anatomical locations of rTMS for Patient 1. T1-weighted MRI. Axial (A) and Left sagittal (B).

Case study #2—cont'd

treatment. Patient 2 reported mild tinnitus symptoms pretreatment with a TRQ rating of 31, and underwent 25 sessions of neuronavigated cTBS targeting Ls6-8, RTE1m, and LPFm at 80% RMT (Fig. 17.2). These sessions were interspersed with support intervals which included Neuro-Emotional Technique (NET), remedial massage, individualized cognitive exercises, and meditation. Patient 2's tinnitus symptoms 6-months following treatment indicated only slight tinnitus symptoms, with a TRQ rating of 5.

Patient 2 reported a pretreatment EQ5D score of 0.276 which was significantly improved following treatment, as seen in the post-treatment EQ5D rating of 0.560. Her general wellbeing was further improved at the 6-month follow-up, with an EQ5D score of 0.647.

Case study #3

Patient 3 was a 73-year-old-male patient who presented with tinnitus, which he had been experiencing for over 30 years. Patient 3 had comorbidities of mild anxiety and depression.

The pretreatment TRQ rating of Patient 3 was 46, which indicated that he had been experiencing moderate tinnitus. He underwent a treatment protocol that alternated between sessions that targeted the auditory regions of both cerebral hemispheres and sessions that stimulated the frontal lobe to aid with rigid thinking. Treatment involved a combination of neuronavigated cTBS and iTBS at 80% of RMT. He received 15 sessions of cTBS on Right A5, 15 sessions of cTBS on Left A4, and 10 sessions of iTBS on 8Av (Fig. 17.3). These sessions were interspersed with mobility-walking, and repeated use of tinnitus iPad applications: audionotch (Teletest, Canada), Whist-tinnitus (Sensimetrics, USA). Patient 3 reported improvements in his tinnitus symptoms post-treatment, indicated by a TRQ rating of 6. A 2-year follow-up reported that the alleviation of his tinnitus symptoms were maintained, with slight tinnitus symptoms as suggested by a TRQ rating of 11.

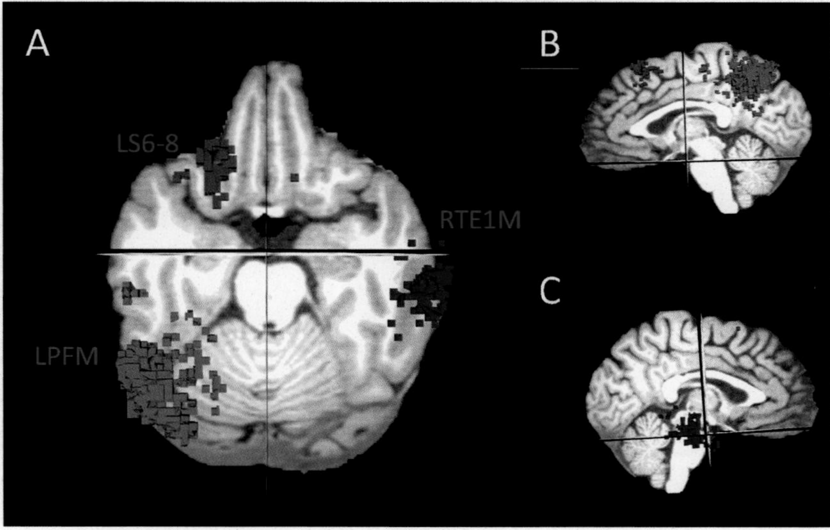

FIGURE 17.2

Anatomical locations of rTMS for Patient 2. T1-weighted MRI. Axial (A), left sagittal (B), and right sagittal (C).

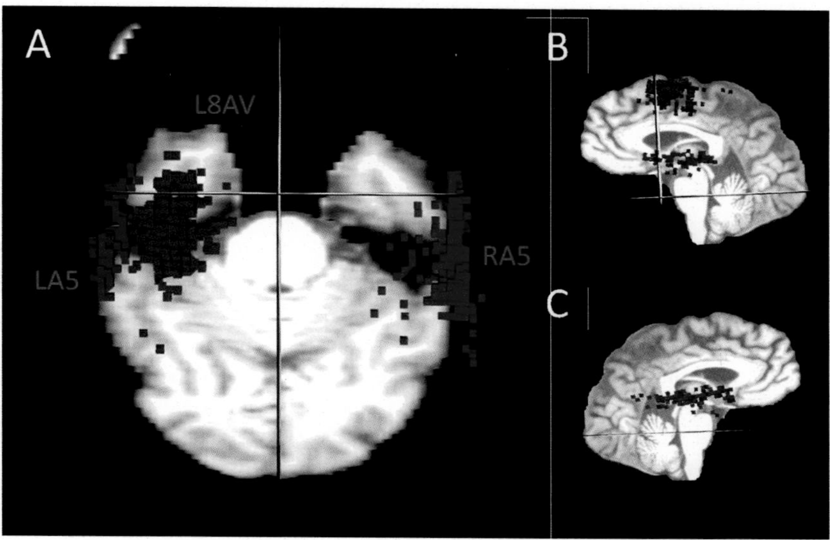

FIGURE 17.3

Anatomical locations of rTMS for Patient 3. T1-weighted MRI. Axial (A), left sagittal (B), and right sagittal (C).

Our experience

Improvement can be attained in patients by studying individual functional connectomes and establishing an off-label rTMS strategy to select individualized targets for stimulation based on the notion that there are other functional networks involved in the pathogenesis of tinnitus.

Treatment and management of tinnitus is often difficult, with limited treatment options that yield variable outcomes. So far, cognitive behavioral therapy (CBT) is the major treatment modality with efficacy in improving distress, quality of life, and depression scores associated with tinnitus (Dalrymple et al., 2021; Zenner et al., 2017). CBT has strong evidence of efficacy, but it does not reduce tinnitus symptoms. Rather, its ultimate therapeutic goal is gradual habituation. Other treatment options include sound therapy and avoiding exposure to noise to impede the progression of tinnitus, but with little evidence to support their effectiveness.

Although pharmacological interventions are administered for tinnitus-associated symptoms such as mood disorders, irritability, and sleep disturbances, there are no currently available medications to reduce tinnitus itself (Landgrebe et al., 2009). With a gap in treatment options to relieve patients from this disturbing affliction, our approach of personalized rTMS could potentially benefit a large population of patients. So far, we have had good success in alleviating patient's tinnitus as measured by established tinnitus inventories as demonstrated by the cases above.

Why we focus on more than the primary auditory circuit

It is reported in the literature that patients with tinnitus display abnormalities in various brain circuits, such as the DMN, salience network, and CEN (Chen et al., 2017; De Ridder et al., 2013; Vanneste et al., 2018). The dorsolateral prefrontal cortex (DLPFC) is a prime example of a brain area that is targeted for the treatment of depression and is implicated in selecting salient auditory signals (Barbas et al., 2011). In addition, the DLPFC can have direct modulation on the primary auditory cortex (Knight et al., 1989). It is reasonable to consider the aforementioned networks in mediating tinnitus pathogenesis since tinnitus itself is associated with various psychiatric comorbidities. We included investigations of other networks with the thought that comorbid psychiatric conditions may exacerbate tinnitus intensity and their modulation may help reduce the perception of and "mentally ignore" the tinnitus (Vanneste & De Ridder, 2013).

The literature suggests that the prevalence of depression and/or anxiety ranges from approximately 25%–75% of patients with chronic tinnitus (Bhatt et al., 2017; Sullivan et al., 1988; Zoger et al., 2006). In our experience, our patients often have depression, anxiety, or similar symptoms as manifestations of previous concussion. Given that depression, anxiety, and other psychiatric conditions are majorly governed by these three major networks, it would be reasonable to investigate alterations in connectivity among these networks as part of the pathogenesis of tinnitus (Li et al., 2021; Liston et al., 2014; Xiong et al., 2020; Yan et al., 2019). A common area that is targeted with comorbid anxiety is L8av, an area that we have found to be consistent in predominantly anxiety presentations and L46, which is a part of the dorsolateral prefrontal cortex that is commonly targeted for depression (Fox et al., 2012). In fact, the use of sertraline, an approved treatment for both depression and anxiety, was shown to significantly improve the symptoms of tinnitus in a randomized controlled trial (Zoger et al., 2006). Of course, sertraline can have effects on multiple brain networks, but this points further to the possibility that internetwork connectivity are implicated in chronic tinnitus (Kim et al., 2020).

Personalized strategy in identification of rTMS targets

The need for individualized treatment plans for patients with chronic tinnitus with psychiatric comorbidities stems from the need to treat the symptoms of multiple diagnoses simultaneously. This was inspired by the transdiagnostic approach to treat emotional disorders that has been shown to be more effective in the treatment of patients with diagnostically heterogeneous conditions, in this case being chronic tinnitus but with various coexisting emotional disorders (Farchione et al., 2012). In order to do so, we had to be able to visualize and quantify functional connectivity of each patient relative to the normal healthy population. Specifically, this approach involved the generation of the hypothesis that chronic tinnitus can be driven by networks outside of the primary auditory network and then evaluating those major networks (DMN, CEN, and salience) to identify any comparable vulnerabilities or connections that can serve as targets for rTMS treatment (Doyen et al., 2021, 2022; Yeung et al., 2021). When we analyse these three major networks together with the primary auditory network, we start to identify connectivity anomalies outside of 3-sigmas of the normal range of rsfMRI data from healthy adults (e.g., estimated against a normative group of 200 healthy rsfMRI data). From these analyses, we can identify and quantitatively demonstrate parcellations with three or more abnormalities among other parcellations and then subsequently attempt to normalize these connections using various forms of rTMS stimulations (iTBS versus cTBS).

Conclusion

Repetitive TMS treatment with target selection using a personalized, agile approach is safe and can provide durable symptomatic relief for patients with chronic tinnitus. In the next chapter, we will discuss how similar treatments can be applied to patients with other multi-network conditions.

References

Barbas, H., Zikopoulos, B., & Timbie, C. (June 15, 2011). Sensory pathways and emotional context for action in primate prefrontal cortex. *Biological Psychiatry, 69*(12), 1133–1139. https://doi.org/10.1016/j.biopsych.2010.08.008

Bhatt, J. M., Bhattacharyya, N., & Lin, H. W. (February 2017). Relationships between tinnitus and the prevalence of anxiety and depression. *The Laryngoscope, 127*(2), 466–469. https://doi.org/10.1002/lary.26107

Chen, Y. C., Bo, F., Xia, W., Liu, S., Wang, P., Su, W., Xu, J. J., Xiong, Z., & Yin, X. (2017). Amygdala functional disconnection with the prefrontal-cingulate-temporal circuit in chronic tinnitus patients with depressive mood. *Progress in Neuro-Psychopharmacology & Biological Psychiatry, 79*(Pt B), 249–257. https://doi.org/10.1016/j.pnpbp.2017.07.001

Dalrymple, S. N., Lewis, S. H., & Philman, S. (June 1, 2021). Tinnitus: Diagnosis and management. *American Family Physician, 103*(11), 663–671.

De Ridder, D., Joos, K., & Vanneste, S. (April 2016). Anterior cingulate implants for tinnitus: Report of 2 cases. *Journal of Neurosurgery, 124*(4), 893–901. https://doi.org/10.3171/2015.3.JNS142880

De Ridder, D., Song, J. J., & Vanneste, S. (May 2013). Frontal cortex TMS for tinnitus. *Brain Stimulation, 6*(3), 355–362. https://doi.org/10.1016/j.brs.2012.07.002

Doyen, S., Nicholas, P., Poologaindran, A., Crawford, L., Young, I. M., Romero-Garcia, R., & Sughrue, M. E. (2022). Connectivity-based parcellation of normal and anatomically distorted human cerebral cortex. *Human Brain Mapping, 43*(4), 1358–1369. https://doi.org/10.1002/hbm.25728. Epub 2021 Nov 26.

Doyen, S., Taylor, H., Nicholas, P., Crawford, L., Young, I., & Sughrue, M. E. (2021). Hollow-tree super: A directional and scalable approach for feature importance in boosted tree models. *PLoS One, 16*(10). https://doi.org/10.1371/journal.pone.0258658. e0258658.

Farchione, T. J., Fairholme, C. P., Ellard, K. K., Boisseau, C. L., Thompson-Hollands, J., Carl, J. R., Gallagher, M. W., & Barlow, D. H. (2012). Unified protocol for transdiagnostic treatment of emotional disorders: A randomized controlled trial. *Behavior Therapy, 43*(3), 666–678. https://doi.org/10.1016/j.beth.2012.01.001

Fox, M. D., Buckner, R. L., White, M. P., Greicius, M. D., & Pascual-Leone, A. (October 1, 2012). Efficacy of transcranial magnetic stimulation targets for depression is related to intrinsic functional connectivity with the subgenual cingulate. *Biological Psychiatry, 72*(7), 595–603. https://doi.org/10.1016/j.biopsych.2012.04.028

Huang, Y. Z., Edwards, M. J., Rounis, E., Bhatia, K. P., & Rothwell, J. C. (January 20, 2005). Theta burst stimulation of the human motor cortex. *Neuron, 45*(2), 201–206. https://doi.org/10.1016/j.neuron.2004.12.033

Kim, M., Jung, W. H., Shim, G., & Kwon, J. S. (November 26, 2020). The effects of selective serotonin reuptake inhibitors on brain functional networks during goal-directed planning in obsessive-compulsive disorder. *Scientific Reports, 10*(1), 20619. https://doi.org/10.1038/s41598-020-77814-4

Kim, H. J., Lee, H. J., An, S. Y., Sim, S., Park, B., Kim, S. W., Lee, J. S., Hong, S. K., & Choi, H. G. (2015). Analysis of the prevalence and associated risk factors of tinnitus in adults. *PLoS One, 10*(5), e0127578. https://doi.org/10.1371/journal.pone.0127578

Knight, R. T., Scabini, D., & Woods, D. L. (December 18, 1989). Prefrontal cortex gating of auditory transmission in humans. *Brain Research, 504*(2), 338–342. https://doi.org/10.1016/0006-8993(89)91381-4

Landgrebe, M., Langguth, B., Rosengarth, K., Braun, S., Koch, A., Kleinjung, T., May, A., de Ridder, D., & Hajak, G. (2009). Structural brain changes in tinnitus: Grey matter decrease in auditory and non-auditory brain areas. *NeuroImage, 46*(1), 213–218. https://doi.org/10.1016/j.neuroimage.2009.01.069

Lanting, C. P., de Kleine, E., & van Dijk, P. (September 2009). Neural activity underlying tinnitus generation: Results from PET and fMRI. *Hearing Research, 255*(1–2), 1–13. https://doi.org/10.1016/j.heares.2009.06.009

Leaver, A. M., Seydell-Greenwald, A., & Rauschecker, J. P. (April 2016). Auditory-limbic interactions in chronic tinnitus: Challenges for neuroimaging research. *Hearing Research, 334*, 49–57. https://doi.org/10.1016/j.heares.2015.08.005

Li, J., Liu, J., Zhong, Y., Wang, H., Yan, B., Zheng, K., Wei, L., Lu, H., & Li, B. (2021). Causal interactions between the default mode network and central executive network in patients with major depression. *Neuroscience, 475*, 93–102. https://doi.org/10.1016/j.neuroscience.2021.08.033

Liston, C., Chen, A. C., Zebley, B. D., Drysdale, A. T., Gordon, R., Leuchter, B., Voss, H. U., Casey, B. J., Etkin, A., & Dubin, M. J. (2014). Default mode network mechanisms of transcranial magnetic stimulation in depression. *Biological Psychiatry, 76*(7), 517–526. https://doi.org/10.1016/j.biopsych.2014.01.023

Mühlau, M., Rauschecker, J. P., Oestreicher, E., Gaser, C., Röttinger, M., Wohlschläger, A. M., Simon, F., Etgen, T., Conrad, B., & Sander, D. (2006). Structural brain changes in tinnitus. *Cerebral Cortex, 16*(9), 1283–1288. https://doi.org/10.1093/cercor/bhj070. New York, N.Y.: 1991.

Rauschecker, J. P., Leaver, A. M., & Muhlau, M. (June 24, 2010). Tuning out the noise: Limbic-auditory interactions in tinnitus. *Neuron, 66*(6), 819–826. https://doi.org/10.1016/j.neuron.2010.04.032

Schlee, W., Weisz, N., Bertrand, O., Hartmann, T., & Elbert, T. (2008). Using auditory steady state responses to outline the functional connectivity in the tinnitus brain. *PLoS One, 3*(11), e3720. https://doi.org/10.1371/journal.pone.0003720

Sonmez, A. I., Camsari, D. D., Nandakumar, A. L., Voort, J. L. V., Kung, S., Lewis, C. P., & Croarkin, P. E. (2019). Accelerated TMS for depression: A systematic review and meta-analysis. *Psychiatry Research, 273*, 770–781. https://doi.org/10.1016/j.psychres.2018.12.041

Sullivan, M. D., Katon, W., Dobie, R., Sakai, C., Russo, J., & Harrop-Griffiths, J. (July 1988). Disabling tinnitus. Association with affective disorder. *General Hospital Psychiatry, 10*(4), 285–291. https://doi.org/10.1016/0163-8343(88)90037-0

Tsagaris, K. Z., Labar, D. R., & Edwards, D. J. (2016). A framework for combining rTMS with behavioral therapy. *Frontiers in Systems Neuroscience, 10*, 82. https://doi.org/10.3389/fnsys.2016.00082

Vanneste, S., & De Ridder, D. (March 2013). Differences between a single session and repeated sessions of 1 Hz TMS by double-cone coil prefrontal stimulation for the improvement of tinnitus. *Brain Stimulation, 6*(2), 155–159. https://doi.org/10.1016/j.brs.2012.03.019

Vanneste, S., Joos, K., Ost, J., & De Ridder, D. (February 2018). Influencing connectivity and cross-frequency coupling by real-time source localized neurofeedback of the posterior cingulate cortex reduces tinnitus related distress. *Neurobiol Stress, 8*, 211–224. https://doi.org/10.1016/j.ynstr.2016.11.003

Williams, L. M. (May 2016). Precision psychiatry: A neural circuit taxonomy for depression and anxiety. *The Lancet Psychiatry, 3*(5), 472–480. https://doi.org/10.1016/S2215-0366(15)00579-9

Xiong, H., Guo, R. J., & Shi, H. W. (2020). Altered default mode network and salience network functional connectivity in patients with generalized anxiety disorders: An ICA-based resting-state fMRI study. *Evid Based Complement Alternat Med*, 4048916. https://doi.org/10.1155/2020/4048916

Yan, C. G., Chen, X., Li, L., Castellanos, F. X., Bai, T. J., Bo, Q. J., Cao, J., Chen, G. M., Chen, N. X., Chen, W., Cheng, C., Cheng, Y. Q., Cui, X. L., Duan, J., Fang, Y. R., Gong, Q. Y., Guo, W. B., Hou, Z. H., Hu, L., … Zang, Y. F. (2019). Reduced default mode network functional connectivity in patients with recurrent major depressive disorder. *Proceedings of the National Academy of Sciences of the United States of America, 116*(18), 9078–9083. https://doi.org/10.1073/pnas.1900390116

Yeung, J. T., Young, I. M., Doyen, S., Teo, C., & Sughrue, M. E. (October 2021). Changes in the brain connectome following repetitive transcranial magnetic stimulation for stroke rehabilitation. *Cureus, 13*(10), e19105. https://doi.org/10.7759/cureus.19105

Zenner, H. P., Delb, W., Kröner-Herwig, B., Jäger, B., Peroz, I., Hesse, G., Mazurek, B., Goebel, G., Gerloff, C., Trollmann, R., Biesinger, E., Seidler, H., & Langguth, B. (2017). A multidisciplinary systematic review of the treatment for chronic idiopathic tinnitus. *European Archives of Oto-Rhino-Laryngology: Official Journal of the European Federation of Oto-Rhino-Laryngological Societies (EUFOS): Affiliated With the German Society for Oto-Rhino-Laryngology - Head and Neck Surgery, 274*(5), 2079–2091. https://doi.org/10.1007/s00405-016-4401-y

Zoger, S., Svedlund, J., & Holgers, K. M. (February 2006). The effects of sertraline on severe tinnitus suffering–a randomized, double-blind, placebo-controlled study. *Journal of Clinical Psychopharmacology, 26*(1), 32–39. https://doi.org/10.1097/01.jcp.0000195111.86650.19

CHAPTER

Connectomic approach to treating everything else 18

Introduction

The literature is full of TMS being used to treat a large variety of things. As reiterated throughout previous chapters, we know it is safe, and often effective when used on the correct patient to stimulate the correct part of the brain. People have treated epilepsy, TBI, Alzheimer disease, tinnitus, schizophrenia, PTSD, ADHD, OCD, and essentially every other brain disease you can imagine (Cocchi et al., 2018; Memon, 2021; Petrosino et al., 2021; Theodore, 2003; Zhou et al., 2021). So where do we start?

If depression has a potentially massive decision tree, then this explodes when we add in numerous other complex problems, many of which are less well understood. This book could have hundreds of more chapters if we decided to exhaustively cover everything else we have treated, heard of someone treating, or seen a study about. The brain is a complex place and there are as many problems as there are patients. Many times, we are confronted with patients presenting with problems which lack an obvious solution, but which we have been able to achieve something positive and life improving. This does not mean we can help everyone, but connectomic medicine is the process of trying to think through the problem based on data and known facts, and coming up with a reasonable approach.

Having said this, instead of trying to outline a paradigm for every possible problem you can treat or we have treated, we will focus on how to think through unusual cases. We will provide some ideas which expand on the previous chapters, as well as integrate some core ideas from the chapters on postsurgical rehab, stroke, and depression/anxiety, and highlight some approaches we have modified from various trials or other clinical experiences. We also will provide a few examples of treatment strategies for unique issues to supplement other disease states. Many of these problems cross multiple disease states, and even medical disciplines, so in lieu of artificially separating problems by causes, we have written this chapter transdiagnostically, focusing on the problem, and how to treat it.

General advice for tackling unique cases

(1) *Maintain a systematic approach*: Our previous chapters were written to focus on outlining fundamentals and best practices. This is more important when you are treating patients with odd, poorly defined symptoms, patients who cross diagnostic boundaries, or patients with problems that lack clear, well-trodden treatment paths. Defining therapeutic goals, focusing on knowing where to look in the connectome, integrating structural and functional information, and forming a solid treatment plan around the TMS plan are key to preventing haphazard approaches, missing

Connectomic Medicine. https://doi.org/10.1016/B978-0-443-19089-6.00009-4

key points, or addressing the wrong problem. Creative strategies do not imply a lack of discipline; in fact, when done well this is quite the opposite.

(2) *Consider depression/anxiety physiology in the disease process*: These fundamental problems are extremely common and they make many problems worse. When reduced to their essence, these diseases involve functional connectivity problems in various circuits of the brain (Young et al., 2023). When those circuits are involved in emotional, reward, or cognitive circuits, the result is the patient has problems in the depression and/or anxiety complex. However, when the physiologic problem occurs in the upper insuloopercular or interoceptive regions, it can cause somatic systems. Other systems can cause over sensitivity to pain, or even stranger problems in other networks that we recognize as manifestations of psychiatric not neurologic causes, though it has always been questionable if such a distinction means anything. Additionally, many problems are made much worse by anxiety or depression complex problems, such as pain, fatigue, tinnitus, and other similar problems (Sheng et al., 2017). We routinely add depression or anxiety core targets to other symptom-specific targets in cases like this. If a patient meets criteria for depression or anxiety, it is likely mandatory to treat these problems when addressing other functional neurologic disorders, whether or not the problems are related. However, we have seen success adding anxiety targets to treatment plans for patients with issues like pain and tinnitus, even when patients do not endorse these symptoms, as it often gets them to pay less attention to their initial symptom which is a great blessing.

The concept of using a core depression target (often a classical target) to address several related symptoms at once is outlined in Fig. 18.1. Here, the common depression circuit does the heavy lifting, and additional targets address things outside the core target.

(3) *Use neuropsychologic principles when appropriate*: While every aspect of cognition is not known, and many higher functions are not simple uni-network, or unifocal problems, a lot of solid neurology and neuropsychology helps guide us to good ideas. We know hemispatial neglect is a right parietal and mainly a right ventral attention network injury pattern (Corbetta & Shulman, 2011). We know what a homonymous hemianopsia means, as well as location of agnosias. We know that working memory problems can result from CEN issues and/or be broken down into transdiagnostic circuits (Champod & Petrides, 2010). We know anterograde amnesia results from problems with the medial temporal structures, the memory circuit of the DMN and cingulum bundle, and/or mamilloforniceal tract problems (De Simoni et al., 2016). Newer data link cognitive and/or attentional decline to newer pathways like the DAN or multiple demand networks (Dadario & Sughrue, 2022). We also know that attention and language problems can cause problems with more complex functions and need to be accounted for. We know ideomotor apraxia is a left frontoparietal dysfunction and Gerstmann's syndrome results from left parietal injury (Shahab et al., 2022). Using these ideas provides useful clues on where to look, how to define prognosis and goals, and how to direct interstimulation therapy.

(4) *Use transdiagnostic tools when appropriate*: Cognition is complex. Much of this can be explained by circuits which handle key components of decision making, emotional control,

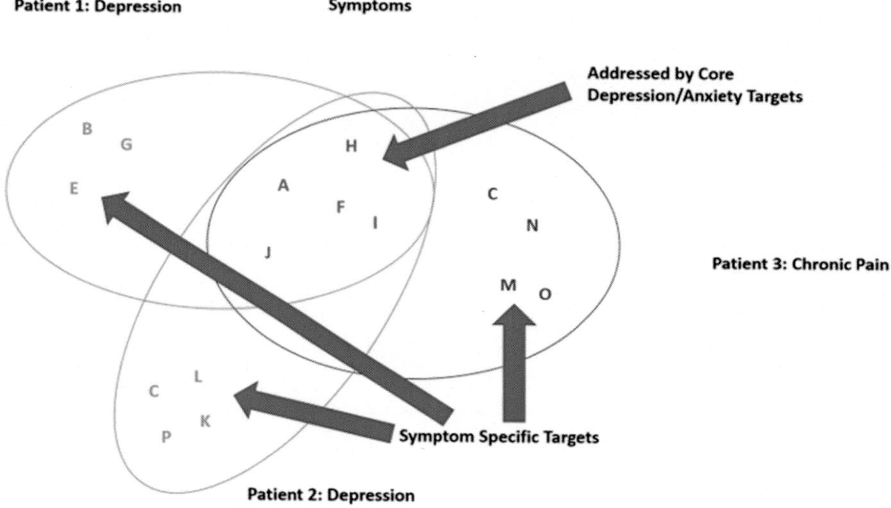

FIGURE 18.1

Schematic representing when a core depression target (center circle) is treated to address several related symptoms.

motivation, and social cognition. Note however, that the RDoC type approach was designed to handle mental illness and not higher cognitive function, though much of this gap can be covered by neuropsychology, neurology, and large-scale brain network anatomy (Insel et al., 2010). At minimum, many problems can be addressed by evaluating these circuits and evaluating the patient in these contexts.

(5) *When in doubt, look at the control networks*: Cognition is complex. Sometimes, we can't nail down a symptom. When this happens, looking at DMN, CEN, and Salience in different combinations usually points to something (Fig. 18.2) (Dadario et al., 2021). We have seen this help in many instances.

A compendium of treatment approaches

We finish this book off by putting all of these principles into practice with a list of strategies based on our current thinking. Many of these we have successfully treated patients with. Others reflect good ideas, at least at the current time. The goal is not to provide a definitive guide to how to treat anything, but to provide you ideas how to think through problems using the connectome. It is likely our approaches will change as we learn more, but the important idea is to outline the concept and how to think through complex cases.

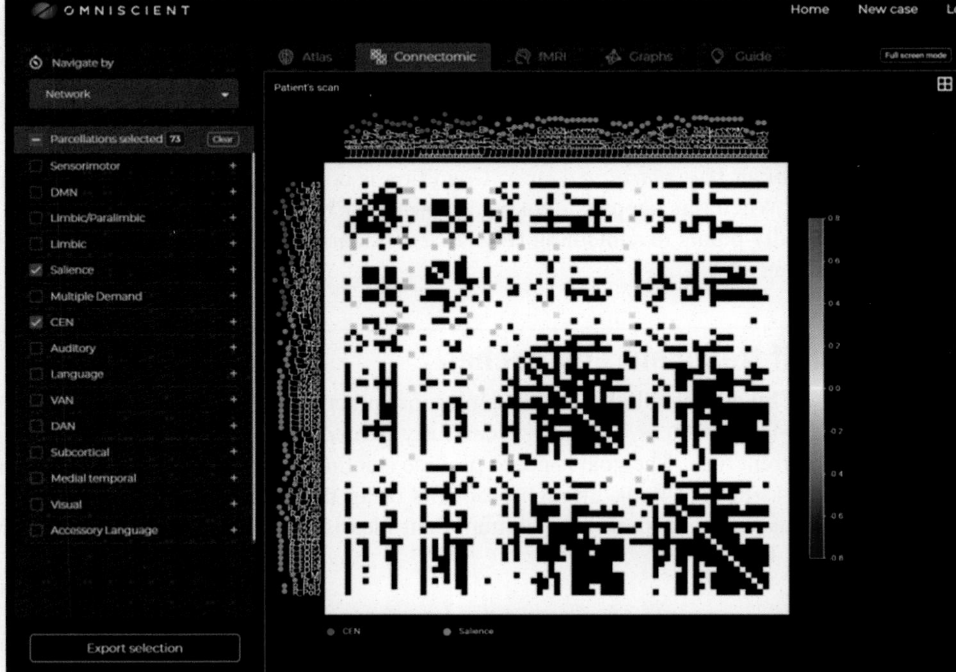

FIGURE 18.2

The cognitive control networks. Anomaly matrices are presented for the (A) DMN and CEN as well as the (B) salience and CEN networks.

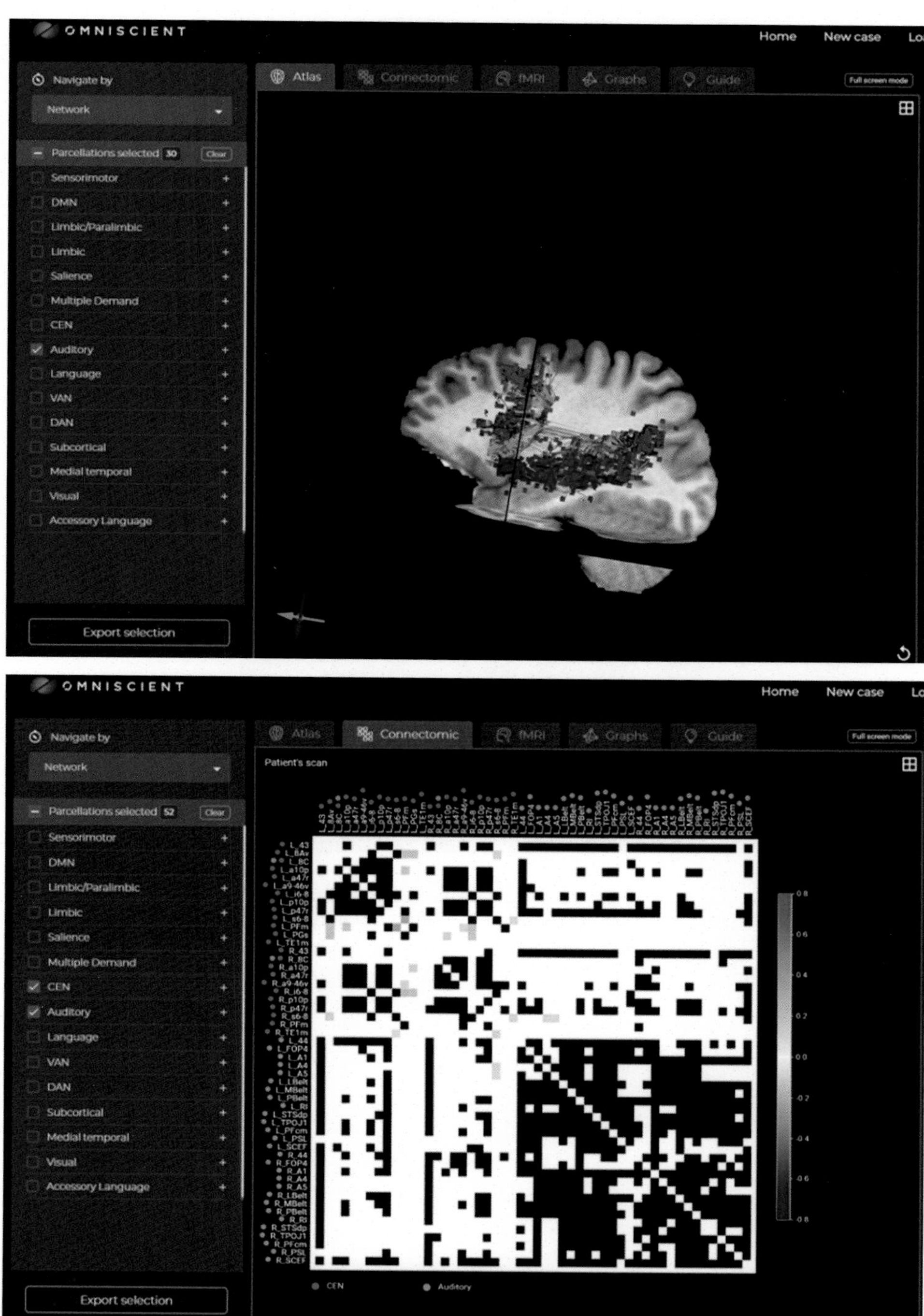

FIGURE 18.3

Relevant networks of tinnitus. The panels shown highlight the (A) auditory network and (B) CEN and auditory networks.

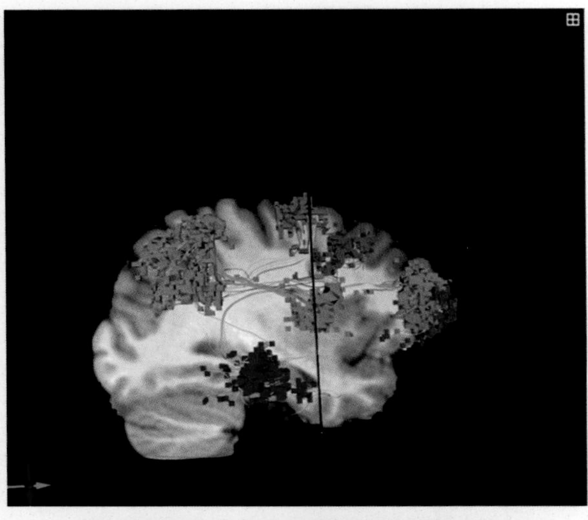

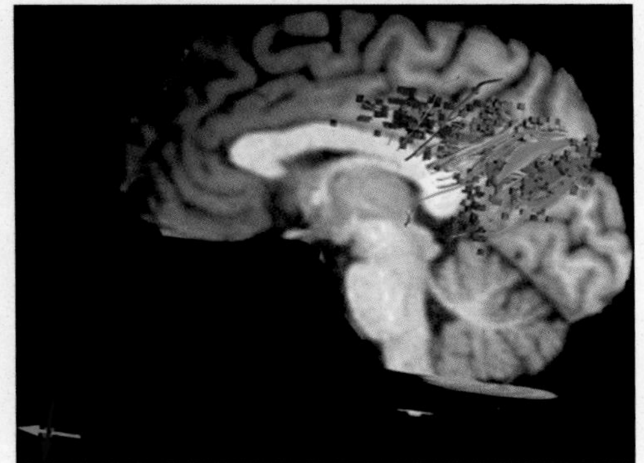

FIGURE 18.4

Relevant structural connections for memory. (A and B) working memory, (C) short-to-long term memory, (D–G) semantic memory including fiber bundles in the medial temporal lobe, language network, and the accessory language system.

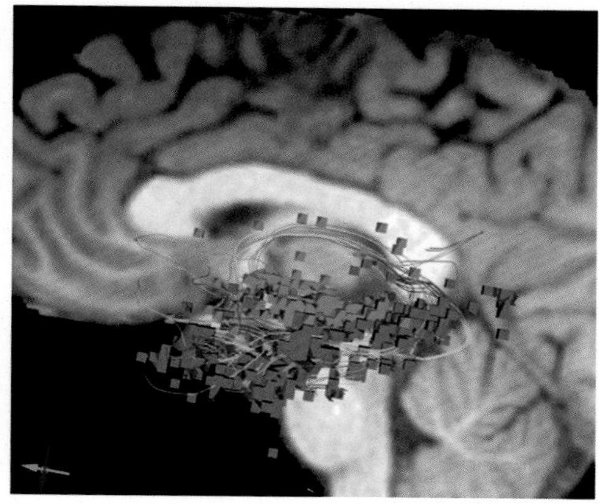

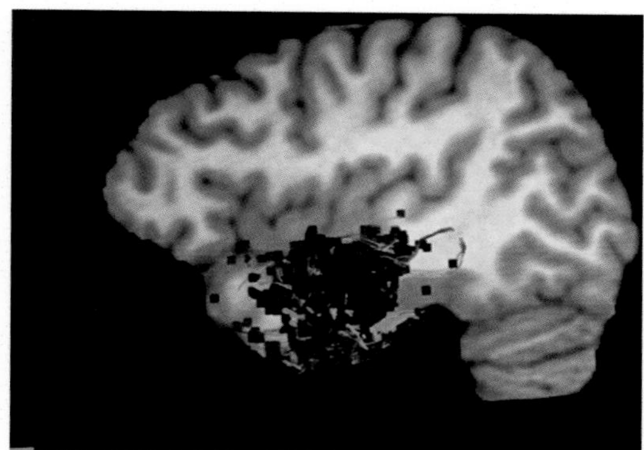

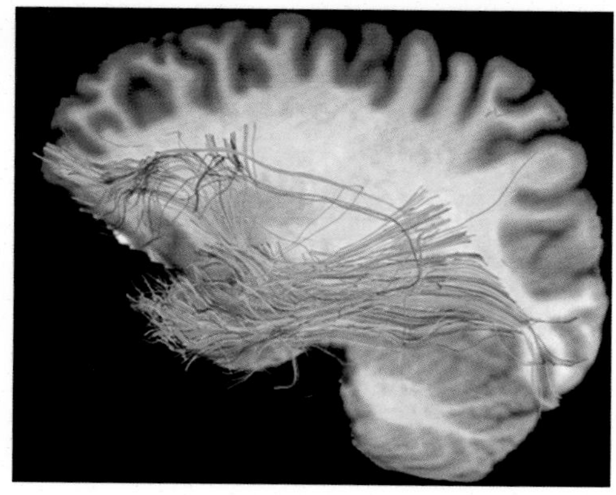

FIGURE 18.4 Cont'd

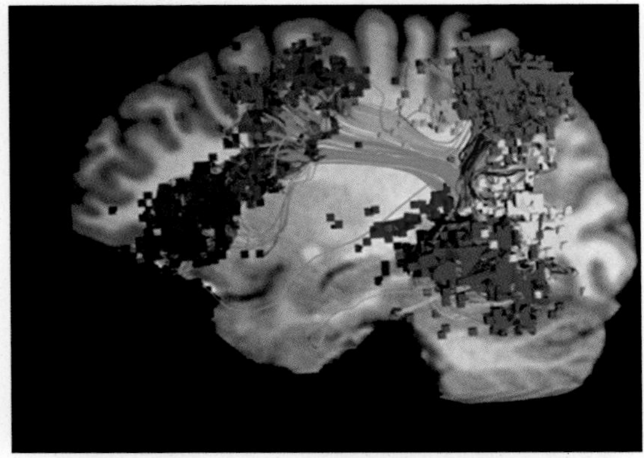

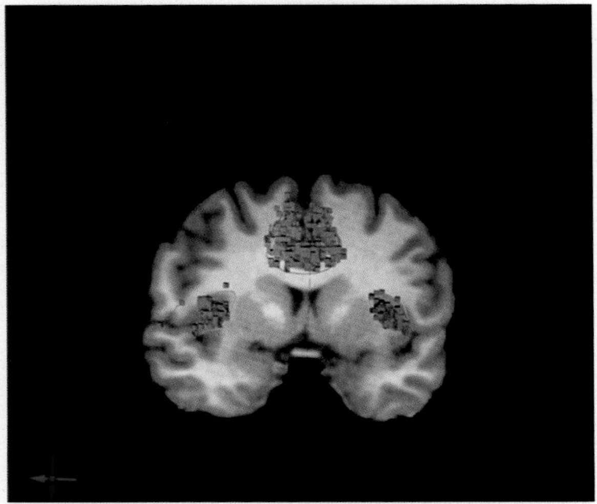

FIGURE 18.4 Cont'd

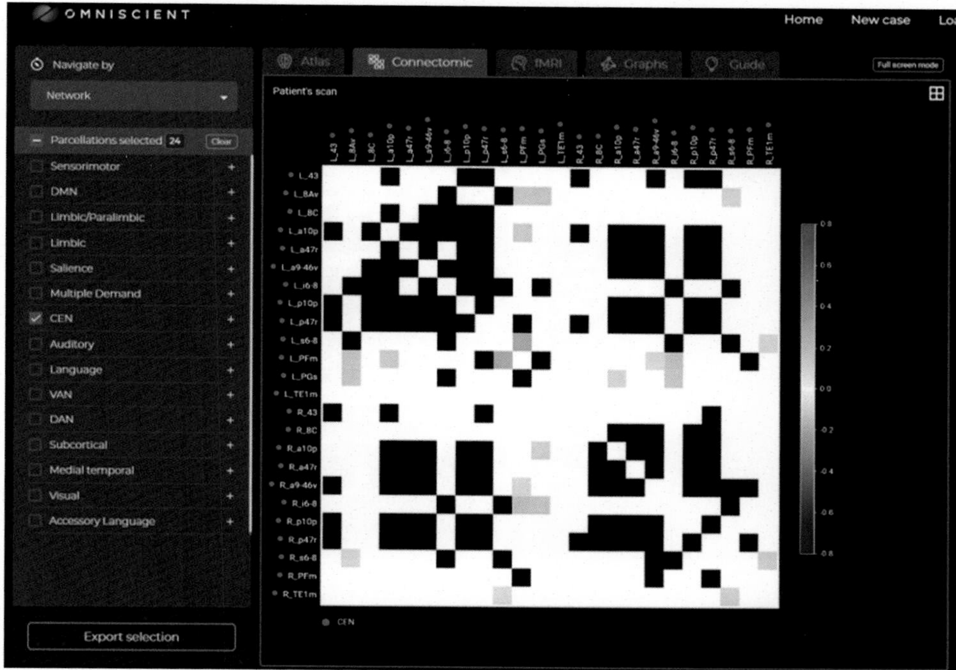

FIGURE 18.5

Relevant functional connectivity for memory. The panels highlight the following networks: (A) CEN, (B) multiple demand, (C) DMN-temporal component, and (D) language, medial temporal, and accessory language networks.

TMS principles for treatment of tinnitus (Fig. 18.3)

Step 1: define the problem

First off, it is critical to ensure that the patient has already undergone a reasonable workup. In our clinic, this is basically always the case (people usually do not seek out strategies like TMS until exhausting other options); however, it is very important you do not treat a vestibular schwannoma, some inner ear problem, or some other serious issue with TMS when they need surgery for a tumor, for example.

Assuming the problem has been worked up and left with no clear option better than "learn to deal with it," the problem is fairly simple: their ears ring, it bothers them, and they want that to stop. However, it is critical to consider that while some of these people have damaged inner ears, or brain stem or CN VIII injury, others have a complex, idiopathic etiology and we assume that these are complex, functional connectivity problems.

Step 2: look at the relevant parts of structural connectome

The auditory network is a mixture of primary sensory cortices around Heschl's gyrus, and portions of the extended DMN like A4 and A5. It has a direct, arcuate fasciculus connection to Broca's area/area 44 also. We do not generally see structural damage in this pathway, but it should be examined.

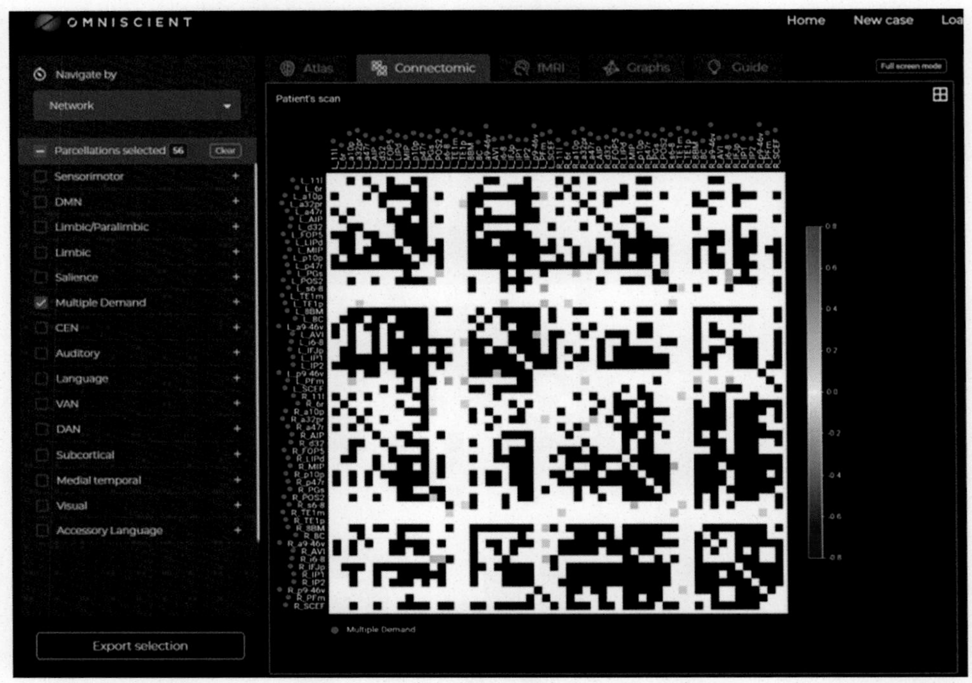

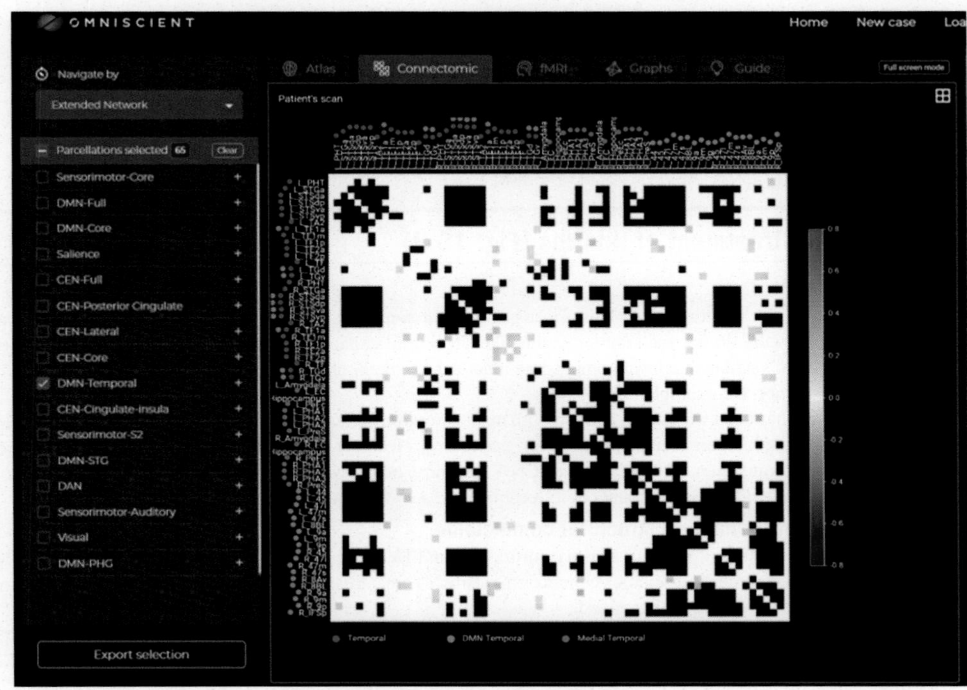

FIGURE 18.5 Cont'd

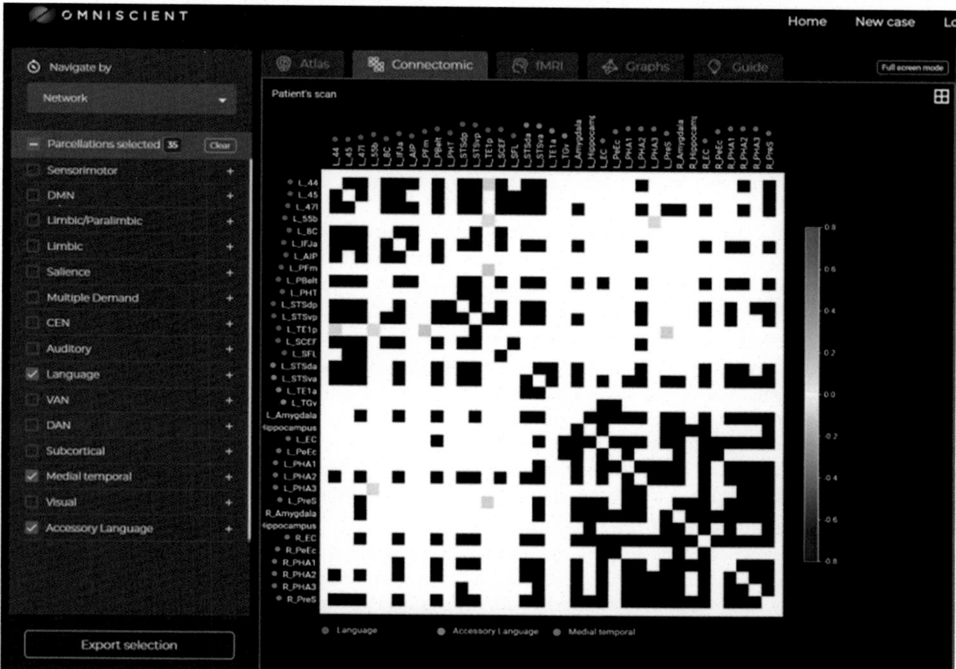

FIGURE 18.5 Cont'd

TMS principles for treatment of tinnitus (Fig. 18.3)—cont'd

Step 3: look at the relevant parts of the functional connectome

The standard targeting approach in the literature involves stimulating the auditory areas bilaterally. We pick the specific point by looking at the interactions of the auditory system and the CEN, in the idea this may be the part of the auditory network most being focused on.

Anxiety core:

While this approach given alone works for some patients, we found a substantial improvement in our outcomes and patient overall sense of well-being when we began adding an anxiety target to the standard approach in every case. This was discovered by chance in a patient who asked for treatment of anxiety midway through a treatment of tinnitus which had only marginally been successful. This approach means that if the ringing does not stop, we look for anomalies in a montage comparing DMN and CEN. The most common thing we see are anomalies in areas Left 8AV, L PGs, and/or either TE1M (Fig. 14.1B). It is important to also look at fear-related transdiagnostic montages if phobia or panic type symptoms are present.

Step 4: determine a treatment which addresses these issues and integrates into the overall care plan

We have used noise canceling headphones in these cases.

TMS principles for treatment of mental illnesses other than depression/anxiety

Step 1: define the problem

This has been the key question in psychiatry for decades. These diseases have multiple symptoms, and often multiple diagnoses. It is key to not only think about symptoms as multiple independent diseases, but also to prioritize the worst and/or most dangerous symptoms first. Also, addressing symptoms which generally do not respond to medical therapy, like negative symptoms in schizophrenia, makes more sense than trying to replace medicines with TMS as we cannot treat all symptoms with a focal therapy.

Additionally, looking for symptoms which overlap into disease clusters (for example potential threat is a problem in anxiety and PTSD, or several symptoms in depression involve problems with the reward system, and might be helped with a core depression target) can often help multiple symptoms with one target. Admittedly, though priorities are the key thing and this involves putting some issues ahead of others.

Step 2: look at the relevant parts of structural connectome

Most of these diseases are idiopathic and do not show overt structural issues. Note though that structural damage can cause mental illness type symptoms, and should be considered in appropriate patients as this may direct your thoughts toward other networks as potential targets. For example, damage to the DMN, salience, or CEN in a patient with symptoms of secondary mental illness might guide you to a nontraditional target which fits that patient better.

Studying smaller scale injury in TBI patients with neuropsychiatric dysfunction can also direct you to which part of a transdiagnostic circuit is the best place to target in that circuit. This is not always possible however.

Step 3: look at the relevant parts of the functional connectome

There are three ways to consider treating patients with mental illnesses, which are not entirely exclusive of each other.

Core target strategy: There is a literature demonstrating utility in treating the standard depression type dlPFC target in other diseases. Often many other disease states coexist with depression and this target may stimulate multiple transdiagnostic circuits. This is not dogma though. None of the trials doing this have 90% success rates, and it is hard to buy that hitting the dlPFC in everyone is a panacea. It does not appear to be. However, depression can exacerbate other problems as well, and treating it, when it exists seems possibly helpful. Anxiety targets may also be useful.

Transdiagnostic strategy: This approach seems likely to be helpful in tackling complex problems. It involves taking a careful transdiagnostic survey, identifying appropriate circuits, and looking for targets inside of a circuit which fits the data seems like a reasonable strategy (Barron et al., 2021). It is unclear that stimulating a region which works abnormally inside of a circuit related to an abnormal function will improve that circuit or relieve the symptom, but this seems more likely that stimulating the same area in everyone, or stimulating a circuit which is not behaving abnormally.

Network-based targeting strategy: When it is not clear, looking at DMN, CEN, and salience can be a good strategy for identifying good patient-specific targets. These networks are abnormal in most mental illness patients, and can carry the core role of tackling several symptoms at once.

Step 4: determine a treatment which addresses these issues and integrates into the overall care plan

There is a lot of room for creativity in this area. Most transdiagnostic circuits have CBT type modalities or other therapies thought to modulate them, and improve outcomes. Alternatively, modalities like cognitive training, meditation, and many other non-psychotherapy-related treatments can also be useful for various patients.

TMS principles for treatment of memory issues (Figs. 18.4 and 18.5)

Step 1: define the problem

Detailed neuropsychologic assessment is key here as memory issues are complex, can coexist with each other, and result from different network problems. Key patterns to be aware of (not exhaustive but common):

Working memory: CEN related, have multiple transdiagnostic subparts, can also be affected by attention problems (Champod & Petrides, 2010).

TMS principles for treatment of memory issues (Figs. 18.4 and 18.5)—cont'd

Short-term to long-term term transfer: Related to the Papez circuit type areas including mamillary bodies, fornix, cingulum bundle, inferior and medial temporal subnetwork of the DMN, medial temporal lobe, and hippocampus (Briggs et al., 2021).

Semantic memory/verbal memory: Generally left temporal pole and accessory language areas (Milton et al., 2021).

Confounders: attentional problems, aphasias, psychomotor retardation, and other neuropsychological dysfunctions (Dadario & Sughrue, 2022).

These patients often present in the context of other diseases, such as degenerative diseases, and it is very, very important to set realistic goals for these patients as TMS is palliative and does not change the natural history of the disease. It is also unidimensional and in its best cases can improve a single domain of a multidomain degeneration. So realistic goals like small gains in memory, improvement of depression, etc., are far more likely than changing every aspect of these patients.

Finally, a search for reversible causes of these issues need to be performed if they have not already been done. In addition to labs, other considerations include normal pressure hydrocephalus, strokes, and medication side effects need to be ruled out before trying TMS.

Step 2: look at the relevant parts of structural connectome (Fig. 18.4)

Structural damage to these circuits is often visible and evaluation should focus on the following:

Bilateral CEN
Posterior cingulate DMN
Bilateral hippocampi and fornix
Left temporal pole and accessory language areas
ILF and IFOF tracts
Language system
Salience network

Diseases like Alzheimer's disease and degenerative conditions often show a global atrophy pattern, but rarely show overt changes in the gross anatomy of networks and tracts. Structural changes in these diseases can be detected with more complex methods employing machine learning, but at present these tools are in early stages.

Step 3: look at the relevant parts of the functional connectome (Fig. 18.5)

First, we focus on language, alertness, or attention issues if they are identified, with emphasis being on the language system, DMN, and salience.

Second, if depression or anxiety exist, we always treat these as they can make everything worse in these patients.

Third, we supplement targets determined from the previous two steps with targets (if any) aimed at the specific type of memory disturbance we noted. This may or may not be informed by structural connectivity data.

Working memory (Fig. 18.4A and B): We look for targets in the CEN, and possibly the multiple demand network.

Short to long-term memory (Fig. 18.4C): We pull up a montage of bilateral hippocampi, memory (inferior) circuit of the DMN, and bilateral temporal lobes in these patients. We cannot hit the hippocampus with TMS, but we select lateral temporal lobe targets which show abnormal connectivity, especially hyperconnectivity, with the hippocampus in hopes of normalizing temporal lobe dysfunction. We have had some success with this strategy.

Semantic memory (Fig. 18.4D–F): Here, we study the connectivity in the medial temporal lobe, language network, and the accessory language system.

Step 4: determine a treatment which addresses these issues and integrates into the overall care plan

We focus on cognitive rehab, both with therapists, and supplementing this with targeted app-based training and other tools. Again, it is critical to focus therapy on the actual type of pathology that the patient has.

Conclusions

In this chapter we discuss the use of connectivity-guided TMS treatment for a wide variety of unique disease states and symptoms beyond anxiety, depression, and stroke. We provide a list of advice for approaching unique cases and a list of strategies when treating these patients based on our previous experience, such as for various mental illnesses, tinnitus, and memory. While many treatments will be updated in due time along with future advancements in our understanding of related pathology, this chapter provides a framework in thinking which will be continuously helpful in future times.

References

Barron, D. S., Gao, S., Dadashkarimi, J., Greene, A. S., Spann, M. N., Noble, S., Lake, E. M. R., Krystal, J. H., Constable, R. T., & Scheinost, D. (2021). Transdiagnostic, connectome-based prediction of memory constructs across psychiatric disorders. *Cerebral Cortex, 31*(5), 2523–2533. https://doi.org/10.1093/cercor/bhaa371. New York, N.Y.: 1991.

Briggs, R. G., Tanglay, O., Dadario, N. B., Young, I. M., Fonseka, R. D., Hormovas, J., Dhanaraj, V., Lin, Y. H., Kim, S. J., Bouvette, A., Chakraborty, A. R., Milligan, T. M., Abraham, C. J., Anderson, C. D., O'Donoghue, D. L., & Sughrue, M. E. (2021). The unique fiber anatomy of middle temporal gyrus default mode connectivity. *Operative Neurosurgery (Hagerstown, Md.), 21*(1), E8–E14. https://doi.org/10.1093/ons/opab109

Champod, A. S., & Petrides, M. (2010). Dissociation within the frontoparietal network in verbal working memory: A parametric functional magnetic resonance imaging study. *Journal of Neuroscience, 30*(10), 3849–3856. https://doi.org/10.1523/jneurosci.0097-10.2010

Cocchi, L., Zalesky, A., Nott, Z., Whybird, G., Fitzgerald, P. B., & Breakspear, M. (2018). Transcranial magnetic stimulation in obsessive-compulsive disorder: A focus on network mechanisms and state dependence. *NeuroImage Clinical, 19*, 661–674. https://doi.org/10.1016/j.nicl.2018.05.029

Corbetta, M., & Shulman, G. L. (2011). Spatial neglect and attention networks. *Annual Review of Neuroscience, 34*, 569–599. https://doi.org/10.1146/annurev-neuro-061010-113731

Dadario, N. B., Brahimaj, B., Yeung, J., & Sughrue, M. E. (2021). Reducing the cognitive footprint of brain tumor surgery. *Frontiers in Neurology, 1342*.

Dadario, N. B., & Sughrue, M. E. (2022). Should neurosurgeons try to preserve non-traditional brain networks? A systematic review of the neuroscientific evidence. *Journal of Personalized Medicine, 12*(4), 587.

De Simoni, S., Grover, P. J., Jenkins, P. O., Honeyfield, L., Quest, R. A., Ross, E., Scott, G., Wilson, M. H., Majewska, P., Waldman, A. P., Patel, M. C., & Sharp, D. J. (2016). Disconnection between the default mode network and medial temporal lobes in post-traumatic amnesia. *Brain, 139*(Pt 12), 3137–3150. https://doi.org/10.1093/brain/aww241

Insel, T., Cuthbert, B., Garvey, M., Heinssen, R., Pine, D. S., Quinn, K., Sanislow, C., & Wang, P. (2010). Research domain criteria (RDoC): Toward a new classification framework for research on mental disorders. *The American Journal of Psychiatry, 167*(7), 748–751. https://doi.org/10.1176/appi.ajp.2010.09091379

Memon, A. M. (2021). Transcranial magnetic stimulation in treatment of adolescent attention deficit/hyperactivity disorder: A narrative review of literature. *Innov Clin Neurosci, 18*(1–3), 43–46.

Milton, C. K., Dhanaraj, V., Young, I. M., Taylor, H. M., Nicholas, P. J., Briggs, R. G., Bai, M. Y., Fonseka, R. D., Hormovas, J., Lin, Y. H., Tanglay, O., Conner, A. K., Glenn, C. A., Teo, C., Doyen, S., & Sughrue, M. E. (2021). Parcellation-based anatomic model of the semantic network. *Brain and Behavior, 11*(4), e02065. https://doi.org/10.1002/brb3.2065

Petrosino, N. J., Cosmo, C., Berlow, Y. A., Zandvakili, A., van't Wout-Frank, M., & Philip, N. S. (2021). Transcranial magnetic stimulation for post-traumatic stress disorder. *Therapeutic Advances in Psychopharmacology, 11*. https://doi.org/10.1177/20451253211049921, 20451253211049921.

Shahab, Q. S., Young, I. M., Dadario, N. B., Tanglay, O., Nicholas, P. J., Lin, Y. H., Fonseka, R. D., Yeung, J. T., Bai, M. Y., Teo, C., Doyen, S., & Sughrue, M. E. (2022). A connectivity model of the anatomic substrates underlying Gerstmann syndrome. *Brain Communications, 4*(3), fcac140. https://doi.org/10.1093/braincomms/fcac140

Sheng, J., Liu, S., Wang, Y., Cui, R., & Zhang, X. (2017). The link between depression and chronic pain: Neural mechanisms in the brain. *Neural Plasticity, 2017*, 9724371. https://doi.org/10.1155/2017/9724371

Theodore, W. H. (2003). Transcranial magnetic stimulation in epilepsy. *EPI - Epilepsy Currents, 3*(6), 191–197. https://doi.org/10.1046/j.1535-7597.2003.03607.x

Young, I. M., Dadario, N. B., Tanglay, O., Chen, E., Cook, B., Taylor, H. M., Crawford, L., Yeung, J., Nicholas, P. J., Doyen, S., & Sughrue, M. E. (2023). Connectivity model of the anatomic substrates and network abnormalities in major depressive disorder: A coordinate meta-analysis of resting-state functional connectivity. *Journal of Affective Disorders Reports, 11*, 100478. https://doi.org/10.1016/j.jadr.2023.100478

Zhou, L., Huang, X., Li, H., Guo, R., Wang, J., Zhang, Y., & Lu, Z. (2021). Rehabilitation effect of rTMS combined with cognitive training on cognitive impairment after traumatic brain injury. *American Journal of Translational Research, 13*(10), 11711–11717.

Index

Note: 'Page numbers followed by "*f*" indicate figures "*t*" indicate tables and "*b*" indicate boxes'.

B

C

R

S

T

V

W

Printed in the United States
by Baker & Taylor Publisher Services